ENVIRONMENTAL POLLUTION ANALYSIS

ENVIRONMENTAL POLLUTION ANALYSIS

SECOND EDITION

S M Khopkar
MSc., PhD
Professor Emeritus
Department of Chemistry
Indian Institute of Technology Bombay
Mumbai

Publishing Globally

NEW AGE INTERNATIONAL (P) LIMITED, PUBLISHERS
LONDON • NEW DELHI • NAIROBI
Bangalore • Chennai • Cochin • Guwahati • Hyderabad • Kolkata • Lucknow • Mumbai
Visit us at www.newagepublishers.com

Published by New Age International (P) Ltd., Publishers
First Edition: 1993
Second Edition: 2011
Reprint: 2020

GLOBAL OFFICES

- **New Delhi** **NEW AGE INTERNATIONAL (P) LIMITED, PUBLISHERS**
  7/30 A, Daryaganj, New Delhi-110002, (INDIA)
  Tel.: (011) 23253771, 23253472, **Telefax:** 23267437, 43551305
  E-mail: contactus@newagepublishers.com • Visit us at www.newagepublishers.com
- **London** **NEW AGE INTERNATIONAL (UK) LTD.**
  27 Old Gloucester Street, London, WC1N 3AX, UK
  E-mail: info@newacademicscience.co.uk • Visit us at www.newacademicscience.co.uk
- **Nairobi** **NEW AGE GOLDEN (EAST AFRICA) LTD.**
  Ground Floor, Westlands Arcade, Chiromo Road (Next to Naivas Supermarket), Westlands, Nairobi, KENYA
  Tel.: 00-254-713848772, 00-254-725700286
  E-mail: kenya@newagepublishers.com

BRANCHES

- **Bangalore** 37/10, 8th Cross (Near Hanuman Temple), Azad Nagar, Chamarajpet, Bangalore- 560 018
  Tel.: (080) 26756823, **Telefax:** 26756820, **E-mail: bangalore@newagepublishers.com**
- **Chennai** 26, Damodaran Street, T. Nagar, Chennai-600 017, **Tel.:** (044) 24353401, **Telefax:** 24351463
  E-mail: chennai@newagepublishers.com
- **Cochin** CC-39/1016, Carrier Station Road, Ernakulam South, Cochin-682 016, **Tel.:** (0484) 2377303, **Telefax:** 4051304
  E-mail: cochin@newagepublishers.com
- **Guwahati** Hemsen Complex, Mohd. Shah Road, Paltan Bazar, Near Starline Hotel, Guwahati-781 008,
  Tel.: (0361) 2513881, **Telefax:** 2543669, **E-mail: guwahati@newagepublishers.com**
- **Hyderabad** 105, 1st Floor, Madhiray Kaveri Tower, 3-2-19, Azam Jahi Road, Near Kumar Theater, Nimboliadda Kachiguda, Hyderabad-500 027, **Tel.:** (040) 24652456, **Telefax:** 24652457
  E-mail: hyderabad@newagepublishers.com
- **Kolkata** RDB Chambers (Formerly Lotus Cinema) 106A, 1st Floor, S N Banerjee Road, Kolkata-700 014
  Tel.: (033) 22273773, **Telefax:** 22275247, **E-mail: kolkata@newagepublishers.com**
- **Lucknow** 16-A, Jopling Road, Lucknow-226 001, **Tel.:** (0522) 2209578, 4045297, **Telefax:** 2204098
  E-mail: lucknow@newagepublishers.com
- **Mumbai** 142C, Victor House, Ground Floor, N.M. Joshi Marg, Lower Parel, Mumbai-400 013
  Tel.: (022) 24927869, **Telefax:** 24915415, **E-mail: mumbai@newagepublishers.com**
- **New Delhi** 22, Golden House, Daryaganj, New Delhi-110 002, **Tel.:** (011) 23262368, 23262370, **Telefax:** 43551305
  E-mail: sales@newagepublishers.com

ISBN: 978-81-224-2747-9
C-20-01-12382

Printed in India at R.K Printers, Delhi.
Typeset at Pagitek Graphics, New Delhi.

NEW AGE INTERNATIONAL (P) LIMITED, PUBLISHERS
7/30 A, Daryaganj, New Delhi-110002
Visit us at **www.newagepublishers.com**
(CIN: U74899DL1966PTC004618)

Dedicated to

My (Late) Mother, Manorama

&

(Late) Father, Moreshwar

Foreword

A delicate balance exists between man and his environment. Lately, this balance has been seriously disturbed by man's direct as well as indirect activities. Thousands of toxic containments of various types, viz. inorganic, organic and organo-metallic, have found their ways to our environment. Many of them are highly toxic even at trace levels. Some have long biological half-lives and, consequently, will remain toxic for a long time. The extent of this contamination is so vast that the *food we eat, air we breathe and water we drink* are suspected to contain a number of toxins.

Extensive research on the detection, identification and measurement of environmental pollutants in recent years has led to an enormous growth in the field of environmental analytical chemistry. With the availability of many sophisticated and sometimes expensive analytical techniques for measuring pollutants, it is often difficult to select one which is not only appropriate but also practical in a given situation.

In a systems analysis approach, it is necessary to identify the sources, pathways and sinks of pollutants. During this translocation, the speciation of a pollutant may very well change leading to a more toxic and bioavailable species. Analytical methods being developed today need also to consider these aspects of environmental pollution.

While teaching analysis of pollutants, it is always interesting to quote actual data of real samples. If the samples happen to be collected from the same region as the student is, the data become more meaningful. Professor Khopkar's efforts to include data from India are commendable.

"Environmental Pollution Analysis" is a well-conceived book which covers many important areas of this vast subject. It should be valuable not only to undergraduate and graduate students as a textbook but also to practicing environmental chemists for reference. It is indeed a pleasure to commend this book.

A. Chatt
Director
Professor of Chemistry

SLOWPOKE-2 Facility
Trace Analysis Research Centre
Department of Chemistry
Dalhousië University
Halifax, Nova Scotia
CANADA

Preface to the Second Edition

Since last two decades, there has been spectacular progress in the analysis of environmental pollution. Newer methods have come to the forefront and classical methods have become outdated. In order to maintain parity for comparison of data in interlaboratory performance, standard methods were developed. Of late, more stress was led on the instrumental methods of analysis as such methods are simple, rapid and give reproducible results. In earlier edition of this book greater stress was given to principles of analysis. In view of this the outline of each method was provided without further experimental details. However, in the course of time several readers felt the need to have detailed experimental procedure for the environmental pollution analysis. Therefore in this revised version of the book, a detailed experimental process is included in several chapters to facilitate carrying out such experiments in the laboratory. Further new chapters have been added to give up-to-date knowledge of pollution analysis in one single volume.

There are new chapters on several topics like sampling methods for air pollution analysis, from ambient and stationary sources. Modern methods like atomic absorption and plasma spectroscopy, anodic stripping voltammetry and capillary electrophoresis are extensively used. Therefore, detailed procedure for such analysis involving instrumental techniques for analysis of suspended particular matter, halogens, ozone and hydrocarbons have been included. The stepwise description for easy analysis of hydrological parameters as DO, BOD, COD have been also included. Anions constitute an important source of water pollution and its characterization attains special importance in the area of the environmental speciation, hence analysis of many anions is also presented.

With growing industrialization, the level of organic pollutants is rising, therefore, methods for the analysis of phenols, poly aromatic hydrocarbons, trihalidemethane and several toxic pollutants encountered in air and the water are discussed. The quality of specially potable water depends upon the presence of microorganisms present in it. Therefore, pertinent description of the microbiological examination of water is thoroughly discussed. It includes analysis of coliforma, streptococci, planktones and fungi. The characterization and determination of these microogranism is provided with all minor details in the text.

All the important features of the previous edition have been retained and modified with up-to-date information on the environmental pollution analysis. With the inclusion of new topics, now it is felt that new version will be very useful to the students of environmental science both at undergraduate and postgraduate classes. It should serve as an ideal laboratory manual for performing all kinds of experiments related to air and water pollution analysis after comprehending basic principles in methodology employed.

I take this opportunity to thank Mr. Saumya Gupta, Managing Director, Mr. V.R. Damodaran and the editorial staff of New Age International Publishers (P) Ltd. for providing new version of this book in its present form. I am also thankful to family of my son Samir, Madhavi, Rohan, as well as that of my daughter Dr. Supriya, Keran and Ruchir, for all kinds of assistance. Finally it is my prime duty to thank my wife Dr. Sucheta for offering wonderful cooperation and encouragement during the formulation of this book.

Professor Emeritus,
Dept. of Chemistry
Indian Institute of Technology, Bombay
Powai, Mumbai - 400076

S.M. KHOPKAR

Preface to the First Edition

Environmental Science is an interdisciplinary science in a real sense. To appreciate nature and its vagaries, one requires an intricate knowledge of chemical, physical and biological sciences. Most of such phenomenon are explained in terms of basic principles of well led theories of these sciences. Sanitary engineering was part of civil engineering which largely had ramification on water and sewage. With its expanding boundaries to environmental science and engineering. It covered broader aspects of air, water, soil biosphere and their interaction. Environmental chemistry plays an important role in the understanding of our environment. Analytical chemistry in particular is considered as the backbone of chemistry of environment. A thorough knowledge of the environmental analytical chemistry greatly helps us to measure extent of environmental pollution and in turn adopt suitable devises to control pollution.

With this objective in mind this book is planned. It is presumed that the reader is quite conversant with the fundamental principles of analytical chemistry and he is primarily interested to understand their applications to solving problems in environment. This book is not intended to serve as exhaustive reference volume for detailed procedure of analysis. For such purpose the reader is advised to consult authorative monographs suggested as the additional reading at the end of each chapter. The main aim here was to explain important aspects of a pollution and immediately dwell upon the principles and outline of methods of analysis. Such of those methods which are practical to conditions in India are only considered.

This book is outcome of fifteen years of my experience of research as well as teaching of environmental pollution analysis to postgraduate students in science as well as engineering. The work has originated from my class notes which were extensively consulted by my students in such course. Perhaps this is the first venture by Indian author in this field, which should prove fruitful for graduate and post-graduate students in chemistry and environmental science majoring in environmental analysis or monitoring. I am quite confident that readers would greatly appreciate my efforts in furnishing data relevant to Indian environment. Finally, this book should be asset to practicing environmental scientist, analytical chemist and serious research workers.

This book covers important areas of environmental pollution analysis. An introductory chapter presents a grim picture of levels of pollution, while second chapter summarises highlights of important modern methods of analysis, while a chapter on toxic pollutants and their analysis largely covers physiological manifestations and analysis with which a chemist by and large is unfamiliar. There are two chapters each on air pollution and water pollution analysis. The important aspects of atmosphere, characterisation, analysis as well as monitoring of air pollutants are then discussed. Water is an important commodity of life hence quality and analysis of potable water, chemistry for water treatment processes is adequately covered along with its analysis. The waste water or industrial effluents as well as domestic effluents are important sources from view point of recycling and recovery of valuable materials, hence their characterisation and quantitative analysis are explained in greater details. Soil or earth is important component of environment, therefore its pollution, the role of micronutrients and impact of pesticide residue is considered in accounting their analysis. Noise as well as odour are important constituents of environment. Their presence is felt only if present in excess, hence measurement of noise or odour is considered in last two chapters. Throughout the text along with method of analysis wherever possible monitoring methodology is also discussed.

I am thankful to Prof. B Nag, Director, Indian Institute of Technology, Bombay and Coordinator, Curriculum Development Programme for providing financial assistance for writing this book. I am grateful to Prof. A Chatt, Director, Slowpoke-2 Facility, Trace Analysis Research Centre, Dalhousie University, Halifax, Canada for sparing his valuable time to write Foreword and offering valuable comments. I have no words to express my thanks to my research scholars specially Dr. Nandu Deorkar and Dr. Mayuri Gandhi who have rendered valuable help in carrying out all kinds of odd jobs for this book. I am thankful to staff of New Age International Publishers (P) Ltd. for the excellent production. Before I conclude it is my privilege and honour to thank my wife Dr. Sucheta, daughter Supriya, son Samir and his wife Madhavi for all encouragement, assistance, patience and utmost sacrifice without which I could not have completed this onerous task.

S.M. KHOPKAR

Contents

CHAPTER 1

Introduction to Environmental Pollution

1.1 INTRODUCTION

Amongst the developing countries, India is the foremost nation to take note of degradation of environment. In 1972, in the world environmental conference held at Stockholm, our government declared that it is their responsibility to control pollution in India. Even during recent global conference held at Kyoto in Japan, India accepted the responsibility of reducing global warming and depletion of ozone layer by banning the use of chloro-fluoro carbons in cryogenic engineering within a stipulated period. Even in the recent conference held in Rio-de-Janeiro, India declared solidarity by conforming to the standards as stipulated by United States Environmental Protection Agencies (US-EPA). Perhaps India was the foremost country in third worlds to implement rigorously pollution legislation both for air and water. We accepted the stipulated levels of the pollutants in act designated as 'maximum permissible level' to be implemented in successive stages in few years.

Within India, Maharashtra state was the front runner to enact Water Pollution and Control Act in 1962. This was followed by passing of Comprehensive Environmental Protection Bill for prevention of air pollution in 1983. Maharashtra, Gujarat, Madhya Pradesh were the first states in Indian union to have pollution control board with attached analytical laboratories. This was followed by the setting of series of laboratories in all remaining states of the nation. A legislation was passed. The grave tragedy of Bhopal in 1984 by accidentally releasing deadly poisonous gas such as methyl isocyanate took a toll of more than 4500 people who were innocent and helpless. With the growing civilisation and with rise in traffic, auto-exhaust emission also takes a toll of people. Therefore government enacted 'PUC' act to check every vehicle periodically and control, release of hazardous gases like NO-NO_x, CO, and hydrocarbons in air.

As a matter of fact, India did much more than any other developing country could do to control pollution. Even in controlling global warming and reduction of the Green House gases we were certainly ahead of developed countries like Canada, USA, Australia. General incentives were given by our government to design and develop indigenously pollution monitoring instruments within the country and make us self-sufficient. At moment, separate ministry looks after all problems related to pollution. In fact, all nations have their own ministry or department to implement effectively legislation related to control of pollution. We have passed act and prevented transportation of hazardous chemicals a big offence on express highways. The abatement of noise pollution surrounding airports, hospitals was taken care by our legislation. Environmental Impact Assessment (EIA) was rigorously followed during setting of new industries and environmental modelling has been used to carry out future foresting of pollution and resorting to effective control measures of pollution.

Basically, our environment is composed of atmosphere, earth, water and space. In absence of pollution, it remains clean and enjoyable. The interaction of the atmosphere, lithosphere, hydrosphere and biosphere is continuing for years together. On account of the various activities of man, the composition and complex nature of environment gets changed. These activities include industrialization, construction, transportation. Such activities, although desirable for human development and welfare, lead to generation and release of objectionable materials into the environment thus turning it foul, and rendering our life miserable.

The natural environment is clean, but due to multiferious activities of man it gets polluted resulting in what is called as environmental pollution. Our main aim should be to keep our environment clean, by curbing industrial activities. However, in order to keep pace with the rapid industrialisation world over, a developing country like India cannot afford to arrest its industrial growth. We can select such industrial processes, which would cause minimum or zero pollution. Further, it is necessary to undertake pollution control measures, so as to enable us to keep our environment as clean as possible.

Our environment is complex. It gets fouled when industrial activities grow. To comprehend this complex nature of environment, we need knowledge of all disciplines of chemical, physical and biological science. To devise control measures, we need knowledge of engineering and technology. Environmental Science and Engineering is inter-disciplinary in nature. To understand atmosphere we need knowledge of physics and meteorology, while for knowing hydrosphere and air we need knowledge of chemistry, the complex behaviour of living organism can be really explained with a knowledge of the life sciences. The interaction in lithosphere can be best understood from principles of earth science, while the entire control mechanism can be devised only with intricate knowledge of engineering and technology. In the process of measurement we collect enormous data, which can be interpreted and understood only if we are conversant with statistical analysis and computer programming and environmental modelling. Some concerted effort must be made to keep our environment clean. It will lead to betterment of our lives and peaceful coexistence on the earth.

1.2 WHAT IS POLLUTION?

Our next task to keep the environment clean is to measure the damage caused to it by pollution. Without understanding the extent of damage caused to our environment due to pollution, it may not be possible for us to clean it. Such cleaning involves development of suitable control measures. Now these suitable control measures can be meticulously planned, provided we know what the level of pollution is? Hence, to understand the level of pollution, we should undertake the analysis or measurement of pollution. This analysis or measurement is termed as *Environmental Pollution Analysis*, which will throw light on exact degree of pollution. It will guide us to recover valuable products from waste and help us to take appropriate control measures to minimise pollution. With this objective in mind we should try to learn environment pollution analysis, so as to undertake the analysis, we must know the type of pollution.

Environmental pollution is classified into various groups. For instance, pollution of air is termed as the atmospheric pollution, the pollution of hydrosphere or water is termed as water pollution, while pollution due to disposal of waste water is termed as industrial effluents pollution. Similarly, indiscriminate dispersal of domestic sewage or sullage is called domestic effluent pollution. In addition to these major sources of pollution we have pollution of lithosphere or land, called soil pollution. For instance pesticide residue contribute towards soil pollution. Urban areas are blessed with the menace of noise, which at times becomes intolerable. It is called noise pollution. Some kind of pollution lead to generation of smell or odour; and as such its measurement becomes important. In this chapter, let us have a glimpse of different kinds of pollution.

1.3 ATMOSPHERIC POLLUTION

Amongst the various types of pollution, air pollution is of greatest importance. It can be readily noticed as it causes immediate health problems. The hazardous gases that lead to pollution are oxides of carbon, nitrogen and sulphur (Table 1.1). Particulate matter also causes air pollution. Table 1.2 indicates level of suspended particulate matter in cities of India. The emission due to auto-exhaust causes serious pollution of air due to formation of oxides of nitrogen and benzopyrene. Table 1.3 indicates the level of BaP *i.e.* Benzopyrene in various towns of the world, and Table 1.4 shows BaP level at various places in Mumbai. BaP is generated by incomplete combustion of petrol. This is a serious air pollutant as it

Table 1.1 Growth of air pollution (1978–1986) in μg/m³

S. No.	*Place/Town*	*SPM*	SO_2	$NO\text{–}NO_x$
(A) Levels of air pollution in Mumbai (1986)				
1.	Chembur	341	65	51
2.	Lalbaug	454	128	–
3.	Khar	349	11	77
4.	Babula Tank	318	56	56
5.	Parel	326	47	70
6.	Dadar	–	60	–
7.	Sewree	–	60	–
8.	Ghatkopar	407	75	–
9.	Sakinaka	–	49	–
(B) Levels of air pollution in India (1978) and abroad				
10.	Bombay (Mumbai)	238	172	50
11.	Calcutta (Kolkata)	500	58	51
12.	Delhi	700	90	35
13.	Kanpur	500	40	30
14.	London	150	87	29
15.	New York	–	135	42
16.	Tokyo	–	59	62

Table 1.2 Suspended particulate matter (SPM) in major cities in India (1978) μg/m³

City	*Urban area*	*Suburban area*	*Residential area*
Bombay (Mumbai)	123	150	221
Calcutta (Kolkata)	352	324	246
Delhi	361	256	391
US (EPA) Standard	75	75	75

Table 1.3 Benzopyrene levels abroad

Country	*Town*	*Level μg/m³*
Germany	Hamburg	17–1863
Hungary	Budapest	32–1060
Iran	Tehran	0.61–63
Japan	Osaka	11.1
Poland	Warsaw	29–133
Spain	Madrid	0–9.4
Sweden	Stockholm	15–27
USA	Urban average	0.5–18.6

Table 1.4 Benzopyrene levels in city of Mumbai

Place	*Range of concentration (μg/1000 m³)*	*Source of emission*
Lalbaug	165.0 to 861.0	Gas plant where coal is burnt
Dadar	7.1 to 16.2	The gas plant was shut down
Sion	15.0 to 36.0	Automobile exhaust
Koliwada	17.4 to 225.1	Kiln for firing pottery
Trombay	2.0 to 13.0	Petrochemical area
Bhandup	0.8 to 2.0	Residential area
Colaba	3.1 to 3.9	Coastal area

Table 1.5 Levels of noise pollution in general

Locality/Source	*Noise level, dB*
Garden	20
Bed rooms	25
Libraries	30
Living rooms	40
Conversation	60
Business office	67–70
Average street traffic	80–85
Heavy truck traffic	90
Pneumatic chipper	100
Pop music	110
Jet take off (at 100 m distance)	125
Jet engine (25 m distance)	130
Threshold of pain	140

has carcinogenic properties causing lung cancer. The figures refer to pollution noted two-three decades ago. The present level of pollution is 4-5 times the figures cited.

Noise is another source of pollution. Noise is measured in decibels *i.e.* dB. Table 1.5 list levels of noise in general while Table 1.6 gives noise in industrial town and Table 1.7 presents noise level in Mumbai in various locations, which is domestic noise. The human being can tolerate on an average 45-60 dB without much discomfort, however, beyond this limit it causes mental fatigue, irritation and other health hazards.

Table 1.6 Noise levels in two typical industrial towns in India

Visakhapatnam (Andhra Pradesh)		*Rourkela (Orissa)*	
Location	*Noise level (dB)*	*Location*	*Noise level (dB)*
Aseelemetta	81	Near Court	85.5
Gopalpatnam	82	Mangal Bhavan	86.1
B.H.P.V. Junction	75	Daily Market	83.2
Gyanapuram	78	Near Nala Road	76.3
Gurudwara Junction	77	Old Bus Stand	80.5
Dondaparthi	76	Madhusudan Market	84.8
Old Kotha Road	73	New Bus Stand	84.5
Ram Nagar	72	Near UCO Bank	76.6
Siddartha Nagar	76		
Thadichetlapalam	75		

Table 1.7 Noise level during festivals and normal times in Mumbai

S. No.	*Locality*	*Noise level (dB)*	
		In festival	*Normal times*
1.	Prabhadevi	101	70
2.	Pedder Road	108	75
3.	Opera House	98	70
4.	Tejpal Ground	90	68
5.	Goregaon (Dindoshi)	81	50
6.	Goregaon (Charkop)	88	55
7.	Colaba	57	40

1.4 SOURCE OF AIR POLLUTION

We have various sources for air pollution. These are burning of coal and coke, industrial emissions, commercial activities, thermal power generation unit and last but not the least is transportation. Table 1.8 lists various sources of atmospheric pollution (Fig. 1.1). Table 1.9 indicates as to how much air pollution is caused by burning sulphur containing fuels. Though wood causes zero pollution, its calorific value is poor, while fuel oil produces pollution due to SO_2 but it has high calorific value.

Transportation is a major source of pollution causing the highest pollution in metropolitan cities in India. Table 1.10 lists the population of two wheelers in various cities of Maharashtra, while Table 1.11

Table 1.8 Source of growing atmospheric pollution in developed countries

Source	*1973*	*1978*	*1983*	*1986*	*1990*
Domestic sources	34.2	43.5	47.12	53.58	55.00
Commercial point	4.92	6.15	6.34	6.95	7.00
Industrial estates	43.8	51.2	56.4	64.4	70.0
Power generation	12.2	12.2	48.94	46.95	40.0
Transportation	39.8	64.6	10.63	13.14	15.6
Others	40.0	64.6	10.64	13.15	15.0

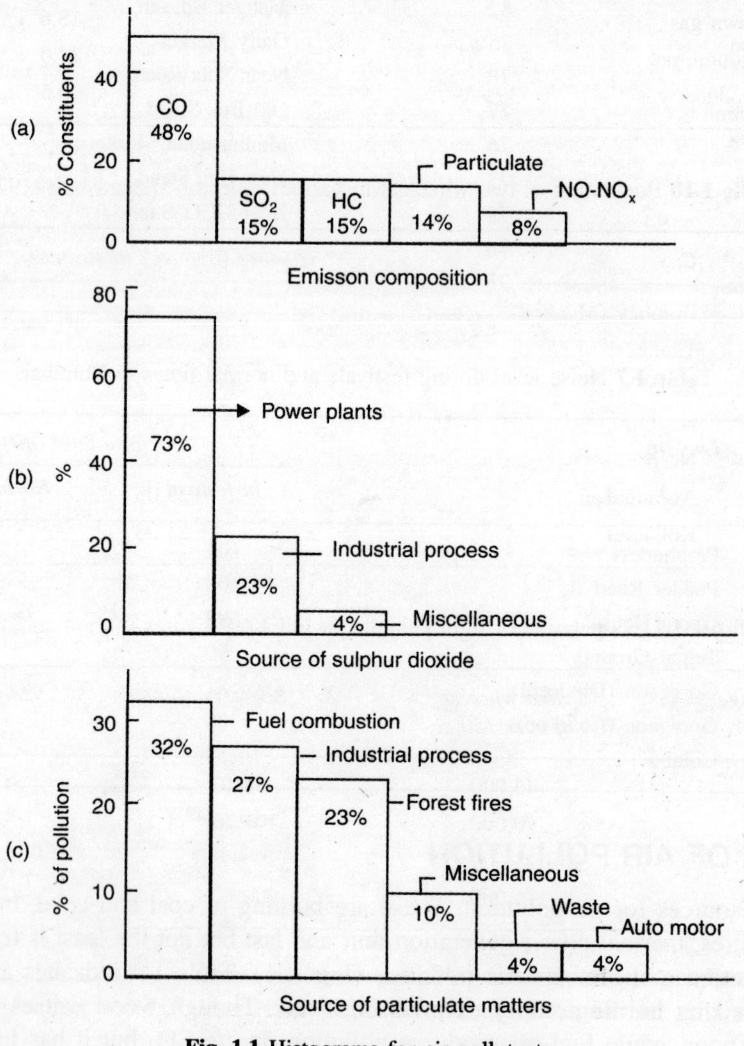

Fig. 1.1 Histograms for air pollutants

Table 1.9 Amounts of sulphur in fuels emitting SO_2 gas

Substance	*% S*	*Calorific value*
Wood	0	18
Peat	1	18
Lignite	1.3	23.2
Bituminous fuel	1.6	25.5
Semibituminous fuel	1.2	36.0
Anthracite	1.0	34.0
Coke	<1.0	29.0
Crude oil	0.5–3.0	–
Fuel oil	2–4	43.5
Coal tar fuel	0.5	37.8
Town gas	–	18.6
Natural gas	Trace	34.0
Producer gas	–	4.5

Table 1.10 Population of two wheelers in various cities of Maharashtra (1970)

City	*Number (approx.) (thousands)*
Bombay (Mumbai)	200.00
Thane	7.00
Pune	150.00
Nagpur	77.00
Nasik	14.00
Aurangabad	29.00
Kolhapur	21.00

Table 1.11 Population of light vehicles in India (all figures in thousands)

Year range	*Two wheeler of over 150 cc*	*Scooters*	*Mopeds*
1970-75	43,000	58.40	11.70
1975-80	70,000	101.70	36.20
1980-85	11,600	210.00	105.00
1985-90	175,000	400.00	425.00

lists overall production of light vehicles in India, which are responsible for vehicle emissions. The growth rate shows that as the activity of production increases, vehicle pollution also rises. The photochemical smog which is obtained by combination of smoke and fog contributes several pollutants to air (Figs. 1.2 and 1.3) e.g. CO, CO_2, NO–NO_x, SO_2, Hydrocarbon (HC), etc., as listed in Table 1.12. A typical petrochemical plant in Gujarat has shown different levels of air borne (SPM) particulate in the atmosphere (Table 1.13). The thermal power plants in India are responsible for causing extensive air pollution. A typical thermal power plant in the capital of India throws daily 80 tons of flyash and several tons of SO_2 in the atmosphere. Amongst Indian cities, Delhi and Kanpur have the highest dust fallout, Mumbai emits highest SO_2 emissions, while Kolkata contributes its share by emitting largest proportion of NO–NO_x in the air.

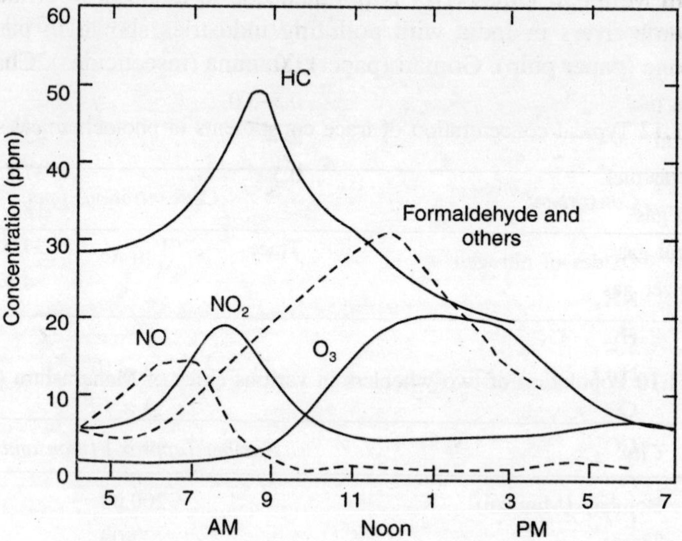

Fig. 1.2 Different pollutants in photochemical smog

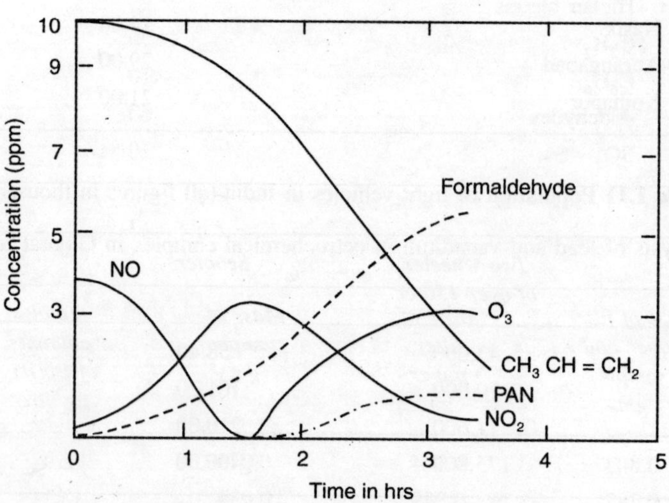

Fig. 1.3 PAN and other pollutants in smog

1.5 WATER POLLUTION

Water pollution has been studied in great details. The indiscriminate disposal of water after use in the form of waste causes water pollution. The tragic incident of Minnamata in Japan is well known. A paper factory using mercury compounds carelessly dumped its waste effluents into the sea, it formed $(CH_3)_2Hg$ and $(CH_3CH_2)_2Hg$, which in turn was consumed by the sea fish. The Japanese people who consumed such fish showed symptoms of mercury poisoning like gingivitis, vomiting, fever, diarrhoea, paralysis of extremities, etc. There were several instances of marine flora and fish dying in the sea, on account of deoxygenation of water, perhaps due to thermal pollution. Most of the rivers in India are polluted due to industrial activity. In Mumbai, Ulhas river is polluted due to disposal of effluents from rayon and dyestuff industries. Some rivers in India with polluting industries shown in parenthesis, are listed: Ganga (jute, sugar), Sone (paper pulp), Gomati (paper), Yamuna (insecticides), Chaliyar (rayon waste),

Table 1.12 Typical concentration of trace constituents in photochemical smog

Constituent	*Concentration, ppm*
Oxides of nitrogen	20
NH_3	2
H_2	50
H_2O	2×10^6
CO	4×10^3
CO_2	4×10^4
O_3	50
CH_4	250
Higher paraffins	25
C_2H_4	25
Higher olefins	25
C_2H_2	90
C_6H_6	10
Aldehydes	60
SO_2	20

Table 1.13 Analysis of lead and vanadium in petrochemical complex in Gujarat refinery campus

Period hr (interval)	*Mass of filter paper and sample gm*	*Mass of filter paper gm*	*Mass of sample gm*	*Total amount of Pb(II) µg/litre*	*Amount of V(V) mg/litre*
18	0.803	0.724	0.079	36.67	0.73
4	0.787	0.749	0.038	13.78	1.93
6	0.789	0.750	0.039	7.63	1.52
16	0.771	0.719	0.052	3.43	1.14

Kaveri (rayon, sugar), Godavari (paper, small-scale industries), Mahi (dyestuff), Mullamutha (antibiotics), Brahmaputra (black liquor), Juhari (fertiliser waste), Patalganga (organic chemicals) and Valdhun (disposal of dye intermediates). It is pardoxial to note that holier is the river, more it is polluted. It is bound to be polluted due to disposal of industrial effluents in the streams that feed the river. The Government of India had constituted the Ganga River Authority to keep the Ganges clean. We need such authority for all principal rivers in India. There exist similar organisations in most developed cities of the world, e.g. Thames River Authority in England.

1.6 INDUSTRIAL EFFLUENTS

The water after it is used once for industrial purpose cannot be reused for the same purpose without treatment. Such water which emerges out after use from industries is called as the industrial effluents. Such effluents have no definite composition, as anything which is not required is carelessly dumped into its stream. Such unwanted disposable material is contributed by chemical firms, food and beverage industry, textile and apparel industries, electronics and electrical material industries, or thermal power plants. The quality of such water is characterised by the study of its various physical, chemical and biological properties. The disposal of arsenic compounds by paper industry, or toxic chemicals by fertiliser plant had caused serious problem in western India. In many places, town authorities have not hesitated to let up sewage into streams or brooks or rivers, presuming that dilution would solve the pollution problem. Unfortunately, biological matter and bacteria such as coliforma, streptococcii, fecal coliforma, crenthronix, are not eliminated by mere dilution. Such water needs chemical treatment, like disinfection. This would also kill pathogenic bacteria.

1.7 POLLUTION DUE TO DOMESTIC EFFLUENTS

Very few cities in India have regular sewage treatment plants. Even major cities like Mumbai, Kolkata do not have plants to treat all sewage and sludge. Further most of our villages are fitted with antiquited treatment plants or septic tanks. The growth of slums in bigger metropolitan cities like Mumbai, Kolkata, Delhi are posing alarming problems of sewage disposal. In fact, the greatest problem that Ganga Water Authorities had faced during their initial working was the tackling of untreated and indiscriminately disposed off sewage. The untreated domestic sewage causes serious health problems. Even the cities in China with largest population are clean, because they have efficient sewage disposal network. The release of untreated sewage in sea causes serious problem with marine life, e.g. a huge amount of sardines in Mumbai sea were killed a decade ago (1958) due to release of sewage from a nearby industrial plant without proper treatment. The oil pollution deteriorates the aquatic marine environment, as several grades of oil are nondegradable. The oil slick caused due to Gulf War had caused enormous problem for India and neighbouring countries.

1.8 POLLUTION DUE TO PESTICIDES

Pesticide residue is the definite source of pollution of land and soil. A list of such pesticides in use is quite lengthy. The most important ones like DDT, malathion, para-malathion, aldrine, dialdrins, cause a serious problem of land pollution. The characterisation and determination is not simple. Some pesticides have deleterious effect on health. Several of them are carcinogenic and cause long term harmful effects upon health and hence need urgent analysis.

1.9 PROBLEM OF SOLID WASTE

The liquid waste was thoroughly investigated by environmentalists, but that is not the case with solid wastes. As a matter of fact with greater industrialisation, the production of colourful cartons, boxes and packings for food, beverages, drugs and edible oil, we have an alarming problem of solid waste disposal. The problem is further aggravated due to nondegradable nature of plastic and polymeric materials. A survey was made decades ago in the city of Mumbai on a typical solid waste composed of garbage, paper, glass, plastic, etc. It was found that animal waste (1500 tons) constitutes the largest share followed by mineral waste (1100 tons), agricultural waste (550 tons), with household waste of (250 tons) per year. In comparison, paper (30 tons) and plastic waste (4 tons) was very less. The methods for disposal of such solid waste are tedious. They are physical compositing incineration and land filling. All these methods available for their disposal are expensive. Table 1.14 gives percentage of solid waste in different towns in India as well as in typical cities of America, England, Japan and Switzerland. Such waste had all components of nature. From Table 1.14 it will be noted that the highest garbage was encountered in city of Pune, while paper and glass was disposed heavily in American cities, rags were extensive in Japanese solid waste, while Mumbai headed the list for having maximum of non-degradable plastic waste.

Table 1.14 Disposal of solid waste (%)

	Garbage	*Paper*	*Glass*	*Rags*	*Plastic*
USA	5.0	54.4	9.1	2.6	1.7
UK	13.0	50.0	6.0	3.0	–
Switzerland	14.5	33.5	8.5	3.0	2.0
Japan	36.9	24.8	3.3	7.1	2.2
India (average)	31.7	0.25	1.0	3.6	0.2
Nagpur	31.4	0.2	0.07	0.3	0.2
Pune	67.6	8.7	0.6	1.6	0.7
Kanpur	35.8	4.5	1.0	5.7	0.8
Kolkata	45.1	3.2	0.4	3.6	0.6
Mumbai	52.3	12.0	–	4.3	2.7

1.10 METAL POLLUTANTS

We have several elements in the periodic classification, of this few are metals. Fortunately, only 17 of these metals are toxic. Of these some are commonly utilised metals, while other are less frequently used and are termed as rare metals. The important toxic metals are: Be, Os, Pt, Hg, T1, V, Te, Co, U, Se, Cd, Pb, As, Sb, Cr, Ni, Zn, Mn. Their toxicity varies with threshold limiting value (TLV) = 2 $\mu g/m^3$ to 5000 $\mu g/m^3$. This is the maximum or permissable tolerable limit for a human. The different metals have different degrees of toxicity. Mercury causes tremor, renal disorder; lead causes plumbism, gastric and colic problems; cadmium is a cumulative poison, it causes anaemia and hepatic disorder; vanadium causes wheezing dyspnea; arsenic causes skin cancer and dermatitis; osmium causes halo around eyes and bronchitis, while tellurium consumption gives garlic odour to breath and sweat. Some of these metals are carcinogenic. They lead to cancer of lung, nose, liver pleura, thyroid. Sources of these metal pollutants are varying, some of these pollutants are encountered in water and air, effluents, solid waste and sewage. A reliable method for their characterisation and quantitative determination is needed. They are emitted out from various sources, e.g. waste water from electroplating industry (Cr, Ni), air emission from fluorescent lamps (Cd, Be), paint pigment waste (Pb, Cr), auto-exhaust emissions (Pb, Te). It is imperative to understand their toxic effects upon health, before devising the method for their analysis and control.

1.11 ENVIRONMENTAL CARCINOGENS

Apart from metals, several of the organic compounds encountered in environment cause cancer. Their chemistry is complex and as such it is really difficult to ascertain the mechanism of interaction. 3-4 Benzopyrene (BaP) is one such deadly compound, which causes cancer of lung in industrial areas. BaP exist in auto-exhaust emissions, incompletely burned tobacco and cigarettes. In Table 1.3 various levels of BaP in different towns in the world are indicated; a similar list for Mumbai in Table 1.4 is also available. In a typical winter, the central Mumbai region showed highest level of BaP as 400 $\mu g/m^3$ due to location of coal gas manufacturing unit in the Lalbaug one time a labour area of Mumbai. The south Mumbai had 3.5 $\mu g/m^3$ while north Mumbai had 1.4 $\mu g/m^3$. The eastern suburbs of Mumbai had varying limit of 15-36 $\mu g/m^3$ due to location of several brick and kiln factories. A consumption of 1000 cigarettes emits 2-122 $\mu g/m^3$ BaP, which is sufficient to kill a large human population. A survey of the workers in England working in gas plant, gas retorts with coal or producer gas, crane operators, printers, chimney sweepers, asphalt workers, street cleaners showed that they were largely affected by cancer of lung. A large number of organic compounds like benzidine, β-napthylamine, 4-aminodiphenyl, 4-dimethylaminobenzene, 4-nitrodiphenyl are all confirmed as carcinogenic compounds.

1.12 CONTROL OF POLLUTION

The environmental pollution is the most burning problem, requiring immediate attention by the society. A survey of basic needs of man in America revealed the order of preference that people wanted consideration of issues such as environment, odour, bus fare, parks, employment, noise, medical care, schools, roads, garden and home. However, they gave top priority to the stoppage of degradation of their environment in preference to other important issues. In England there was greater awareness since 1952. With the enforcement of clean air act (1968), air was cleaned, while in the USA, Environmental Protection Agency (USEPA) was established. In India since 1962 Maharashtra was the first state to enact Water Pollution Act, followed by preparation of the draft proposal for Air Pollution Control in 1980. Fortunately the Central Government is aware of the problems. It enacted comprehensive environmental protection bill in 1986 to include legislation on abatement of noise pollution. Mumbai city was the first to introduce a bill for transportation of hazardous chemicals (1988). In spite of this we have tragedy of spilling of chemicals on roads. We have now boards for abatement and control of air and water pollution in every state of India, including union territories. However, these control measures shall not be effective, unless environmental education is imparted compulsarily from primary school to college and university level. We have the Department of Environment and Forest of the Central Government under the Ministry of Environment. Agencies to control environmental pollution will be effective only if we properly monitor pollution and implement laws governing pollution strictly.

1.13 ENVIRONMENTAL POLLUTION ANALYSIS

What is the need for monitoring environmental pollution and analyse the data thus collected? This is a vexing question. One cannot think of control measures unless one knows what is the extent of pollution?

A lot of awareness can be created about pollution, a great awareness is already there in urban society in large cities on pollution, but we cannot effectively control pollution in the absence of authentic data about air pollution, water pollution, water treatment, industrial effluents, sewage management, land and pesticides pollution or abatement of noise pollution. Such measures can be effective only if we have reliable data on exact levels of pollution. Such reliable data can be obtained by monitoring of environmental pollution at site. The knowledge of levels of pollution will enable us to use and adopt effective, reliable and economic measures for their control. To control pollution, which is growing in alarming proportion, the most effective way is to analyse environmental pollution and then adopt a strategy for its control. Therefore, in this book it is envisaged to consider ways and means to analyse environmental pollution. The modern methods for analysing environment pollution shall be first considered. This will be followed by overall consideration of industrial toxicology and hygiene due to inorganic and organic pollutants. It will be followed by discussion on atmospheric pollution and its measurement, potable water treatment and analysis of water as well as the analysis of industrial waste effluent. Finally, the methods for characterisation and analysis of sewage and sludge, pesticide residue and noise pollution shall be considered. A detailed account is presented in few chapters for the analysis of air pollutants as well as water polluting materials with all minor details on experimental procedure in the separate chapters.

ADDITIONAL READINGS

1.1 R.A. Horne, *Chemistry of Environment*, Wiley-Interscience (1978).

1.2 A.S. Boughey, *Man and Environment*, 2nd Ed., MacMillan Publishing Co. Inc., New York (1975).

1.3 E.T. Chanlett, *Environmental Protection*, 2nd Ed., McGraw-Hill Kogakusha Ltd. (1979).

1.4 I.L. Marr and M.S. Cresser, *Environmental Chemical Analysis*, International Textbook Co., New York (1983).

1.5 C.N. Sawyer and P.L. McCarty, *Chemistry for Environmental Engineers*, 3rd Ed. McGraw-Hill Book Co., (1978).

1.6 A.K. Sen, *Environmental Management and Planning*, New Age International Publishers (P) Ltd., New Delhi (1988).

1.7 S.M. Khopkar, *Environmental Pollution, Monitoring and Control*, New Age International Publishers (P) Ltd., New Delhi (2006).

1.8 K.C. Sahu, *Proceedings of Symposium on Role of Earth Sciences in Environment* (1987). I.I.T. Press.

1.9 *Water Treatment Handbook*, 5th Ed., Degremont (1979).

1.10 A.K. De, *Environmental Chemistry*, 3rd Ed., New Age International Publishers (P) Ltd., New Delhi (2005).

1.11 R. Nash, *Environment and Americans—a problem of priorities*, Holt Rinehart and Winston (1972).

1.12 H.B. Elkins and L.D. Pagnotto, *Industrial Health Hazards in Treatise of Analytical Chemistry*, Ed. I.M. Kolthoff, P.J. Elving, Part III, Vol. 4, Section B, Wiley-Interscience (1980).

1.13 J.W. Moore and E.A. Moore, *Environmental Chemistry*, Academic Press, New York (1976).

1.14 K.J. Irgolic and A.E. Martell, *Environmental Inorganic Chemistry*, VCH Publishers Inc., USA (1985).

CHAPTER 2

Modern Methods of Pollution Analysis

2.1 INTRODUCTION

With the advent of Science & Technology, newer instrumental methods have come to forefront for analysis of pollutants. In earlier years when someone was interested in analysing SO_2 from air, he had to follow classical lead candle method involving gravimetric analysis. Even the analysis of dissolved oxygen had to be carried out by Winkler's method involving redox titration. The hardness of water was ascertained by complexometric titrations. The main disadvantage of these methods, although they were good, was long time of analysis, level of analysis and method restricting only single component at a time for quantitative analysis. These methods were therefore unsuitable for continuous monitoring and characterisation of pollutant.

However with the discovery of absorption spectroscopy utilising UV, visible, IR spectroscopy, it was possible to carry out analysis of sample at microgram concentrations. Further with the discovery of atomic absorption spectroscopy as well as plasma spectroscopy one could analyse as many as six constituents at a time in one single operation. Further, these methods needed no special procedure to remove interfering ions. With discovery of NMR, ESR it facilitated the characterisation of organic pollutants while both the process of separation as well as of the analysis was accomplished in single operation by hyphenated GC-MS, GC-ICP-AES or MS-MS i.e., tandem mass spectrometry. With the advent of time electroanalytical methods such as Conductometry or Polarography not only facilitated the quantitative analysis but also the identification of the pollutant from the knowledge of ($E_{1/2}$) half-wave potential. Within chromatography technique, GC, HPLC, facilitated separation but new methods like ion chromatography techniques, capillary electrophoresis or supercritical fluid chromatography facilitated the task of ion-speciation as well as quantitative analysis at very low concentration. Radioanalytical methods were not of much use.

However with the need of analysing pollutants in outdoor environment, portable GC, Flow injection analysis, continuous monitoring of conductivity came into forefront. These methods facilitated the in situ or outdoor analysis with greater accuracy and speed. The technique of nephelometry was good for analysis of SPM or SO_4^{2-} while molecular luminescence spectroscopy involving chemiluminescence promoted excellent analysis of $NO\text{-}NO_x$, SO_2. The CO from atmosphere could be best analysed by (ND-IR) non-dispersive IR spectroscopy. The most interesting part was in spite of progress in instrumental analysis the classical methods like volumetry (for DO) or the gravimetry (SPM) still continued to receive patronage from environmental analyst.

The control measures for the abatement of pollution cannot be properly devised unless we are aware of the extent or level of pollution. Such level can be ascertained by precise measurement and analysis of the environmental pollution. There are numerous methods for the measurement of pollution. They consist of classical methods of analysis like gravimetric or volumetric methods. These methods are generally applicable at milligram concentrations. Several of the pollutants are present in environment at microgram levels, e.g. dust and particulate matter, while few metal pollutants like beryllium are having their low threshold limiting value as 2 $\mu g/m^3$. As such spectral methods or electro analytical methods are best suited in such analysis. Many a times the pollutants are found in association with one another. Under these circumstances determination becomes more complicated. One has to resort to various separation techniques, like chromatography for the isolation of these pollutants. In this chapter, an endeavour is made to present an overview of various methods utilised for the analysis of the various pollutants. It is not envisaged here to describe what the said method is? It is presumed that the reader is quite conversant with the basis of these methods. Only, an attempt would be made to highlight the applications of these methods in environmental pollution analysis.

2.2 METHODS FOR QUANTITATIVE ANALYSIS

We have several methods for quantitative analysis of pollutants. These methods are applicable when the pollutant is present at milligram (10^{-3} g) to femtogram concentration (10^{-15} g). Table 2.1 lists some methods. The classical gravimetric and volumetric methods can analyse a pollutant up to 10^{-4}g concentration, while analytical absorption or emission spectroscopic method can analyse pollutants to the level of nano gram (*i.e.* 10^{-9} g) concentration. Other methods involving light scattering or molecular luminescence spectroscopy operate up to microgram levels. The radioanalytical or electroanalytical methods can work in the range 10^{-7}–10^{-9} concentration. At this stage, methods used for pollution analysis shall be reviewed. No attempt is made to provide details of these instrumental methods of analysis. In fact it is assumed that reader is quite conversant with these modern methods of analysis. Only endeavour is made to pinpoint the specific method for analysis of particular pollutant with due emphasis on the basic principles involved in such analysis.

Table 2.1 Methods of quantitative analysis

Names of method	*Level of analysis*	*Range of analysis (gm)*
Gravimetry methods	mg	10^{-1}–10^{-3}
Volumetry methods	mg	10^{-1}–10^{-4}
Molecular absorption spectroscopy (UV/visible/infrared)	μg	10^{-3}–10^{-6}
Emission spectroscopy	ηg	10^{-6}–10^{-9}
Atomic absorption spectroscopy	ηg	10^{-6}–10^{-9}
Molecular luminescence methods	ηg	10^{-3}–10^{-9}
Light scattering methods	μg	10^{-3}–10^{-6}
Radioanalytical techniques	ηg	10^{-6}–10^{-9}
Electroanalytical methods	μg	10^{-3}–10^{-6}

2.3 GRAVIMETRIC METHODS

When the pollutant is present in the environment at milligram levels, this method is used e.g., the lead candle method for analysis of SO_2 from air is based upon this principle. A candle of PbO_2 on exposure to SO_2 forms $PbSO_4$, which is weighed to find out SO_2 in the air. The method involves simple weighing of PbO_2 and $PbSO_4$ before and after exposure respectively, to find the amount of SO_2 present. The other example is the determination of suspended particulate matter from air. We resort to weighing of particulates deposited on millipore filter paper. This residue is collected by sampling procedure involving high volume sampler, however in analysis of water pollution we rarely use gravimetric methods.

2.4 VOLUMETRIC METHODS

These methods are classified into four groups involving acid-base, redox reactions, precipitation and complexometric titrations. These are used for monitoring of air as well as water pollution, e.g., the acid-base titration with conductance measurement is used for the analysis of SO_2 from air or analysis of boron from sample of soil. The simple acid-base titration is also used for ascertaining the acidity or alkalinity of sample of water. The application of redox titration is used in the Winkler's method for the determination of dissolved oxygen (DO) from water. It is based upon redox reaction of manganese salt which is reacted in the presence of oxygen with KI to liberate iodine which is titrated with standard $Na_2S_2O_3$. Many of the organic pollutants from air can be determined by redox titrations. In all these cases, analysis is restricted to milligram level. The precipitation titration has wide ranging applications in water analysis, e.g., the determination of the dissolved chloride from sample of water can be accomplished by Volhard's method or Mohr's method. The complexometric titrations has maximum utility in pollution analysis. For instance, the hardness of water can be ascertained by titration of calcium or magnesium with EDTA (disodium salt), with suitable metallochromic indicators like Patterson Reed indicator. Similarly, level of iron or manganese from polluted sample of water can be ascertained by complexometric titrations. In comparison to water analysis, the complexometry technique has limited applications in analysis of air pollution. In spite of great advancement of instrumentation and with development of DO (dissolved oxygen) analysers, the Winkler's method with redox titration is still considered to be the best and reliable method for analysis of DO, BOD (biological oxygen demand) and COD. In fact chemical oxygen demand (COD) for water can be ascertained by redox titrations involving $K_2Cr_2O_7$, H_2SO_4, KI and $Na_2S_2O_3$.

2.5 ULTRAVIOLET AND VISIBLE ABSORPTION SPECTROPHOTOMETRY

This is perhaps the most popular technique, which has maximum utility in the pollution analysis. The principal reason for such popularity is the easy availability of colorimeters or spectrophotometers. The principle of visual colorimetry also forms the basis of Ring Oven technique. Table 2.2 indicates how UV Spectroscopy can be used for the characterisation of organic pollutants. The task is complicated, when one has to characterise organic compounds in waste water. A peak at 280 nm indicates presence of azo group; a peak at 186 nm shows presence of carbonyl group; peaks at 240–280 nm shows possibility of the presence of aromatic compounds like napthalene or styrene, respectively. Visible spectrophotometry has, in fact, made the great revolution. In air pollution analysis, SO_2 is determined with p-roseaniline (PRA) method or $NO–NO_2$ from air is determined by Saltzman method; H_2S is analysed spectrophotometrically by molybdenum-blue method; in water pollution analysis. The metal pollutants like iron (1-10 phenanthroline), Chromium (sym diphenylcarbazide), Manganese (formaldioxime); nickel (Dimethylglyoxime), zinc (dithizone) methods are commonly used, where a

Table 2.2 Ultraviolet absorption of pollutant with chromophores

Chromophore	*λ max nm*	*Molar absorptivity (ε)*	*Nature of transitions*
Alkene	177	1.3×10^4	$\pi \rightarrow \pi^*$
Carbonyl	186	1.0×10^3	$n \rightarrow \alpha$ or $n \rightarrow \pi^*$
Amide	214	60	$n \rightarrow \pi^*$
Azo	339	5	$n \rightarrow \pi^*$
Nitro	280	22	$n \rightarrow \pi^*$
Olefin	184	1.0×10^4	delocalise π^*
Ketone	280	27	$n \rightarrow \pi^*$
Methanol	184	1.5×10	$n \rightarrow \alpha^*$
Methyl chloride	173	2×10^2	$n \rightarrow \alpha^*$
Benzene	204	7.9×10^3	$\pi \rightarrow \pi^*$
Toluene	207	7.9×10^3	$\pi \rightarrow \pi^*$
Phenol	211	6.2×10^3	$\pi \rightarrow \pi^*$
Aniline	230	8.6×10^3	$\pi \rightarrow \pi^*$
Naphthalene	286	93×10^3	$\pi \rightarrow \pi^*$
Styrene	244	1.2×10^4	$\pi \rightarrow \pi^*$

chromogenic ligand is reacted with metal pollutant to form coloured complex. So long as the system adhers to Bear's law (viz. $A = abc$, where A = absorbance, a = absorptivity, b = optical path, c = concentration), one can use this technique for analysis of majority of the metal pollutants in air and water. There are numerous examples which would be explained in subsequent chapters (Fig. 2.1) along with details of experimental conditions.

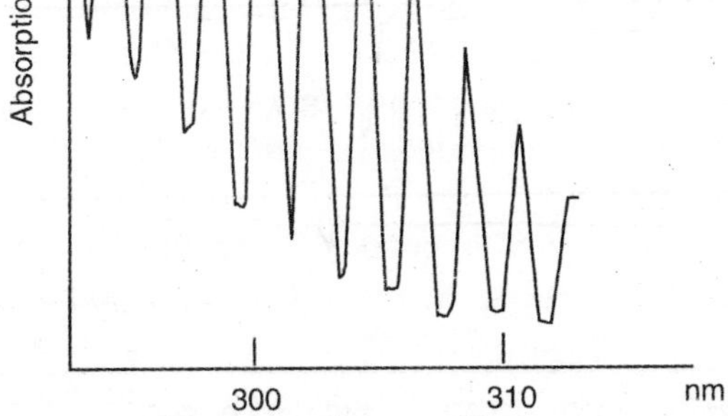

Fig. 2.1 UV Spectra of SO_2

2.6 INFRARED AND RAMAN SPECTROSCOPY

While characterising the organic pollutant from air or water, only one technique may not furnish definite information. At a time, it is necessary to use more than two methods to ascertain the presence of certain pollutants. Infrared spectra originates on account of vibrational frequency of functional group

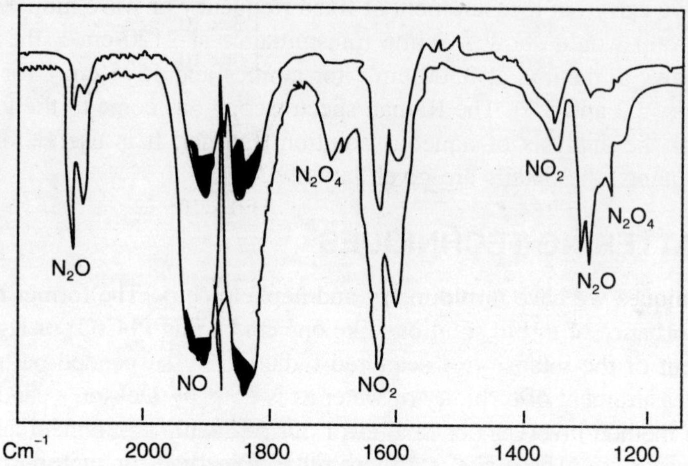

Fig. 2.2 IR spectra of NO–NO$_x$

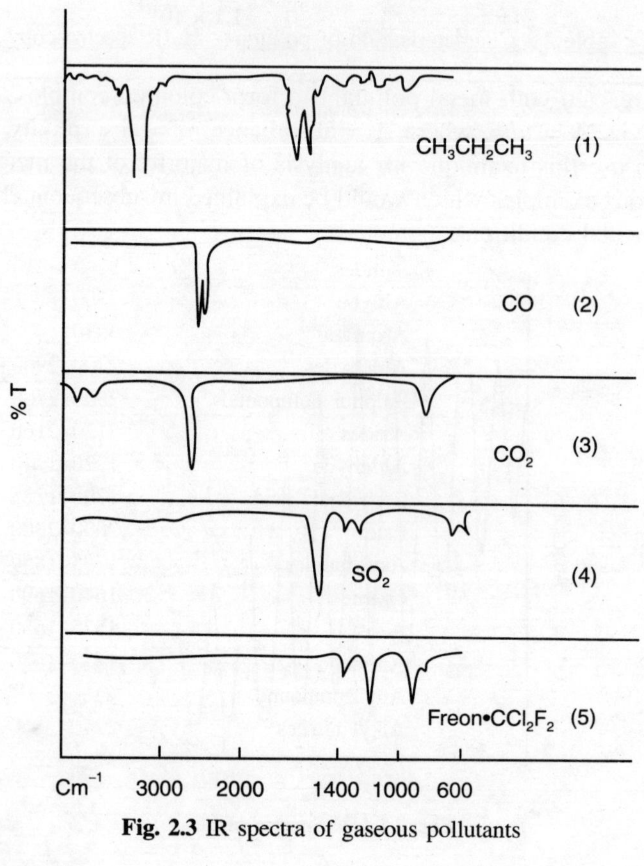

Fig. 2.3 IR spectra of gaseous pollutants

in a molecule. As such every functional group has characteristic frequency. A typical group frequency can identify a molecule in aquatic environment or air. Table 2.3 lists such characteristic frequency of various functional groups. e.g. a peak at 1720–1740 cm^{-1} indicates the presence of >CO group. In fact, this had formed the basis for analysis of carbon monoxide from air by non-dispersive infrared spectroscopic technique (NDIR). It is discussed in subsequent chapters. We call this technique as non-dispersive, because we carry out measurement at fixed frequency or wave number, e.g. 1800 cm^{-1}. The presence of azide group would show maxima transmittance at 2120 cm^{-1}, the effluent sample from dyestuff industry shows a maxima at 3500 cm^{-1} for amines and 1575 cm^{-1} for azo group present in sample of water (Figs. 2.2 and 2.3). The Raman spectroscopy has come to the forefront only recently. It is more useful for the analysis of aqueous solution. Further, it is useful for remote sensing and detection of air pollutants, the details are given latter.

2.7 LIGHT SCATTERING TECHNIQUES

Amongst these techniques we have turbidimetry and nephelometry. The former technique involves the measurement of absorbance of turbid solutions like one containing $PbCrO_4$ or $HgCrO_4$, while the latter involves measurement of the intensity of scattered radiation by suspended particles e.g. the classical method involving measurement of turbidity of water as is done by Jackson's candle technique. The unit involved is JU. This method involves personal error during such measurement and hence is unreliable. Instead a nephelometry with NTU units is a more reliable method for such purpose. The best method for the analysis of SO_4^{2-} from water is by measurement of $BaSO_4$ by nephelometry technique.

Table 2.3 Characterisation of pollutants by IR spectroscopy

Group	*Environment*	*Frequency (cm^{-1})*
OH	Alcohols	3580–3650
	Acid	2500–2700
NH	Amines	3500
	Amides	3140–3320
CH	Alkyne	3300
	Aromatic	3030
	Aldehydes	2700–2900
SH	Sulphur compounds	2500–2700
$-N_3$	Azides	2120–2160
>CO	Aldehyde	1720–1740
COOH	Carboxylic acid	1700–1725
	Ester	2000–2300
	Acyl halides	1795–1850
NOH	Oximes	1640–1690
–N = N–N	Azo	1575–1630
N–H	Amine	1575–1650
$-(CH_2)n$	Any compound	722
CN	Alkyl nitriles	2240–2260
	Isocyanates	2240–2275

2.8 MOLECULAR LUMINESCENCE SPECTROSCOPY

These methods are termed as chemiluminescence techniques. They have made a spectacular progress in the analysis of gaseous pollutants like SO_2 or NO–NO_x from atmosphere. These are most reliable methods. The analysis of Be (morin, *i.e.*, pentahydroxy-flavanol) Th (Alizarin), Al (quinanzarene) by fluorescence technique is well known. The analysis of organic compounds, e.g. riboflavin, thiamine hydrochloride, quinine from waste water emanating from pharmaceutical industry, is best carried out by fluorescence spectroscopy. In comparison, the phosphorescence spectroscopy has not become popular in environmental analytical chemistry. A relatively cheap set of fluorimeter in India will make this technique more popular in future.

2.9 ATOMIC ABSORPTION SPECTROSCOPY

This technique has made a radical difference in quantitative chemical analysis. The main reason for popularity of this method is the possibility of analysing large number of samples without doing any separation. This is useful for multi-element analysis. It is applicable for micro and nanogram concentration of the metal pollutant analysis. Certain critical analysis of metals like As, Sb, Sn, Se, Te, Bi need hydride generator and one must know what kind of thermal stability the hydrides have. Table 2.4 lists prospective metal pollutants which can be determined by AAS. In spite of the discovery of cold vapour AAS technique for mercury, most of the state pollution control laboratories in India prefer Bindscheiderr's Green for Spectrophotometric method of analysis. A look at sensitivity (Table 2.4) indicates certain metals like mercury and uranium have poor sensitivity. The details of such method are spelled out in great details in subsequent chapters.

Table 2.4 Atomic absorption spectroscopic methods for metal pollutant analysis

Toxic metal	*Wavelength* $(\lambda)_{max}$ *nm*	*Sensitivity* *μg/ml*	*Type of flame used*
Zn	213	0.009	AA
Pb	217	0.11	AA
As	197	0.60	AH
Be	234	0.016	NA
Cd	228	0.011	AA
Cr	357	0.55	AA
Hg	253	2.2	AA
Mn	279	0.21	AA
U	358	113	NA

AA – air acetylene, AH – air hydrogen, NA – nitrous oxide acetylene.

2.10 ATOMIC EMISSION SPECTROSCOPY

This technique includes Flame Emission Spectroscopy, Inductively Coupled Plasma-Atomic Emission spectroscopy (ICP-AES) and conventional emission spectrography. Of these methods, flame photometry is relatively cheap and dependable. Table 2.5 lists metal pollutants which can be analysed by flame photometry. The greatest advantage of this technique, is that, it is relatively a cheap instrument if appropriate filters are available in the market. All determinations can be carried out in visible region of the spectra. A greater accuracy can be sought by standard addition and the internal standard method. Only metals like Rb or Cs can be determined at near IR region, for which appropriate light filters are not easily available.

Table 2.5 Flame emission spectroscopic analysis of toxic metal pollutants

Element	*Wavelength* (λ) *nm*	*Sensitivity* *μg/ml*	*Source of pollution*
Li	670	0.017	Glass, photocells
Na	589	0.003	Water, soap, alkali
K	766	0.009	Soap, photocells, pharmacy
Ca	422	0.013	Milk, plants, water, vegetation
Sr	460	0.041	Fission reaction, soil, leaf
Ba	554	0.20	Vegetables, glass, X-ray

The ICP-AES technique is of recent origin. It is quite expensive but very rapid and reliable. It is useful for the multi-element determination of large number of samples containing metal pollutants. The accuracy obtained is quite satisfactory. An argon torch is used for the purpose of excitation with radio frequency generator. In developed countries, it is used as the hyphenated technique by coupling it with HPLC or a similar suitable technique, no doubt it is expensive tool.

X-ray fluorescence also provides the same result with greater advantage.

2.11 RADIOANALYTICAL METHODS

Amongst nuclear methods of the analysis, neutron activation analysis and isotope dilution methods are most popular. However, the greatest limitations of these methods is non-availability of radioactive isotope. The neutron source with high flux is available only with Bhabha Atomic Research Centres, with restricted access to Government agencies. In more advanced countries like Canada and France, such facility is readily available freely through private enterprises. In India private agencies may be permitted to handle radio active isotopes with the recent signing of nuclear treaty with United States of America.

In neutron activation analysis, a sample to be analysed is irradiated with slow neutron source to form unstable isotope, with mass number greater by one unit. Such unstable isotope undergoes transmutation (n, γ) reaction to revert to its original isotope, with emission of γ-radiation. These emitted gamma (γ)-radiation are proportional to the amount of metal irradiated. By comparing known weight and the activity, it is possible for us to ascertain weight of the unknown by measuring its activity on irradiation. Many metal pollutants like Fe, Mn, Cu, Sb, Bi, Cd, Zn, V can be easily determined at nanogram concentrations by neutron activation analysis technique (NAA).

The second largely used technique is the isotope dilution method (IDM). This method involves the dilution of active isotope within active material. Consequently, the reduction in activity is measured and from this indirectly the concentration of material is evaluated.

2.12 ELECTROANALYTICAL TECHNIQUES

These techniques comprise of four major groups, viz. potentiometry, conductometry, coulometry and polarography (Table 2.6). Potentiometric method involves measurement of the potential. The indirect application of this technique in pollution analysis is the measurement of pH of solution. The technique of chronopotentiometry can be also used easily for the analysis of the metal pollutants. The potentiometric method can be used for indirect determination of the metal pollutant. The technique of conductometry is used to ascertain quality of water. Apart from measuring purity of water, conductometric technique is used for the continuous measurement of SO_2 from air. This technique is described in details in the subsequent chapters. The coulometry technique is based upon the Faraday's Law of Electrolysis. The electrical generation of bromine is compared with its generation of bromine from reaction of KBr with SO_2 by reduction. This led to monitoring of SO_2 by coulometry. The anodic stripping voltametry (ASV) is an instance of the specialised category of the coulometric method. In recent years, ASV has become very popular. This method involves two cells for reduction, while ion selective electrode (ISE) is an instance of potentiometric technique. With the easy availability of specific ion electrodes, the analysis of anions from water is a matter of minutes. A great details on ASV is provided in latter chapters. With progress in solid state and enzyme electrode chemistry as well as electrodes fitted with liquid ion exchangers, or crown ethers the analysis of cationic species is reduced to single measurement.

Table 2.6 Limits of analysis by electroanalytical methods

Technique	*Limit of analysis*
AC polarography	$10^{-4} - 10^{-5}$
Classical polarography	$10^{-5} - 10^{-6}$
Square wave polarography	$10^{-6} - 10^{-7}$
Pulse polarography	$10^{-7} - 10^{-8}$
Anodic stripping voltametry	$10^{-7} - 10^{-9}$

Amongst electroanalytical techniques, polarography or voltammetry has the greatest advantage over other methods. Table 2.7 summarises important metal pollutants with their half-wave potentials ($E_{1/2}$), which can be analysed by polarography. The distinct advantage of such a technique is the possibility of identification of metal pollutant from the effluent sample. $E_{1/2}$ is the characteristic property of a metal under a set of supporting electrolyte, which helps one to identify a metal. The polarography has distinct advantage for analysing the organic pollutants, e.g. the pesticide residues can be best analysed by this technique with a suitable electrolyte like lithium hydroxide.

Table 2.7 Polarographic characterisation of toxic metals

Metal pollutant	*Half-wave potential ($E_{1/2}$)*
Cd	–0.76 V
Co	–1.46 V
Cu	–0.47 V
Fe	–1.12 V
Pb	–0.76 V
Zn	–1.50 V

2.13 RING-OVEN TECHNIQUE

This technique was largely recommended by SCOPE for the analysis of metal pollutants. This is a relatively cheap method, easy to operate and dependable and useful for semiquantitative analysis. It is best suited for outdoor work, for field analysis and *in situ* analysis, by comparison of rings with standard rings obtained by interaction of chromophoric group with the metal pollutant. It is quite possible to qualitatively analyse and get semiquantitative information of the metal to be analysed. Tables 2.8 and 2.9 list typical results of analysis of metal pollutants from air and water by ring oven technique along with well-known quantitative methods.

Table 2.8 Ring-oven technique for analysis of metal pollutants in Mumbai

Metal ion present	*Found by ring-oven μg/m³*	*Found by comparison method μg/m³*
Pb (II)	340	360
–	220	230
V (V)	78	73
–	62	68
Se (IV)	26	27
–	24	28
Cd (II)	13.5	14
–	16	19

Table 2.9 Ring-oven technique for metal analysis in waste water

S. No.	*Metal*	*Amount taken μg/litre*	*Amount found by ring-oven technique μg/10 ml*					*Amount found μg/litre*
			A	*B*	*C*	*D*	*Average*	
1.	Cd	100.0	0.90	1.00	1.10	0.90	0.975	97.5
		120.0	1.20	1.10	1.10	1.30	1.175	117.5
		130.0	1.30	1.20	1.20	–	1.233	123.3
		140.0	1.40	1.20	1.30	1.40	1.325	132.5
2.	Cr(VI)	25.0	0.20	0.30	0.20	0.20	0.225	22.5
		50.0	0.50	0.40	0.50	0.40	0.45	45.0
		100.0	0.80	1.10	1.00	1.00	0.975	97.5
3.	Hg	75.0	0.70	0.80	0.70	0.60	0.70	70.0
		100.0	1.00	1.10	1.00	1.10	1.05	105.0
		125.0	1.20	1.30	1.10	1.20	1.20	120.0
4.	Pb	10.0	0.10	0.10	0.10	–	0.10	10.0
		50.0	0.50	0.60	0.50	0.50	0.525	52.5
		75.0	0.80	0.80	–	0.80	0.80	80.0
		100.0	1.00	0.90	0.90	1.10	0.975	97.5
		125.0	1.30	1.30	1.20	1.20	1.25	125.0

(Contd.)

S. No.	Metal	Amount taken μg/litre	Amount found by ring-oven technique μg/10 ml					Amount found μg/litre
			A	B	C	D	Average	
5.	V	15.0	0.10	0.10	0.10	0.10	0.10	10.0
		50.0	0.40	0.40	0.50	0.40	0.425	42.5
		60.0	0.50	0.60	0.60	0.60	0.575	57.5
		75.0	0.80	0.80	0.80	0.80	0.80	80.0
		100.0	1.00	1.00	1.10	1.00	1.025	102.5

2.14 SEPARATION METHODS

Before concluding this chapter, it is worthwhile to consider the review of prominent separation methods utilised before quantitative analysis. Amongst separation methods, one thinks of chromatographic techniques. Within these methods HPLC and GC have become most popular with advancement of instrumentation. However, it is worthwhile to consider them on the basis of physical properties, on which such techniques are based.

(a) Adsorption Chromatography

This largely consists of columnar method and gas-solid chromatography. These are based on the phenomenon of adsorption but both techniques have limited applications in environmental separations.

(b) Partition Chromatography

This technique is based upon exploitation of difference in partition coefficients of solute for promoting differential migrations. The important methods under these class are paper chromatography, thin-layer chromatography, gas-liquid chromatography, liquid-liquid partition and reversed phase extraction chromatography. All these methods have extensive applications. Paper chromatography (both ascending and descending) is used for the separation of inorganic as well as organic species, e.g. separation of Hg, As, Pb with ethylacetate butanol mixture as stationary phase and hydrochloric acid as the mobile phases. The TLC is used for characterisation and separation of pesticides residue. The GLC has extensive applications for organic pollutant analysis. GC-MS *i.e.* hyphenated gas chromatography mass spectrometry is very useful for the separation of volatile and thermally stable organic pollutants, specially from air. All hydrocarbons, polymers, etc., from air can be characterised by GC with appropriate detectors like TCD, FID and ECD. Of these flame ionisation detector finds extensive applications in the environmental pollution analysis. Mass spectrometer can have various kinds of analysers like single channel, double channel, time of flight and quadrupole analysers. Of these, the last one is best for analysis of pollutants.

(c) Ion Exchange Chromatography

Both cation as well as anion exchange chromatographic methods are very useful in environmental analysis. The well-known process of deionisation of water is based upon the principle of ion exchange methods. Apart from this, any kinds of separations of anions can be done by anion exchange chromatography. All cationic impurities can be isolated by cation exchange chromatography. Liquid anion exchangers are more or less based upon similar principles of anion exchange chromatography. In addition to the separation of inorganic compounds, the organic compounds like proteins, carbohydrates, amino acids can best be separated by this technique.

(d) Exclusion Chromatography

This technique is based upon the separation of molecules depending upon their geometry. Gel filtration is an example of this technique. But this technique has limited use in environmental analysis, excepting for the separation of biological products.

(e) Electrochromatography

Electrophoresis or electrochromatography has limited applications in environmental pollution analysis. It is commonly used for separation of metals from effluents.

(f) Ion Chromatography, HPLC and SFC

The ion chromatography has come to the forefront only recently. It has substantially helped in characterisation, identification and determination of anions from the samples of water. It is being used presently for the interesting work of chemical speciation of metal pollutants. The chemical speciation is having greater ramification, as the nature of species in the environment will throw light on the toxicity of such metals on human beings. As a matter of fact, composition of Hg, Cr or Fe in water was ascertained by ion chromatography. Sodium salts are used for the elution of ions taken up by ion exchange column. This eluent is subsequently passed on anion exchange column, which in turn generates corresponding acid to indicate rise in conductivity. Such change in conductivity is related to quantum of anions present in the solution.

HPLC or high performance liquid chromatography is a technique and not a separate class of chromatography. Any kind of chromatographic technique discussed above can be carried out with HPLC. What one needs is a good pump with pressure of 1000–3000 psi and a suitable detectors like thermal, ultraviolet or refractive index. This technique has accelerated the pace of the work and facilitated quick separation of organic pollutants. Finally supercritical fluid chromatography is gradually getting popular in pollutant analysis. This technique is of recent origin, which utilises CO_2 as supercritical liquid.

2.15 SOLVENT EXTRACTION SEPARATIONS

This is the most powerful tool for separation. It does not need sophisticated instruments, excepting a good separatory funnel. The separations are largely based upon exploiting the difference in distribution coefficient of solute in two solvents, which are immiscible with each other. Various kinds of solvating oxygenated solvents like ethers, ketones, aldehydes and alcohols have been used for the purpose. In many instances the macrocyclic polyethers like crown ethers and high molecular weight amines, *i.e.* liquid anion exchangers have been used for separation of metal complexes. When the distribution ratio is exceedingly small, one resorts to extraction chromatographic technique, wherein hydrophobic extractant is coated on stationary support and interesting separations are carried out with mineral acids, as the mobile phase. The success of such a method depends upon the availability of large number of theoretical plates in the chromatography.

Finally, the important chelating extractants are β-diketones, oxine, oximes cupferron, dithizone, naphthol and azonaphthol, and xanthates and sodium diethyl-dithiocarbamates (cupral). These extractant not only permit separation at microgram concentration, but also facilitates simultaneous extraction and spectrophotometric determination in one single step. Subsequent chapter shall list several extractive photometric analysis of metal pollutants.

In conclusion, it can be seen that not all the methods available in analytical chemistry are always useful in the environmental pollution analysis. At the same time, one cannot all the time depend upon sophisticated instrumental methods, as on several occasions simple classical methods have furnished valuable data on the extent of pollution. What is more important is the level of analysis and background of analyte sample? We cannot all the time use instrumental methods specially in fields, we need a method suitable for the continuous measurements or monitoring. We will have to adopt radically different approach while selecting a method for monitoring. In a developing country like India, another factor is the cost-benefit analysis. A method no doubt should be sensitive, but must also be economical. Such criteria shall be taken into account while selecting a method for the air and water pollution monitoring. A good amount of information is provided with experimental details for pollutant analysis with modern instruments in subsequent chapters.

ADDITIONAL READINGS

2.1 L. Erdey, *Gravimetric Analysis*, Pergamon Press (1963).

2.2 A.I. Vogel, *Textbook of Quantitative Inorganic Analysis*, International Publishers, 4th Ed. (1980).

2.3 H. Freiser and Q. Fernendo, *Ionic Equilibria in Analytical Chemistry*, Wiley, New York (1963).

2.4 F.J. Welcher, *Analytical Uses of Ethylenediamine Tetraacetic Acid,* Van Nostrand (1958).

2.5 A. Ringbom, *Complexation in Analytical Chemistry*, Interscience, New York (1963).

2.6 C. Schwarzenbach and H. Flaschka, *Complexometric Titrations*, Metheuen and Co. Ltd., London (1969).

2.7 A.K. De, S.M. Khopkar and R.A. Chalmers, *Solvent Extraction of Metals*, Van Nostrand Reinhold (1970).

2.8 O. Samulson, *Ion Exchange Separations in Analytical Chemistry*, John Wiley, 2nd Ed. (1963).

2.9 E. Lederer and M. Lederer, *Chromatography—Theory and Practice,* 2nd Ed., Van Nostrand (1957).

2.10 O.C. Smith, *Inorganic Chromatography,* Van Nostrand (1953).

2.11 J.M. Bobbitt, *Thin Layer Chromatography,* Reinhold, New York (1963).

2.12 J.A. Dean, *Chemical Separation Methods*, Van Nostrand Reinhold, London (1970).

2.13 D.W. Grant, *Gas Liquid Chromatography*, Van Nostrand Reinhold (1971).

2.14 M. Bier, *Electrophoresis*, Academic Press, New York (1959).

2.15 M.G. Mellon, *Analytical Absorption Spectroscopy,* Wiley, New York (1950).

2.16 D.M. Hercules, *Fluorescence and Phosphorescence Analysis*, Wiley-Interscience, New York (1966).

2.17 R.M. Silverstein, G.C. Bassler and T.C. Morrill, *Spectroscopic Identification of Organic Compounds,* Wiley, New York, 4th Ed. (1981).

2.18 I. Fleming and D.H. Williams, *Spectroscopic Methods in Organic Chemistry*, McGraw-Hill, New York, 3rd Ed. (1980).

2.19 J.A. Dean, *Flame Photometry,* McGraw-Hill, New York (1960).

2.20 G.F. Krikbright, *Atomic Absorption Spectroscopy and Related Methods*, Marcell Dekker (1977).

2.21 H.J. Arnikar, *Essentials of Nuclear Chemistry,* New Age International Publishers (P) Ltd., New Delhi, 2nd Ed. (1990).

2.22 L. Meites, *Handbook of Analytical Chemistry*, McGraw-Hill, New York (1962).

2.23 L. Meites, *Polarography Technique*, Interscience, New York (1952).

2.24 H.T.S. Britton, *Conductometric Analysis,* Van Nostrand (1934).

2.25 J.H. Beynon, *Mass Spectrometry and its Application in Organic Chemistry*, Elsevier (1960).

2.26 S.M. Khopkar, *Basic Concepts of Analytical Chemistry,* New Age International Publishers (P) Ltd., New Delhi, 3rd Ed. (2008).

2.27 R.C. Kapoor and S. Agarwal, *Polarography*, New Age International Publishers (P) Ltd., New Delhi (1991).

2.28 S.M. Khopkar, *Environmental Pollution, Monitoring and Control,* New Age International Publishers (P) Ltd., New Delhi (2006).

CHAPTER 3

Toxic Pollutants and Their Analysis

3.1 INTRODUCTION

The term 'toxic' is very loosely used in the Literature. A person intoxicated by alcohol consumption falls under the category of 'inebriation'. The food poisoning amongst children is case of 'poisoning'. The snake or scorpion bite is also not an example of toxicity. However certain pollutants if ingested in human system beyond specified limit cause adverse effects on health. They are called 'Toxic' e.g. CO, As, Hg, Tik-20 (a pesticide) are termed as toxic. They have certain level up to which they can be tolerated beyond which this tolerance limit also called (TLV) threshold limiting value causes deleterious effect on health.

In environmental pollutants several of pollutants are toxic and are injurious to health e.g. amongst metal pollutants we have Be, Cd, Hg, Pb, V, As, Cr and Mn are toxic pollutants. Amongst gases we have CO, $COCl_2$, PH_3, halogens like fluorine are toxic. Amongst organic pollutants we have array of compounds like C_6H_6, hexane, heptane, acetone, ethanol if present beyond certain levels and if ingested in body are injurious to health.

The toxicity is revealed by certain toxicological manifestations on human health. Such of the effects include asphyxiation, pneumonia, hepatic and renal failure, dermatitis, and the carcinogenicity. Some compounds have cumulative effect e.g. the compounds of cadmium initially reside in human system and start showing symptoms of poisoning after several years if the body resistance and the immunity fails to protect. Such of the consequences which can be ascertained by various kinds of test e.g. breath and urine sample show toxicity of metal pollutants, blood analysis reveals the ill effects of organic compounds, hair analysis shows as poisoning by arsenic while the tissue analysis in biopsy as well as in autopsy indicates extent of damage caused to tissues in body.

The main source of these toxicants is chemical, leather, drug, industry. The level of tolerance for few is too small e.g. $(C_2H_5)_4$ Pb, $Ni(CO)_4$, be less than $<100\ \mu g/cm^2$ in air is toxic and we need sensitive methods of characterisation and analysis. AAS, ICP-AES and conventional spectrophotometry serves the purpose. Analysis of volatile organic compounds is best carried out by GLC, HPLC, SFC and Mass spectrometry. The electroanalytical methods at times are used like ASV, polarography and ion-selective electrodes. Most of the toxic pollutants are measured in indoor environment like laboratory with sophisticated instruments.

The common incidents of toxic effects is poisoning due to mercury in fish in Minamata in Japan or cadmium intoxication causing diseases as 'itai-itai', the poisoning due to arsenics in Bengal and Bangladesh; through water or air. Even DDT, Malathion, 2, 4-D causes several adverse effects. So attempts are made to take overview of toxic inorganic as well as organic pollutants and suggest method for their quantitative analysis at microgram concentrations.

Most of the pollutants in air, water or waste water are toxic. Therefore it is worthwhile to consider them in groups like toxic metal pollutants, toxic gases and the toxic organic pollutants. We would consider the source of pollution, toxicological manifestations, methods of analysis and measures to control them. Detailed procedure for experimental set-up for analysis is provided in the respective chapters.

3.2 KINDS OF TOXIC POLLUTANTS

The compounds used in industry are not toxic or injurious to the health of the workers, but few of them cause serious health hazards. Majority of toxic materials exist in gaseous or vapour form, while a few of them exist in the form of fluids or solids. Amongst toxic gaseous pollutants, one can easily name: AsH_3, SbH_3, H_2S, CO or Freon (CCl_2F_2). They are absorbed through skin and eyes by the process of inhalation and are ingested in the body. Several of the organic solvents, specially those having boiling point less than 120-150°C are toxic in nature. They affect skin and cause problem of inhalation. Dust fumes and suspended particulate matter are hazardous, as they are ingested into lungs.

Table 3.1 Source of toxic pollutants in environment

Source	*Toxic pollutant*	*Average level ppm*
Sulphuric acid factory	SO_2	10–30
Steel industry	Dust and particulates	10–1000
	CO	40–400
	Aromatic hydrocarbons	0–100
Chemical factory	NO–NO_x	10–200
	$HCrO_4$	18 mg/m^3
	Al_2O_3	30 mg/m^3
	NH_3	2–50
	SO_2	1–40
Pesticide industry	DDT, parathion,	18 mg/m^3
Drug industry	Talcum powder	–
	C_2H_5OH	800
	C_6H_6	150
	$COCl_2$ (Phosgene)	1
	Fe_2O_3	60 mg/m^3
Heavy engineering works	Al_2O_3	–
	SiO_2	–
	C-particles	–
	Fe-dust	–
	Asbestos	–
Rayon industry	CS_2	40

Similarly, minerals containing Pb, Mn, Be, silica, asbestos cause serious lung diseases. They can produce cancer of lung, and impaired respiratory function. Many pesticides and polymers are harmful to health. Pesticide residue causes intoxication. They are considered under next few chapter. Since plastic and polymer are non-degradable, their disposal poses a serious problem. Their products of decomposition or combustion are very dangerous. There are several skin irritants, like primary irritants or sensitizers. The latter do not cause any irritation on first contact to skin, but later cause dermatitis. Dyes, soaps, oil, resins, explosive plasticizers, cosmetics, coal (Table 3.1) derivatives are irritants by nature. Acids, alkalies, salts and organic solvents belong to this category of primary irritants. Therefore, it is worthwhile to consider the common physiological manifestations of toxic pollutants.

3.3 COMMON TOXIC MANIFESTATIONS

As a rule brain, heart, lungs, kidney and liver are most important organs of our body. These vital organs get affected by toxic pollutants. Any problem with heart is termed as cardiac disfunction. Disorder of lung is termed as pulmonary oedema, whereas that of kidney as renal disorder and with liver we term it as hepatic disorder. The central nervous system, operated mainly by brain also gets affected. We also have problems associated with blood vascular system, digestive system and skin ailments.

(a) Asphyxiation

The absence of oxygen causes cessation of breathing and may result in death. Several gases react chemically and produce asphyxiation, e.g. CO, HCN.

(b) Irritation

The corrosive acids and caustic alkalies produce irritation of eyes, skin and respiratory passage. Those causing harm to lower respiratory passage are more dangerous, such as NO_2, PH_3, CdO, BeO. Organic compounds like aldehydes, alcohols or ketones are equally injurious to health. Alkali injury is more serious than acid injury of the skin.

(c) Pneumoconiosis

This is a problem related to lung, caused due to dust or other similar substances. Compounds like quartz, asbestos, beryllium compounds cause pneumoconiosis. Dust also promotes pneumoconiosis.

(d) Narcosis

Most of the organic solvents are narcotic in nature, e.g. halogenated hydrocarbons, esters, ketones, ethers. Even benzene, carbon tetrachloride causes intoxication. Such effects are not of permanent in nature, but at the same time many effects are irreversible.

(e) Systemic Intoxication

This causes irreversible injury with slow recovery of vital organs, nervous system or of hemopoietic system. Hemotoxic efforts are caused by C_6H_6 and can even lead to leukemia. Anemia is commonly caused by lead-poisoning. CO converts hemoglobin from blood to methemoglobin, which in turn leads to death due to asphyxiation. Aromatic and organic nitro compounds are also bad for the health (Table 3.2).

(f) Hepatic and Renal Disorder

All problems of liver and kidney fall under this category. Compounds of Hg, Se, U, As, TNT, halogenated hydrocarbons cause damage to liver and kidney. The primary function of the liver is to form blood, while kidney is an efficient filtration unit to remove toxins from the body. Neurotoxic effects are indirect fall-out of renal or hepatic disorder. Person may get tremor, paralysis, mental disturbance due to CH_3Br, CS_2, Mn or Hg compounds. A simple problem of bile fluid getting mixed with blood causes jaundice, excessive drinking of wine causes cirrhosis of liver, urinary infection causes cystitis and nephritis.

Table 3.2 Health hazards encountered in various chemical processes

Health hazards	*Chemical process*	*Incidence (%)*
Pneumoconiosis	Coal and zinc	34
Toxicity due to CS_2, Zn	Rayon industry	27
Toxicity of Mn and V	Ferromanganese, steel plants	73
	Processing and refining	16
Ulceration of wounds	Dichromate exposure	20
Insecticide poisoning	DDT manufacturing units	30
	Parathion	21
Byssionsis	Cotton fibre dust	8
Chronic bronchitis	Textile and jute mill process	10
Lead toxicity	Printing press workers	25
	Accumulator	11
Dye interaction	Dyestuff, carcinogenic compounds like benzidine, β-naphthylamine	7

(g) Dermatitis

This is an occupational disease, primarily caused by contact of skin with corrosive chemicals. Dermatological examination reveals cause of such problems.

(h) Carcinogenesis

The cause of cancer is still not known in spite of continuous investigations. It leads to uncontrolled multiplication of cells. Several compounds like β-naphthylamine, benzidine, 3–4 benzopyrene, $Ni(CO)_4$, asbestos, Cr(VI) and As-(III) compounds cause cancer. Such compounds are discussed under individual groups in this chapter.

3.4 BIOLOGICAL SAMPLING

Several problems of body are accurately diagnosed by the sampling process. One can use air, blood, urine for such purposes. Some poisons like cadmium compounds which remain in body for as long as 10 years. We call them cumulative toxins, while a few others are excreted instantaneously, like compounds of vanadium. Very useful and specific information is obtained from biological sampling.

(a) Breath Sampling

Volatile and unreactive substances can be measured in breath sample. Chlorinated hydrocarbons, CO, radioactive gases can be easily identified in breath sample.

(b) Urine Analysis

Just as a physician's diagnosis depends upon analysis of urine, so also an environmental scientist can detect 90% of toxic materials in urine, e.g. Pb. Hg., As, Sb, Cd, Se, Te, U, fluorides are easily detected in urine. Few are detected after late stage. Organic substances are excreted as such. DDT resides in the body for a longer time. A large presence of say proteinuria or coproporphyrine in urine indicates presence of Cd or Pb in the body.

(c) Blood Analysis

Some substances are easily detected in blood, e.g. Pb, Mn, heavy metals, are easily identified in the body by blood test. We can also measure chemiluminescence effect on blood, e.g. poisoning of organophosphorus pesticides like Tik-20 can be detected due to the formation of carboxyhemoglobin in blood. CH_3Br is detected instantaneously in blood.

(d) Nail and Hair Analysis

Certain compounds like As, V, Sb are easily detected in hair and nails. The cystine contents of nail gets affected by their presence. Several criminal cases were detected by analysis of nail and hair right from ashes in the grave by forensic scientists.

(e) Tissue Analysis

This is a sure test of detecting malignant growth. The autopsy is carried out for dead tissue, the biopsy is carried out for living tissue. It gives important information on lesion in the tissue.

3.5 THRESHOLD LIMITING VALUE (TLV)

This is an unique measure of toxicity. This refers to airborne concentrations of substances and indicates conditions under which a worker is continuously exposed day after day without adverse effects, however, there is a wide variation in the individual susceptibility due to immunity. It is interesting to note that on an average, a person of Indian origin can tolerate more dose of a toxin due to intrinsic immunity in comparison with a person from the abroad. Thus TLV represents weighed time average concentration for 7-8 hours work day, or 40 hrs per week. It provides a guide to the control of health hazards. It permits an exposure above the stipulated limit, only if there is compensation by equivalent exposure below the limit during work day or work week. TLV are based on best available information from human or animal studies of exposures. TLV are not always used for knowing relative index of hazards, evaluation of air pollution nuisances, estimation of toxic potential or proof of disease. Most important factor is that, the working conditions differ from place to place e.g. those in the USA cannot be compared with those living in India. So it is not right to overstretch this information to draw conclusions under existing conditions.

3.6 TOXIC METAL POLLUTANTS AND THEIR CHARACTERISATION

There are only 18 important metal pollutants. In periodic classification of elements, we have more than 104 elements, of these approximately 80 are metals; fortunately out of these 80 metals only 18 metals are toxic. The important source for these 18 metal pollutants is indicated in Table 3.3, while the toxic effects of very important metal pollutants are listed in Table 3.4, the general toxic manifestations caused by metals are listed in Table 3.5 and finally the threshold limiting values of only important metals are listed in Table 3.6, while Table 3.7 gives TLV of several other elements, which are also considered to be toxic. The source, toxic effects, important analytical methods for these toxic metal pollutants shall also be considered. The metals shall be discussed according to their position in the periodic classification, since there is some resemblance in toxicity of similar metals. We have groups. These groups contain elements from both A and B sub-groups. The Group I has beryllium while Group II has zinc, cadmium, mercury, while Group III includes thallium, Group IV has lead, Group V has vanadium, arsenic, antimony, Group VI has chromium, uranium, selenium, tellurium, Group VII has manganese, Group VIII includes cobalt, nickel, osmium and platinum. The details of analysis by specific methods or instrumental techniques are provided in chapter 12 and chapter 13 in great details.

Table 3.3 Sources of metal pollutant

Metal	*Source*	*Metal*	*Source*
Be	Fluorescent lamps	Se	Glass, photocells
Os	Catalyst, staining agent	Cd	Cladding alloy, alkali cells
Pt	Catalyst, alloy, thermocouple	Pb	Auto-exhaust emission, water pipes
Hg	Chloroalkali industry, pesticides	As	Paper pulp industry pesticides
Tl	Rodenticide; pesticides	Sb	Alloys, pharmaceuticals preparations
V	Furnace oil, metal mills	Cr	Plating industry, paints, alloys
Te	Auto-exhaust emission	Ni	Alkaline cell, alloys, catalyst
Co	Ink pigment, magnets	Zn	Galvanisation, alloys
U	Nuclear reactors	Mn	Mines, minerals

Table 3.4 Toxic effects of metal pollutants

Metal	*Effects*
Hg	Tremor, Gingivitis, Renal disorder asphyxiation
Pb	Plumbism, gastric problems, colic, anaemia and irritation of lungs, pneumoconiosis
Cd	Shortness of breath, anaemia, liver trouble and narcosis, hepatic and renal disorder
V	Catarrh, cough, wheezing, dyspnea and dermatitis, sore throat
As	Cancer of skin and nasal septum, jaundice
Cr	Dermatitis, ulcer, cancer
Os	Halo around eyes, bronchitis
Te	Garlic smell to sweat and breath

Table 3.5 Toxicological effects of metal pollutants

Toxic effects	*Name of compound*
Asphyxiation	CO, HCN
Irritation	CdO, BeO, NO_2
Pneumoconiosis	Be compounds
Narcosis	$CHCl_3$, CCl_4
Systemic intoxication	Organic compounds
Hepatic and renal disorders	As, AsH_3
Hemotoxic effects	Pb, organolead compounds
Neurotoxic effects	Hg, Mn
Dermatitis	Alkalis, phenols
Carcinogenic effects	Cr(VI). $Ni(CO)_4$, Pt, As

Table 3.6 Threshold limits of selected toxic metals

Periodic group	*Metal pollutant*	*TLV μg/m³*
Group I	Be (II)	2
Group II	Zn (II)	500
	Cd (II)	200
	Hg (II)	10
Group III	Tl (I) (III)	100
Group IV	Pb (II)	200
Group V	V (V)	100
	As (III)	500
	Sb (III)	500
Group VI	Cr (III) (VI)	500
	U (VI)	200
	Se (IV)	200
	Te (IV)	100
Group VII	Mn (II)	5000
Group VIII	Co (III)	100
	Ni (II)	1000
	Os (VIII)	2
	Pt (II)(IV)	2

(a) Beryllium

The main source of pollution is emissions from fluorescent lamp industry. It causes dermatitis and pneumoconiosis and severe respiratory disorder. Fine particles are hazardous, it can cause lung problem for a person exposed in area 1–2 km radius of the factory. It is determined by spectrophotometry with eriochrome cyanine R or fluorimetrically with morine (*i.e.* pentahydroxy flavanol) or 8-hydroxyquinoline. It can also be determined by atomic absorption spectroscopy at 235 nm with nitrous oxide-acetylene flame with sensitivity 0.016 μg/m^3. As small as 0.005 μg/m^3 can be detected in air or tissue. Other effective chromogenic reagents for spectrophotometry of beryllium are p-nitrophenylazoresorcinol, chromoazurol S (see section 13.2).

(b) Zinc

In comparison with Be it is less toxic. The main source is galvanisation and alloy industries. It causes chills, nausea, vomitting and metal fume fever. Fortunately, patient recovers in 48 hours as a person can acquire (temporary) immunity on long exposure, but immunity is also lost quickly. The common methods of analysis are polarography and atomic absorption spectroscopy. The AAS is carried with air-acetylene flame at 214 nm with sensitivity of 0.009 μg/ml of zinc. The spectrophotometry with zincon as the chromogenic reagent is also very effective (also see section 12.13).

(c) Cadmium

The main source is cladding industry, where it is used in protective metal coating, nuclear reactor, alkaline cells; and alloy industry. It is a cumulative poison and can reside in the body for more than 10 years. It causes injury to lungs, death results from pulmonary oedema, shortness of breath, increase in proteinuria in urine, it is excreted through urine very slowly. It affects liver, kidney and pancreas, it is determined spectrophotometrically with dithizone chloroform extraction, by measuring absorbance at 520 nm. It can also be analysed polarographically or AAS at 229 nm in air-acetylene flame, where the sensitivity is 0.0004 μg/ml, with limit of analysis as 0.5–2 μg/ml (see section 13.3).

Table 3.7 Threshold limiting values of metals

S. No.	*Metal pollutant*	*TLV μg/m³*
1	Beryllium	2
2	Barium	500
3	Indium	100
4	Thallium	100
5	Tin	2000
6	Lead	200
7	Tetraethyl lead	75
8	Arsenic	500
9	Selenium	200
10	Tellurium	100
11	Yttrium	1000
12	Hafnium	500
13	Vanadium	500
14	Manganese	5000
15	Cobalt	100
16	Rhodium	100
17	Nickel metal	1000
18	Nickel carbonyl	7
19	Copper	100
20	Silver	10
21	Zinc oxide fume	5000
22	Mercury	100
23	Molybdenum	5000
24	Tungsten	1000
25	Uranium	200

(d) Mercury

It is the king of toxic pollutants and has created history but was thoroughly investigated. It caused greater awareness of pollution in the world due to Minamata Tragedy in Japan, where a paper factory discharged its compounds in sea, which on degradation formed $(CH_3)_2$ Hg or $(CH_3\ CH_2)_2$ Hg. They were consumed by fish; through food chain which in turn was consumed by man and several persons died of mercury poisoning. The compound PMA, *i.e.* phenyl mercury acetate (C_6H_5) Hg $(CH_3\ COO^-)_2$ was used for seed dressing. The birds consumed these seeds and they died in greater number in Denmark. When mercury was used in dental formulations, it caused health problem in Europe. It was discharged from rayon industry, which was using classical Caster-Kellenor cell for sodium hydroxide manufacture in Mumbai and large number of cattle was affected. It was used in felt industry and its discharge in waste water caused severe health problems, it was used there as a carroting agent. It is used in discharge tubes, fungicides, antifouling agents. The main health problems are tremor of extremities, mental disturbance, gingivitis, renal disorder, and impaired central nervous system. About 0.01 μg/litre can be detected in urine. Organic mercury, *i.e.* alkyl or aryl mercury is more harmful than inorganic mercury. Figure 3.1 depicts mercury cycle in total environment. It mainly shows the transport of mercury in atmosphere, lithosphere, biosphere and hydrosphere. There are numerous methods for analysis of mercury, viz. spectrophotometry with dithizone, malachite green; by polarography or best by cold vapour AAS. Since it is volatile, the atomisation of mercury is carried out in vacuum. On test papers, it is detected by ($KI + I_2$) turning red on exposure. SeS paper also indicates darkening. As small as 0.001 μg/ml can be easily determined by AAS at 253 nm, with air-acetylene flame, with limit of analysis 100–400 μg/ml and sensitivity of 2.2 μg/ml (See chapter 12 & 13, Section 13.7).

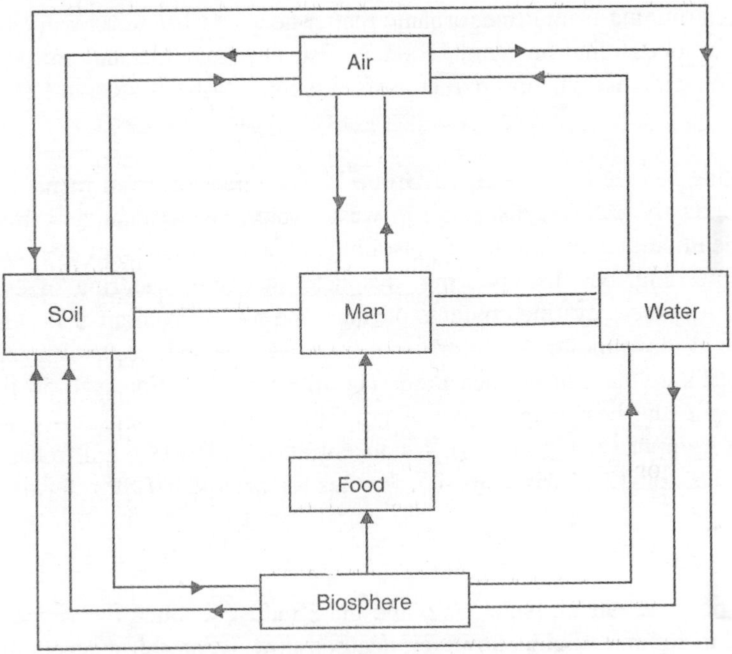

Fig. 3.1 Mercury cycle in the universe

(e) Thallium

This is the only toxic metal from the third group. The main source of pollution is pesticide and cosmetic industry. It is largely used as the depilatory agent. It causes chronic poisoning. It is usually extracted in amylacetate and is determined spectrophotometrically with methyl violet as colour forming ligand. It is best analysed by polarography. AAS is not very useful for thallium analysis. Also see section 13.1 for details of the analysis.

(f) Lead

Lead is the most toxic metal pollutant. The important source is auto-exhaust emission, as $(C_2H_5)_4$ Pb forms an essential ingredient for petrol as the antiknocking agent. Other sources are glass industry, paint industry, low fusible alloy like solder, accumulator and storage batteries, printing industry (type metal) and water pipes. It can enter into lungs through inhalation and into skin through mere contact.

The common disease is plumbism; which causes intoxication, colic gastrointestinal disturbance, anemia, weakness in wrist and legs and impaired functioning of central nervous system. In children, the part of brain *i.e.* medulla oblongata gets affected and it causes meningitis and encephalomyelitis diseases of the brain. On interaction with blood, the presence of high levels of erythrocytes is detected. In urine less than 0.4 mg/litre can be detected. Like mercury, organic lead is more poisonous than inorganic lead.The intake causes increase in urinary coproporphyrine and arrest porphyrine synthesis. δ-ALA, *i.e.* delta-aminolevulinic acid is formed in urine as the intermediate product. As regards analysis, it is collected by electrostatic precipitations (ESP-methods are described in Chapter 4 under sampling) or impingers. Volumetrically it can be determined by iodometry by interaction of $PbCrO_4$, KI and $Na_2S_2O_3$. The dithizone is used for extractive spectrophotometric analysis, but polarography is the preferred method, with AAS it is determined at 217 nm with air-acetylene flame, when sensitivity of analysis is 0.11 µg/ml. While determining from urine, organic matter needs destruction by dry ashning with HNO_3, $HClO_4$ and HCl. The determination from blood is also cumbersome and initial separation by ion exchange or extraction with methylisopropylketone is essential (Also see section 12.8 for detail analysis).

(g) Vanadium

Three metals including vanadium are toxic in Group V. The emission from furnace oil boilers, milling operations in steel industry, catalyst discharge in waste water, heat exchangers fired with residual oil causes pollution. The important symptoms of poisoning are acute irritation of eyes and respiratory tract, nasal catarrh, nosebleeding, cough, wheezing, dyspnea, sneezing, watering of eye, sore throat and blackening of teeth. It reduces cystine contents of hairs and nails. From air it is sampled by ESP. It is determined with 8-hydroxyquinoline in chloroform by extractive spectrophotometer. In aqueous solution, the magenta black colour can be measured. The limits of detection is 1–50 µg/ml. In AAS it is determined at 318 nm, the best way is to use graphite furnace, as sensitivity with nitrous oxide-acetylene flame is very poor (> 0.75 µg/ml). A concentration of 0.003 mg/litre can be easily detected in urine (Also refer Section 12.12 for analysis). Further for details of ESP refer to section 4.2 of next chapter.

(h) Arsenic

The main source is pharmaceutical, paper pulp and glass industry. Since its compound has no taste, it is a popular material to commit suicide, however, death is slow. It can cause dermatitis and the irritation of upper respiratory passage, ulceration and perforation of nasal septum, cutaneous cancer, AsH_3 causes anuria, jaundice and anemia. It is rapidly excreted. Lobster can tolerate 1 mg/litre of As and nails can store 0.1–0.35 mg/100 g of As. The best method for analysis is Gutzeit method, where AsH_3 decomposes to form As mirror in tube. In AAS it is determined by hydride generator, as AsH_3 easily volatilises in air-hydrogen flame. It is determined at 197 nm with a sensitivity of 0.60 µg/ml, with detection limit of 0.26 µg/ml of As. (See section 13.7 for details of the analysis by AAS with hydride generator.)

(i) Antimony

The main source is safety match industry, drug and alloy industry. It causes violent and continuous vomiting, profuse diarrhoea, muscle weakness and damage to kidney and liver. The fatty degeneration of spleen and liver is accelerated by its compounds. Sb_2S_3 is main component of safety match which causes loss of appetite. It affects the heart, causes myocardial infraction; the metal is more dangerous than its compounds, which is used in fire works. SbH_3 dissolves readily in gastrointestinal track, and causes damage to liver. It is determined spectrophotometrically at 565 mm, with Rhodamine B as the chromogenic ligand. A concentration of 0.2 mg/ml of Sb can be detected by this method. It can be determined by AAS at 218 nm in air-acetylene flame, with limits of 10–40 μg/ml of Sb. Also see section 13.7.

(j) Chromium

This is the most toxic metal from Group VI of periodic classification. Cr (III) is not as toxic as Cr(VI). Main source of pollution is paint industry, alloy (stainless steel); glass industry, electroplating effluents and tanneries waste, where $K_2Cr_2O_7$ is used for hide cleaning and leather treatment. Waste is alkaline. It causes dermatitis, ulceration and perforation of nasal septum, chromic mist causes chest trouble. Hexavalent compounds cause lung cancer, to workers working in paints, pigment and tannery industry. A concentration of 3.8 μg/litre is detected in urine of workers. It is determined spectrophotometrically with s-diphenylcarbazide. It can also be determined by polarography. In AAS air-acetylene flame is used to analyse it at 357 nm, with a sensitivity of 0.55 μg/ml. [For detail see section 12.4 and section 13.3 for details of the analysis by spectroscopy as well as by AAS.]

(k) Uranium

Amongst radioactive elements, this is highly toxic. The main source is atomic reactors. It causes histological damage of kidney and is slowly absorbed in the body. U^{235} amongst all isotopes causes radiation hazards. A concentration of 0.03–0.3 μg/litre is detected in urine. It is analysed by α-counting. In spectrophotometric analysis, dibenzoylmethane, 8-hydroxyquinoline and 2-thenoyltrifluoroacetone (TTA) are used as extracting agents to give coloured complex for photometric determination. It can be determined by polarography. Due to poor sensitivity (viz. 113 μg/ml), AAS is not a good method, even if one uses a flame of nitrous oxide-acetylene or graphite furnace at 358 nm. [Kindly refer to section 13.16 for details of the analysis by AAS technique.]

(l) Selenium and Tellurium

The main source is glass and photocell industry, also it is present in auto-exhaust emission, as $(C_6H_5)_6$–Te is used to combat the ill effect of residual lead from antiknocking action. Selenium compounds are irritating and toxic. SeO_2 causes bronchitis, dyspnea, rales, wheezing, headache, sleeplessness, nervous and cardiac disorders, loss of appetite, H_2Se(<3 ppm) causes damage to liver and spleen; $SeOCl_2$, SeO_2, H_2Se are all harmful. In comparison, tellurium is more toxic, it causes fatal intoxication, epigastric distress, metallic taste, and bad odour to breath and sweat (in fact it is a good symptom of toxicity). Se is determined spectrophotometrically with 3 3′ diaminobenzidine, while Te blue solution reduced by $SnCl_2$ is measured directly by nephelometer. It can be also extracted in n-amylalcohol and analysed by colour measurement. AAS need air-hydrogen flame for Se or air-acetylene flame for Te. They are measured at 196 and 214 nm respectively, with sensitivity of 0.47 and 0.26 μg/ml, respectively. [Refer to section 12.10 for details of analysis by AAS technique.]

(m) Manganese

It is the only element in Group VII which is toxic. The main source of manganese is milling factory, alloy industry, mines and mineral depots, smelting operations, dry-cell battery factories, and glass industry. It affects the central nervous system, causes gait and speech disturbance tremors, mask-like face, uncontrollable laughter, lung troubles; symptoms are visible after 2–3 months of exposure. Blood test is most dependable. MnO_2 causes pharyngitis, pneumonia. It is carried into bile gastrointestinal tract. It is determined polarographically, or spectrophotometrically with formaldioxime at 470 nm with AAS, absorption maxima is of 279 nm, with good sensitivity of 0.21 μg/ml for Mn in air-acetylene flame with limit of analysis as 1–4 μg/ml Mn. [See section 12.9 for the details of analysis by AAS and section 13.8 for graphite furnace AAS.]

(n) Cobalt and Nickel

Main source of cobalt is magnet making industry, while Ni come from hydrogenation of oil industry and paint factories. Cobalt carbonyl is most toxic and causes lung problem. It causes pulmonary oedema, while nickel causes dermatitis, pneumonia, lung and nose cancer (due to $Ni(CO)_4$, giddiness, headache, nausea, vomiting. Cobalt is determined spectrophotometrically with Nitroso R-salt or with cupral extraction in chloroform or with α-nitroso β-napthol. Nickel is determined spectrophotometrically with α-dimethylglyoxime or α-furildioxime in chloroform. It is not accurately measured by AAS, as sensitivity is not satisfactory. In AAS cobalt is determined at 240 nm, while nickel at 232 nm, with limits of analysis is 3–12 μg/ml for both. [See section 13.2 for details of analysis by AAS and section 13.10 for plasma spectroscopy.]

(o) Osmium and Platinum

Source of osmium is staining agent, a catalyst in organic synthesis (OsO_4 introduce –C–C– linkage); while platinum is used in thermocouples and as catalyst. OsO_4 is volatile (m.p. < 45°C; b.p > 100°C), causes irritation of eyes, halo around eyes and heaviness of eyelids and ultimate impairment of vision. It causes headache and bronchitis, pulmonary oedema, while platinum causes respiratory diseases, nasal irritation, asthama, dermtitis and eczema and photophobia. Osmium is determined spectrophotometrically with thiourea, while platinum is determined by the same method with p-dimethylaminobenzoyl rhodamine as chromogenic ligand. For AAS sensitivity is not good below a concentration of 1.2 μg/ml, analysed at 295 and 495 nm for Osmium and platinum respectively. [Also see section 13.3 and 13.4 for details of AAS analysis of these metals.]

3.7 TOXIC GASEOUS POLLUTANTS AND NON-METALS

Amongst non-metals, we have boron, carbon, nitrogen, phosphorus, oxygen, sulphur and halogen compounds, which are toxic in nature and they cause atmospheric pollution (Table 3.8). Let us consider them in groups.

Table 3.8 TLV of gaseous pollutants

S. No.	*Gas pollutants*	*Formula*	*Gas ppm*	*Concentration (× 10^5) μg/m³*
1	Carbon monoxide	CO	50	0.55
2	Cyanide (liquid)	CN	–	0.05
3	Cyanogen	CN	10	–
4	Phosgene	$COCl_2$	0.1	0.004
5	Ammonia	NH_3	50	0.0003
6	Nitrogen oxides	$NO–NO_2$	5	0.09
7	Phosphine	PH_3	0.3	0.004
8	Phosphorus (solid)	P_4	–	0.001
9	Ozone	O_3	0.1	0.002
10	Sulphur dioxide	SO_2	5	0.13
11	Fluorine	F	0.1	0.00
12	Chlorine	Cl	1	–

(Contd.)

13	Bromine	Br	0.1	0.007
14	Hydrogen sulphide	H_2S	10	0.15
15	Hydrogen fluoride	HF	3	0.02
16	Hydrogen chloride	HCl	5	0.07
17	Hydrogen bromide	HBr	3	0.10
18	Nitric acid	HNO_3	2	0.05

(a) Boron Compounds

H_3BO_3 is a dressing material. It is less toxic in comparison with various hydrides like B_2H_6, B_3H_8, etc. Of this diborane is a respiratory irritant and a nerve poison, B is estimated by non-aqueous conductometric titration using manitol, or by spectrophotometry with caramine red.

(b) Carbon Compounds

Amongst these CO, CO_2 are environmental pollutants. CO is very dangerous, as it is colourless, odourless and tasteless gas. It is detected in tube as $I_2O_5 + CO \rightarrow CO_2 + I_2$ (violet vapour). Other methods of detection are $HgO + CO \rightarrow Hg$ (luster) $+ CO_2$, $PdCl_2 + CO \rightarrow COCl_2 + Pd$ (luster). CO readily reacts with haemoglobin of blood to form methemoglobin, which causes coagulation of blood, asphyxiation and death. CO is formed by incomplete combustion, CO causes headache, dizziness, nausea impaired brain function, weakness and death. In blood it is tested with pyrotannic acid. The methods for analysis are many. Carbon dioxide is not a pollutant in the real sense. Excess amounts in air causes rapid breathing and headache. [See section 14.13 for details of analysis in water and see also sections 6.13, 6.14, 6.15, 6.16 for analysis by NDIR.]

(c) Nitrogen Compounds

NH_3, NH_2–NH_2, NO–NO_x and cyanides are the main pollutants. Ammonia causes irritation of nose, throat and eyes. A concentration greater than 50 ppm is not tolerable. Hydrazine (NH_2–NH_2) which is used as the rocket propellant causes damage to liver and kidney. It can also cause skin trouble. It is determined spectrophotometrically with p-dimethylaminobenzaldehyde. NO–NO_x contains several gases like NO, NO_2, N_2O_3, N_2O_4 and N_2O_5, of which NO-NO_2 are more toxic. They irritate the eyes, cause problems with lower respiratory passage, vertigo, vomiting, asphyxiation, inhibit enzyme activity in the body. The methods for their analysis are described in Chapter 6. Metal cyanide are deadly poisonous (limit 100 ppm is fatal). [See sections 14.7, 14.8, 14.9 for analysis of oxycompounds of nitrogen in water while sections 6.7, 6.8, 6.9, 6.10 for the analysis of gases in atmosphere.]

(d) Phosphorous Compounds

Important compounds in this group are PH_3, P_4O_6, P_4O_{10} and metal phosphides, which are used as rodenticides and pesticides. Organophosphorus compounds are more toxic. They cause necrosis and phossy jaw for workers, specially in the match industry using white or red phosphorus. PH_3 is toxic, causes diarrhoea, nausea, tightness of chest and epigastric pain. PH_3 is used for fumigation of wheat and corn. Phosphorous compounds are generally determined by spectrophotometry. [For analysis of pesticide see section 16.5.]

(e) Oxygen and Sulphur Compounds

The gases O_3, SO_2 and H_2S, are well known environmental pollutants. High pressure oxygen or pure O_2 is harmful, but not diluted oxygen, with inert gas or nitrogen, as it exists in the environment. O_3 formed by silent electric discharge in air, is toxic, causing irritation, headache and coughing. Also, O_2 by interaction with UV radiation, or welding or high voltage electric discharge forms ozone. Amongst sulphur compounds, SO_2 and H_2S are harmful. More than 10 ppm of SO_2 causes health hazards while H_2S is an odorous gas which can paralyse respiratory tract, can cause conjuctivitis and eye irritation. A dose > 500–1000 ppm is fatal for animals; 10-20 ppm of H_2S causes eye trouble called keratitis. The methods for analysis for them are described in Chapter 4. [Kindly refer to sections 14.10, 14.11 & 14.13 for details of analysis in water and refer to sections 6.1, 6.2, 6.3, 6.4 for analysis of gases.]

(f) Halogens

Most of them have toxic effects on health. The famous experiment of Moseley's during synthesis of fluorine causing irrepairable damage of tissues and incurable wounds are well known. For this calcium gluconate injection is the best antidote. Chlorine gas causes burning of eyes, cough, substernal pain and respiratory distress. In tube railway in England, several persons fainted due to leakage of chlorine in duct. A concentration of $ClO_2 < 0.1$ ppm is not harmful, a greater dosage causes bubling in chest *i.e.* Rales. Bromine is corrosive but less toxic. Iodine in true sense is not toxic. Phosgene ($COCl_2$) a gas used in biological warfare, causes pulmonary oedema. Methylene chloride formed by burning of kerosene is bad for health. They are all analysed by gas chromatography-mass spectrometric methods. [See sections 14.4, 14.5, 14.6 for details of analysis in water and sections 7.2, 7.3, 7.4 for analysis in the air by chemiluminescence.]

3.8 TOXIC MINERALS AND DUST

Table 3.9 and 3.10 indicate dispersion of metal dust, grit and particulate matter in atmosphere, while Table 3.11 indicates typical levels of benzopyrene in atmosphere in 1970. In 1990, the level was much higher with growth of automobiles and two wheelers. Thus minerals and solid mixtures of organic matter can also cause toxicity. Their inhalation causes mechanical loading of lung and fibrosis of lung cells. Of the materials tested silica and asbestos are highly objectionable. SiO_2 causes silicosis, shortness of breath, decreased chest expansion and tuberculosis of lung. The toxic level is 2.5 million particles per cubic foot. While asbestos causes fibrosis of lung and tuberculosis, silica can cause lung cancer. It absorbs protein from human body. Talcum, soapstone, mica, limestone, dolomite, coal, alumina are also injurious to health. Amongst organic dust, we have cotton dust or jute dust, which cause

Table 3.9 Metal concentration in common flyash

Metal pollutant	*Level*	
	25-12.5 μm	*1.5 μm or less*
Al(III)	54,300	59,300
Cd(II)	5	100
Cr(III)	130	300
Fe(III)	59,000	32,000
Mn(II)	240	500
Ni(II)	200	300
Pb(II)	200	500

Table 3.10 Grit and dust in different industrial locations

Type of site/area	*Grit and dust fall out (mg/m²d)*
Industrial	139
Fashioned housing	116
Low density housing	82
Commercial town	112
Town centre	90
Outskirts of a village	70
Countryside, express highways	38

Table 3.11 Benzopyrene in various cities of world

Nation	*Town*	*BaP in air* $\mu g \times 10^{-2}/m^3$
USA	Chicago	0.4–3.9
USA	Detroit	1.10–5.4
USA	New York	0.5–4.1
USA	Philadelphia	0.6–5.9
USA	Los Angeles	0.16–10
Switzerland	Zurich	4.1–7.7
South Japan	Kyushu	5–10
England	London	5–10
Iran	Tehran	6–6.3
Norway	Oslo	0.6–2.6
India	Mumbai	2.6–6.0

byssinosis on exposer to the workers. This is prevalent in mill areas. All dust particulate are best analysed by nephelometry, *i.e.* light scattering methods.

3.9 TOXIC ORGANIC COMPOUNDS

Most of the organic compounds have effect on central nervous system. It is difficult to consider their toxicity in terms of individual compound. It is worthwhile to consider them in groups, on the basis of (a) compounds containing carbon and hydrogen, (b) compounds containing carbon, hydrogen, oxygen, (c) compounds containing carbon, hydrogen, nitrogen and finally (d) compounds containing carbon, hydrogen, sulphur and phosphorus.

(a) Hydrocarbons and their Derivatives

Table 3.12 lists such compounds. They have in general narcotic effect on human beings. Butadiene isoprene affects bone marrow. Benzene (C_6H_6) causes respiratory disorder and narcosis. It causes blood dyscrasia and blood anemia. It is determined spectrophotometrically at 244, 249, 255 or 261 nm, as its dinitro derivative. Toluene ($C_6H_5–CH_3$) in comparison with benzene is less toxic. It causes lymphocytosis. It has bad effect, if the dose exceeds 200 ppm. It causes drowsiness, mild headache, impaired coordination. Fortunately, it cannot penetrate through the skin. It is excreted through urine as hippuric acid. Its dinitro derivative is determined spectrophotometrically at 530 nm. Amongst aromatic hydrocarbon, xylene ($C_6H_4(CH_3)_2$) has least toxicity. A high dose can produce liver and kidney damage and

Table 3.12 Toxic level of organic compounds containing carbon and hydrogen

S. No.	*Compounds*	*ppm*	$\times 10^5$ *µg/m³*
1	Benzene	25	0.8
2	Butadiene	1000	2.2
3	Cumene	50	2.45
4	Cyclopentadiene	75	2.0
5	LPG (liquefied petroleum gas)	1000	18.0
6	Styrene	100	4.2
7	Benzene	25	0.8
8	Toluene	200	7.5
9	Xylene	100	4.35
10	Hexane	500	18.0
11	Heptane	500	20.0
12	Octane	500	23.5
13	Cyclohexane	300	10.5

hyperplasia of bone marrow. It is a narcotic. Ethylbenzene ($(C_2H_5)C_6H_5$) is a substitute for antiknock for aviation fuel. It causes irritation of eyes and skin. Terpentine is a mixture of hydrocarbons, a universal solvent which can cause lung irritation and paralysis of central nervous system. Terpentine should therefore be carefully handled in houses.

Chloroform ($CHCl_3$) was a classical anaesthetic agent, but is not useful now as it affects heart and causes cardiac sensitization. Carbon tetrachloride (CCl_4) is harmful, as it has depressing action on central nervous system, it causes gastrointestinal disturbance and causes liver, kidney and lungs problems. It causes loss of appetite, vomiting. Tetrachloroethylene is a substituent for CCl_4 for the purpose of dry cleaning. Freon (CCl_2F_2) is a gas used for filling compressor in refrigerators. It affects the lung. Methyl (CH_3Cl) chloride is a substitute refrigerant, which also causes symptoms of drunkenness and inebriation; pulmonary congestion, cardic failure. [For details of analysis of CH_4 please see section 16.1, so also for P A-H analysis see in section 16.4.]

(b) Compounds Containing Carbon Hydrogen and Oxygen

The important compounds are listed in Table 3.13, of these methanol (CH_3OH) is one which causes blindness, diminution of vision, and enlargement of liver. In crude form, country wine contains methanol. It is analysed spectrophotometrically with chromotropic acid. Ethanol is not toxic, but it causes inebriation and temporary drowsiness and affects liver. n-Butanol irritates eyes and can cause corneal inflammation. Acetone (CH_3COCH_3) is less toxic, is narcotic in nature and it causes irritation of mucous membrane. MEK, MBK, MIBK are less toxic, as they cause minor problems like throat irritation, nausea, headache. Mesityloxide (4-methyl-3-pentane-2-one) is an efficient extractant for metal separation, but it is toxic if present in concentration more than 12 ppm. It irritates eyes, lung, and throat. A concentration of 25 ppm causes discomfort to lungs and kidneys, however, dioxane is not very toxic, so also tetrahydrofuran (THF). [For analysis of TOC refer to the section 15.8.]

Table 3.13 Threshold limit values of organic pollutants containing carbon, hydrogen and oxygen

S. No.	*Compound*	*Concentration*	
		ppm	*× 10^5 μg/m^3*
1	Acetic acid	10	0.25
2	Acetone	1000	24.0
3	Amyl-acetate	100	5.25
4	Cresol	5	0.22
5	Dioxane	100	3.6
6	Ethyl alcohol	1000	19.0
7	Formaldehyde	5	0.08
8	Formic acid	5	0.9
9	Phenol	5	0.19
10	Tetrahydrofuran	200	5.9
11	Terpentine	100	5.6
12	Butyl alcohol	100	3.0
13	Isoamyl alcohol	100	3.6
14	Ethyl acetate	400	14.0
15	Butyl acetate	150	7.1
16	Diethyl ether	400	12.0
17	Mesityl oxide	25	1.0

(c) Organic Compounds Containing Sulphur, Phosphorus or Nitrogen

Aniline is toxic and causes cynosis (Table 3.14). It forms methemoglobin in blood. Benzidine causes cancer and bladder tumour, also β-napthylamine is carcinogenic. These are dye intermediates. Pyridine has bad odour but is less toxic. TNT and nitrobenzene causes cynosis of blood, blue colouration of skin, nails and lips and death due to respiratory failure. It is absorbed readily through skin. TNT is a warfare material and causes aplastic anaemia and jaundice. Acrylonitrile ($CH_2 = CH–CN$) is as bad as KCN. It causes anorexia and respiratory paralysis, liver damage, vomiting and headache, smoking enhances toxicity. Organonitrogen compounds are carcinogenic and hence one must handle them with care. Amongst organosulphur compounds, mention must be made of CS_2 which causes hysteria and mental fatigue. CS_2 react with amino-acids and proteins in the body to form chelates with heavy metal.

(d) Carcinogenic Compounds

Some organic compounds are confirmed as carcinogenic agents. They are 3–4 benzopyrene, 2-acetylaminofluorene, β-napthylamine, 4-nitrodiphenyl, 4-aminodiphenyl, benzidine, dichlorobenzidine, N-nitrodimethylamine, β-proiplactent and 4-dimethylaminobenzene. The list is not comprehensive, but is indication of hazards caused by organic chemicals. As a matter of fact, extensive research for their characterisation is in progress.

Table 3.14 Threshold limiting value of organic compounds containing nitrogen, sulphur and phosphorus

S. No.	*Compound*	*Concentration*	
		ppm	*× 10⁵ μg/m³*
1	Acetonitrile	40	0.7
2	Acrolein	0.1	0.002
3	Aniline	5	0.19
4	Benzidine	–	A′
5	Hydrazine	1	0.013
6	Methylamine	10	0.12
7	β-naphthylamine	–	A′
8	Toludine	5	0.22
9	Trinitrotoluene	–	0.015
10	Dinitrobenzene	–	0.01
11	2, 4–D	–	0.1
12	DDT	–	0.01
13	2, 4, 5T	–	0.1
14	Lindane	–	0.005
15	Chloroform	50	2.4
16	Carbon tetrachloride	10	0.65
17	Malathion	–	0.15
18	Parathion	–	0.001
19	Dibutyl phosphate	1	0.05

ADDITIONAL READINGS

3.1 F.A. Patty, *Industrial Hygiene and Toxicology*, Vol. 1 and 2, Interscience, New York (1963).
3.2 E.A. Brown, *Toxicity of Metals*, Butterworths Pub. (1968).
3.3 A.K. De, *Environmental Chemistry*, 3rd Ed. New Age International Publishers (P) Ltd., New Delhi (2008).
3.4 M. Sitting, *Toxic Metals*, Noyes Data Corpn., New Jersey (1976).
3.5 R.W. Allen, M.D. Ells and A.W. Hart, *Industrial Hygiene*, Prentice Hall Inc., New Jersey (1976).
3.6 K. Anderson and R. Scott, *Fundamentals of Industrial Toxicology*, Ann Arbor (1981).
3.7 H.G. Seiler and H. Sigel, *Handbook on Toxicity of Inorganic Compounds*, Marcel Dekker, New York (1988).
3.8 M. Sitting, *Handbook of Toxic and Hazardous Chemicals,* Park Ridge, Noyes (1981).
3.9 S.M. Khopkar, *Basic Concepts of Analytical Chemistry*, 3rd Ed., New Age International Publishers (P) Ltd., New Delhi (2008).
3.10 S.M. Khopkar, *Environmental Pollution, Monitoring and Control*, New Age International Publishers (P) Ltd., New Delhi (2007).

CHAPTER 4

Sampling and Analysis of Air Pollutants

4.1 INTRODUCTION

Before undertaking a discussion on the analysis of air pollutants it is essential to study procedures of the sampling of the pollutants. The technique of sampling is as important as the technique of the characterisation of the air pollutants. In fact if by chance it goes wrong with the sampling technique, the entire process of analysis is futile as one would not get reproducible results. The sampling of the atmospheric pollutant is more complicated as compared to the sampling of water. This is because water is having state of homogeneity in liquid state. There is no variation from sample to sample during the analysis on account of its homogeneous nature.

The air pollution sampling is critical. This is because the constituents and the kind of air pollution continues to change from time to time and place to place. Thus it is almost impossible to get what is called as the representative sample. For the variation of the composition of the air, several factors are responsible. They are related to temperature inversions, wind direction, wind velocity and several other meteorological parameters. Apart from these meteorological parameters, what is most important is what is the purpose of the sampling? Further one should know for what duration sampling should be carried out? Then comes the final question of what is the rate of sampling of air?

There has been greater revolution in the field of this sampling of air. Various sampling devices were developed. Such air sampling equipments consisted of high volume sampler, various kinds of filters, including membrane filters, different types of cyclones, the sedimentation chambers, the cascade impactors, and midget impingers. The electrostatic and thermal precipitators were also important. In relation to sampling of suspended particulate matter or aerosols the sampling of gaseous pollutants is relatively simple and clean. In order to understand complexities of sampling we would take extensive discussion on various sampling equipments and procedures. Although the methods and equipments used for the analysis of the pollutants are reliable, the results so obtained for analysis will not be dependable unless the sampling technique gives representative results. The quality of the sample for all purposes should be same for the bulk of material. The important stages in sampling includes collecting the instrument and techniques refinement of sample and analysis of sample and finally the evaluation of pollutant. The size of the sample should be sufficient to accommodate the evaluation technique. For sampling the liquids 1–10 litres volume is more than adequate. Then overall rate of sampling depends upon the kind of equipment used for the collection. The efficiency of collection is of prime importance to get reproducible results. The duration of sampling and the shortest sampling time provide most dependable results is important. The time of sampling vary from 1-hour to 24 hours depending upon the purpose of the analysis as well as what is the objective of sampling? To determine the quality of air and to find out the influence of specific emission, the information on sources must be provided. This is a valuable information in planning control measures in environmental planning and future modelling.

4.2 METEOROLOGICAL PARAMETERS

This is important specially with reference to air pollution before sampling therefore the meteorological factors must be first considered prior to sampling. Such meteorological factors include wind direction and wind velocity. The diurnal and seasonal fluctuations occur during the source of emission. The atmospheric stability, wind speed, wind direction and the phenomena of the precipitation largely influences the quality of air. The rate of change in temperature is called as 'lapse rate' is also important during sampling as it does influence the air stability while wind direction and speed directly influences the sampling process. Ambient pollution is inversely proportional to wind speed. The cup type anemometer can measure rate of wind direction and wind velocity. The other aspect is precipitation or raining which can be measured by the standard rain gauge. The humidity and quality of sunlight must be also considered. Low humidity increases the formation of suspended particulates. The photochemical reactions and generation of secondary pollutants depends upon the kind of solar radiation emitted in a atmosphere.

4.3 LOCATION OF SAMPLING SITE, FREQUENCY AND SIZE OF SAMPLE

The location of sampling site is of considerable importance. The criteria for proper siting of sampler should provide true representation of receptor. A site should provide wide data. The sampling inlet should be at least six feet above ground level. The sampling site should not be located in vicinity of the major obstructions (*i.e.*, tall building). There should not be contamination from neighbourhood of sample (*i.e.*, chimney). The site should be safe from tampering. The electric power to run equipment must be available. The location of site of sampling should be on concentric circular lines around source. Such background of the location is needed during comparative assessment of pollution. If peak level of pollutant is to be ascertained then the sampling time should be small but if the sampling is for average concentration it needs week's sampling. The time should be adequate enough to collect sampled material. Finally size of the sample depends upon its concentration. The relative analytical sensitivity of the method is of equal importance. The sampling rate should fulfil need of the size and time of sampling. The rate should be such that the sufficient material is collected for the purpose of subsequent analysis e.g. in high volume sampling if the method of the analysis is gravimetry, the rate should be rapid. The rates are indicated in Table 4.1 for typical sampling rates for different sampling units. The sampling efficiency is hardly 100%. A qualitative information is needed.

4.4 SOURCES OF SAMPLING

Primarily we have two main sources which we term as "stationary source" and "mobile source". The first includes the emission from stationary industrial plants while the second includes moving objects as petrol, diesel and CNG engine vehicles or aircraft. The source gives better idea for selecting controlling devices. The stationary source needs good planning. The source should be uniform and steady. The stack parameters vary with time. The ideal sampling location rarely exists. While sampling from mobile source e.g. exhaust gas we need a sophisticated equipment. The operating cycle of running engine decides the quality of the sample. Aircraft emission cycles tend to be dependent upon the steady state of operating conditions.

4.5 AIR SAMPLING EQUIPMENT

It comprises of two kinds of the equipment. One for measurement of flow while other is for collection of sample. The first category of equipment consists of air mover, meters like spirometer, wet test meter, dry test meter, rotameters, pitot tubes, orifices and deflecting vane anemometer. The sample collection devices includes filters (membrane filters, paper filters, glass fiber filters) sedimentation chamber, cyclones, impact devices (like cascade impactor, impingers) for electrostatic precipitation (ESP) and thermal precipitation.

Air mover is required for the good sampling of air. The dust fall for collection does not need an air mover or pump however it should properly control the flow rate of the pollutant. The collector is on the vacuum side of the air mover. The carbon vane pump is popularly used, which has rotor with four cabon vanes held by the centrifugal force. The vane in turn separates into chamber leading into different compartments. The change in volume results change in pressure to move the air. Such pump can be operated for long hours. About 18–20 cf/mt air is pulled by pump. Volume of the air moved is constant. The needle valve arrangement controls the flow of air. The pressure is 15 pounds per square inch. Most pumps are motor driven. Their capacity is 2-4 cf/inch. The hand operated pumps are also available for field work. Other kind of pumps are diaphragm pump, and rotary pumps. In former midget sampler is used to absorb air. The flow rate being 1 Lt/mt. In these pumps, diaphragm pumps are popular. The rotary pumps are used for low concentration of the pollutant for short sampling time. They have graphite vanes. Such vanes need periodic replacement. They can provide 20-30 L/mt of air. For collecting very small volume, hypodermic syringes can be used. Other kinds of pumps are screw type pumps. The centrifugal pumps are best and quite efficient. High volume pump can displace 200 cf/mt. The brush motors need periodic servicing. Sometimes ejector pump is also used. The battery operated pumps are also available. They are made mostly of alkaline batteries. They are preferred in outdoor site. The mine safety pumps are four cylinder and cranker pump used with midget impinger. Cranking at 50 rpm produce a vacuum of 12" w.g. and provides a flow of 0.1 cc/mt on midget impinger. Automatic flow rate adjusters are used to control flow rate specially in high volume sampler. Apart from flow meter, the exact measurement of volume of air is crucial. This can be done by single displacement with spirometer. The measurement of sample by continuous displacement is also feasible with dry gas meter. There are also indirect methods of air volume measurement. Spirometer works on the principle of volume displacement of exact air passed through sampling unit. It is well suited for indoor work. The flow rate does not guarantee measurement of total sample volume. The use of the siphon bottle to standardise exact volume of air passed is recommended. The most important part of air sampling devices is air mover. One prefers a carbon vane pump having rotor with four carbon vanes moving under centrifugal force. The vanes divide chamber into the four components which change volume. Such change can produce a change in pressure. Such pumps can be used for several hours (with 2000-6000 rpm). The pumps can adjust the flow in sampling train. The motor driven pumps are commonly used. The centrifugal pumps are used for moving large volume of air.

4.6 HIGH VOLUME SAMPLER

It has capacity to draw 200 cf/mt of air. The proper maintenance is very important. A high flow rate blower (1.1–1.7 m^3/mt) draws the sample of air into a covered housing through 20 × 25 cm rectangular glass fiber filter (Fig. 4.1). The covered housing protects high volume sampler from rain. This is operated for all 24 hrs but this period can be changed depending upon the level of the pollutants. The glass fiber filter extends life of the instrument. A timer can also be attached as to collect samples at regular interval of time. Rate of sampling must be recorded. The sampler is rarely used at low flow rates of 0.85 m^3/mt. The device must be calibrated once in six months. The deviation should not be more than > 4.5%. The second most important component is flow meters. We have spirometer, wet test meter, dry test meter, Rotameters, Pitot tube, orifices and vane anemometer. We would consider special feature of these flowmeters.

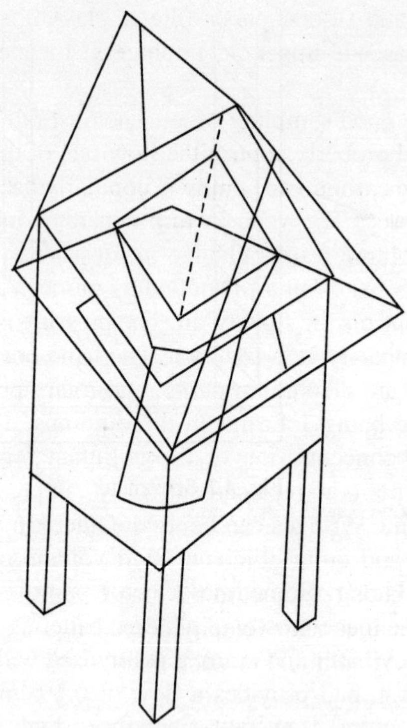

Fig. 4.1 High volume sampler

4.7 VARIOUS FLOWMETERS

We have six kinds of flowmeters.

(a) Spirometer

Spirometer is very accurate volume measuring device which is used for calibration. It can use direct positive displacement of a calibrated cylinder which moves with water scale (A) (Fig. 4.2). A constant ratio of flow as controlled by counterweights (B). Simultaneously the effect of buoyancy on cylinder wall is removed by small weight with a variable moment arm.

(b) Wet Test Meter

It also uses water displacement to rotate segmented drum, gears to indicating dial to count revolution of the drum. The maximum flow ratio is 1 cu ft/mt. It is unsuitable for outdoor work.

(c) Dry Test Meter

The enlargement of bellows of diaphragm is used to drive dials of indicator. Thus for volume measurement two bellows gets empty while other bellow gets filled. However it is useless for outdoor measurements.

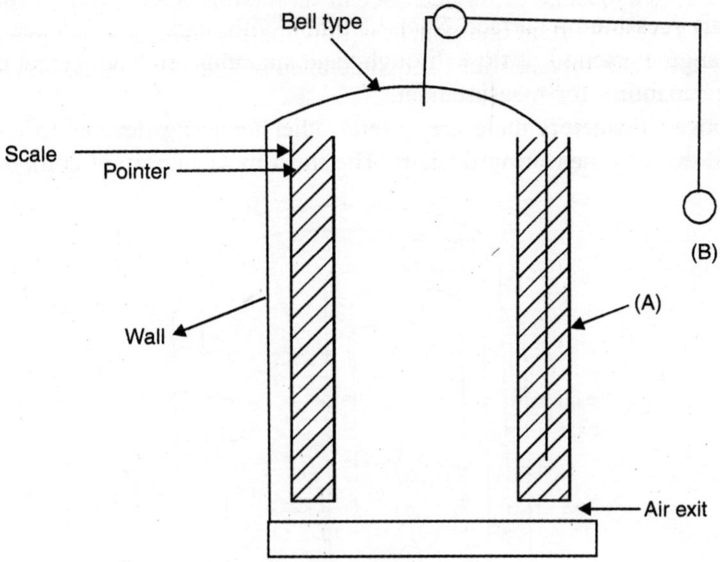

Fig. 4.2 Spirometer

(d) Rotameter

A rotameter has float (1) which is driven upward till drag force balances a weight of float bell cone or other suitable shape. The area changes with velocity of the drag. They need calibration. The oscillation of float must be uniform, but by gentling flow such oscillations can be damped.

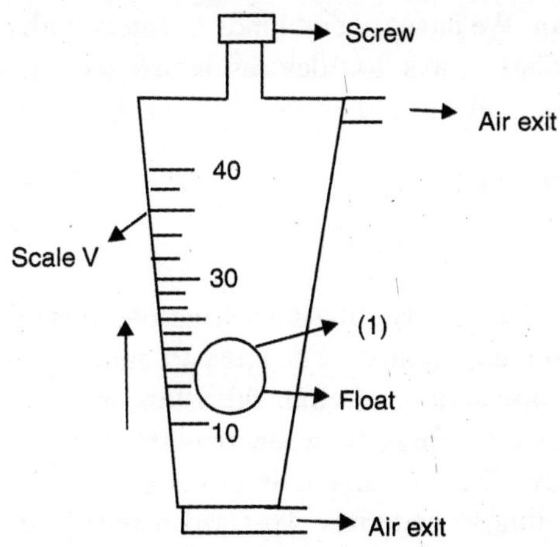

Fig. 4.3 Rotameter

(e) Pitot Tube

It is designed to change velocity head to pressure head which can be measured by manometer (Fig. 4.4). The stream velocity is reduced to zero at tip of tube *i.e.*, pressure changed so obtained is then measured.

(f) Vane Anemometer

It uses wind exerting pressure on surface. Such variation with angle of incidence of the air stream and probes, permits changing fraction of flow through vane anemometer. The weight of vane can be varied to get wider range capability for measurement.

Apart from above flowmeters there are several other metering devices to measure flow rate e.g. Pumps can be used even as measuring devices. The third most important component in air pollution measurement is sample collector. We have several kinds of filters with enormous variations, as we have, sedimentation chambers, cyclones, impaction devices which include cascade impactor, ESP, thermal precipitator. We have also other collecting devices like absorber and condensation devices.

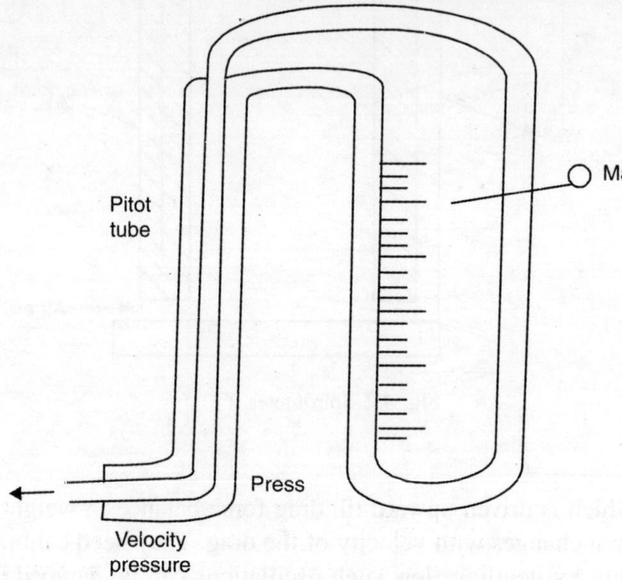

Fig. 4.4 Pitot's tube

4.8 SAMPLE COLLECTORS

(a) Filters

As regards filters, there are variety of the filters such as membrane filters, paper filters, glass fiber filters. Of these last one are more commonly used. The particulate sampling needs some kind of filters. The particles are removed by impaction, sieving, diffusion or brownian movement. The membrane filters consist of maze of branched fibers between two flat surfaces having number of pores. The collection occurs at surface only. The pore size is important. The commonly used pore size is 0.8 μm. These filters are made from cellulose materials. The transparent filters are used in microscopic work. They are quite rapid. The paper filters should stand air pressure and high speed of sampler. Whatman filter paper No. 41 is most commonly used. Glass fiber filters have good efficiency as they do not need impaction velocities. They are good for solvent extraction.

(b) Sedimentation Chamber

We now consider important collection device e.g., It is effective dust collection sampler. It is cubical from 3-18 inches on side. Top, bottom axis are aligned during collection. The small particles are removed by Brownian motion in the random directions and large particles settle down by process of sedimentation.

(c) Cyclones

These are centrifugal collectors. They are specially useful for the collection of particles of large size. They are used for the collection of ash also and respirable dust samples. There are various types of collectors with many variation.

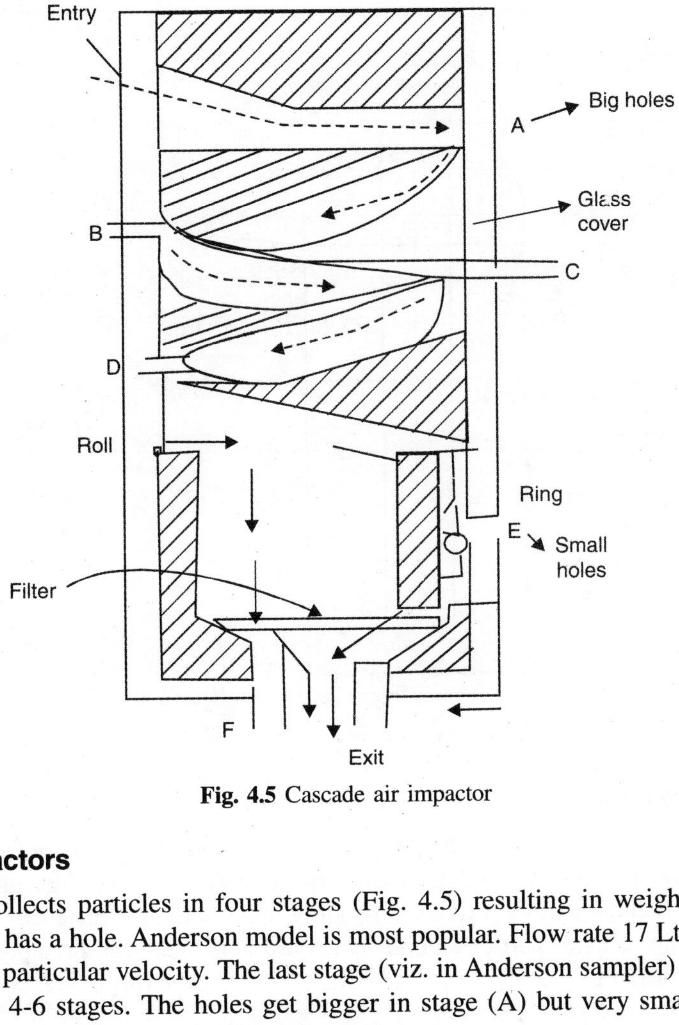

Fig. 4.5 Cascade air impactor

(d) Cascade Impactors

Cascade impactor collects particles in four stages (Fig. 4.5) resulting in weight wise distribution of particles. Each stage has a hole. Anderson model is most popular. Flow rate 17 Lt/mt. Each stage (A → E) has slot to impart particular velocity. The last stage (viz. in Anderson sampler) has 400 holes per one cubic meter through 4-6 stages. The holes get bigger in stage (A) but very small in stage (D) or (E) which pushes particle to orifice at point (F) (Fig. 4.5).

(e) Midget Impingers

The role of impingers (Fig. 4.6) is to collect particles by impaction on the surface below inlet nozzle and by process of scrubbing. The midget impinger is good for flow of 0.1 cft/mt. It collects about 85%

of particles but if particle size is < 1 μm it fails to carry out its job of collection. The movements can be mechanical in nature or those originated by wind flow. Apart from one described above. We have following two important collectors which are extensively used.

(f) Electrostatic Precipitator (ESP)

It imparts negative charge to particles, rendering movement of particles towards oppositely charged electrode where they are collected. Such collector is good for solids or liquids. It has very high collection efficiency. The particles remain intact in shape and size. It has many aluminium tubes to collect particulates. One tube acts as anode, while wire at center of the cylinder behaves as cathode. 12-16 kV potential is applied. It has been used for the collection of sample on grid for electron microscopy. Only limitation is we do not get uniform level deposits of particulates as they are not collected by this device.

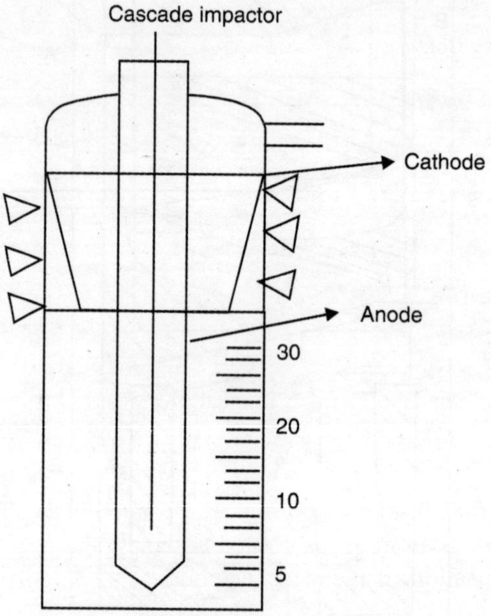

Fig. 4.6 Midget impinger

(g) Thermal Precipitator

It collects particles by thermal force on the particle in thermal gradient a force from hot to cold. A thin wire is heated (A) (Fig. 4.7) air is passed through orifice (o). The air passes through (A) on hot and cold (B) surfaces. There is deposition of particles depending upon mass, for instance, lighter particles deposition zone is (A) while heavier particles deposits toward zone (B). The precipitator is ideal for small sized particles. In addition to these collectors we have three other forms viz. absorber, adsorber, condensation devices which are useful mainly for aerosols.

(h) Absorber

Air sample can be bubbled through solution to collect gaseous sample. This is best accomplished by Midget impinger. If efficiency of collection is low, more impingers can be added in train impingers for the gases. The evaporation of absorbed solution must be avoided, further we have to saturate air before passing through channels.

(i) Adsorber

The organic gases can be thus collected by process of adsorption. Activated carbon is used to adsorb odour and colour of the sample. Polar adsorbers are used in wet form. The stripping is done to remove adsorbed material from the adsorbent Gas-mask canisters are used in high volume samplers. Activated coal can be impregnated with adsorbing liquid to perpetuate contaminants.

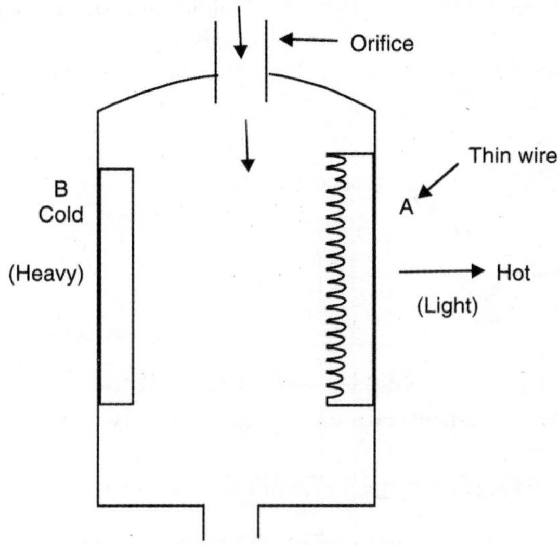

Fig. 4.7 Thermal precipitator

(j) Condensation

The fraction of air can be condensed by cooling at appropriate temperature, but it needs removal of water vapour before the condensation. Traps cooled with salt and ice or dry ice (–78°C) are employed before the process of condensation of the water vapour.

(k) Automated Samplers

Automated samplers are commercially available or are fabricated. They are time controlled or volume controlled sequential samplers. They change from one port to another port. Paper tape sampler does the same job. The periodic sampling by automated sampler is done to relate time with type of sample. The principle of collection of the gases in Orsat Gas analyser depends upon the phenomena of absorption of gaseous pollutants by specific adsorbing liquids stored in gadgets. The Table 4.1 indicates rate of sampling by various sampling units.

So far we have considered sampling of dust/suspended particulates. Before concluding this section, let us consider one important form of dust samplers:

Respirable dust sampler: The small tiny particles entering though the respiratory system to lungs leads to serious health problems. Therefore, one prefers to use standard impingers. Silica, quartz are toxic dust. We determine the threshold limit values it is given as:

$$\text{TLV for silica} = \frac{300}{\%\ \text{Quartz} + 10} \text{ million particles per cubic foot}$$

$$\text{TLV for total dust} = \frac{300}{\%\ \text{Quartz} + 3}\ \text{mq/m}^3$$

Table 4.1 Efficiency of sample collectors

S. No.	*Kind of collector*	*Rates of sampling*	*Remarks*
1.	High volume sampler	45–70 cf/mt	High efficiency
2.	Electrostatic precipitator	0.1–5 cf/mt	Intermediate efficiency
3.	Whatman no. 45 filter paper	10–120 L/mt	Reasonable efficiency
4.	Anderson sampler	1 cf/mt	Average efficiency
5.	Thermal precipitator	0.02–0.1 L/mt	Low efficiency
6.	Smith Greenberg sampler	1 cf/mt	Average efficiency
7.	Midget impinger	0.1 cf/mt	Reasonable efficiency
8.	Cascade impactor	10–40 L/mt	Good efficiency
9.	Paper tape sampler	5–30 L/mt	Fairly good efficiency
10.	Membrane filter	40–50 L/mt	Very good efficiency.

The particles are routed through cyclones (size selection) to analyse them in terms of mass filter medium must be non-hygroscopic in nature e.g. glass fiber, silver membranes or polyvinyl chloride.

4.9 SAMPLING GASEOUS POLLUTANTS

SO_2, NO-NO_x, SO_3, O_3, H_2S, HF, HCl are classified as gaseous pollutants. The technique utilised for collection includes:

(*a*) Absorb by chemical reaction
(*b*) Adsorb on finely divided solid surface
(*c*) Gas is converted to solid/liquid state
(*d*) Collect air contaminated with pollutant

These techniques are identical to the one utilised for the sampling of aerosols described in earlier section. Some details are provided in subsequent sections.

The most common technique is absorption of gases through reagent solution. Such collection depends upon solubility of the gas as well as its what chemical reaction is shown? Some factors influence sorption of gases or size of bubbles formed, contact time and concentration of the absorbing solution. Few processes involve special chemical reaction e.g. absorption of SO_2 in water leads to formation of H_2SO_4. The choice of absorbing liquid is important. Water is best solvent showing maximum solubility. NO or CO_2 are collected over NaOH. Lime water is best for CO_2 absorption leading to formation of calcium carbonate. The efficiency of the collection is also important. Good pump and flow meter must be available. One can employ various absorption devices e.g. glass bottle, absorbing media or bubbler. Flow rate should be around 0.1–1 ft^3/mt. One can use erylmer flask, fritted glass absorber or suitable impingers. The impinger can be easily cleaned. Gases can be also adsorbed on finely divided solids like carbonised coal/coke. Their efficiency depends upon amount of adsorbent used, physical state of adsorbent, its affinity for gas and low temperature promotes adsorption. The cryogenic or freezing technique is also used to collect gaseous pollutant however grab sampling is the most popular method used. It uses evacuated container, or gas is collected by downward displacement of liquid. The inflation of flexible containers like balloons (rubber) is most suitable. Mylar bag can be also used for the purpose of collection of the gas.

4.10 EXPERIMENT DETAILS OF SAMPLING

In these methods good calibration programme provides reliable results. Spirometer can be used to calibrate other metering devices for flow. High volume air sampler (a calibrated version) is commonly used. The low head pumps e.g. ESP are used for calibration. The efficiency of collection should be 100%. One needs the standard for gas absorption calibration. The standard particle distribution method can be also used for verifying collection efficiency. The sampling can be done on the spot followed by quick analysis e.g. measurement of CO by non-dispersive infrared spectroscopy (NDIR) method after sampling. Gas chromatography can be used for same purpose. SO_2 need midget collector (impinger). The storage of sample is also critical. The samples like ozone needs immediate analysis as they cannot be easily stored. One uses mylar bags for storage. Isokinetic sampling is also useful to procure definite information. The stack sampling is used for dust. Pitot tube is used to measure pressure during the process of sampling.

4.11 PROCEDURE FOR SAMPLING OF DUST BY GRAVITY AND FILTRATION TECHNIQUE

Let us first consider sampling of dust by gravity settling. For this purpose the site selection is important. It should be free from overhead obstructions like chimney or incinerator. The dust fall collector should be from earth at distance of 8'–50' above ground level.

Dust fall collector. A juice container of 1–2 gallon capacity, 20 mesh brass, buchner funnel, filtration flask. Steam water bath as evaporating dishes, drying oven, weighing bottle, aspirator.

Clean collecting jar. Add 1 mg/L copper sulphate to prevent algae deposition. Then expose it for one month. Remove unwanted object from sampler. Scrap and collect dust in evaporating dish and heat till 105°C to get dried product. As usual weigh insoluble solids on filtration on filter paper. Also ascertain soluble salts present in solution.

Calculations

We express dust fall for 30 days as gm/m^3. The weight of total solid (w_1) be equal to sum of weights of soluble (w_s) and insoluble (w_{si}) solids

$$\text{Total dust fall } D_g = \frac{w_1 \times 30 \text{ days}}{A_c \times N_d} \quad \text{if } A_c = \text{area of collecting jar mouth}$$

N_d = Number of days exposed

One can carry out chemical analysis to detect the presence of SO_2Cl^- metals bituminous and combustible materials.

4.12 SAMPLING OF SPM BY HIGH VOLUME SAMPLER

Experiment: The sampling of SPM by high volume sampler wherein Stoke's law is applicable. The sampler also called high volume sampler is vacuum cleaner type to draw air. The filter must have diameter of 4" in dimension of 8 × 10". Air is collected at a rate of 40–60 cfm. As much as 0.5 g dust can be scrapped after sampling. Motor of high volume sample works on timer. As such midnight to midnight collection is done (24 hrs) with one sampler filter can be manually changed during day time. Rotameter is used to know rate of flow of air passing during the process of the sampling.

One should install single sheet of tared glass fiber filter in high volume sampler. Measure the flow of air by rotameter. Adjust on/off time switch at partial hour of day. Minimum 24 hrs. sampling is required. Remove filter paper. Lab filter paper may not be used. A glass fiber filter of size (8 × 10") collects as much as 1–1.5 g of SPM. The particles cannot be easily dislodged from glass. Use fiber filter, or dry filter at 110°C and then weigh the amount of suspended particles deposited on the filter (Fig. 4.8).

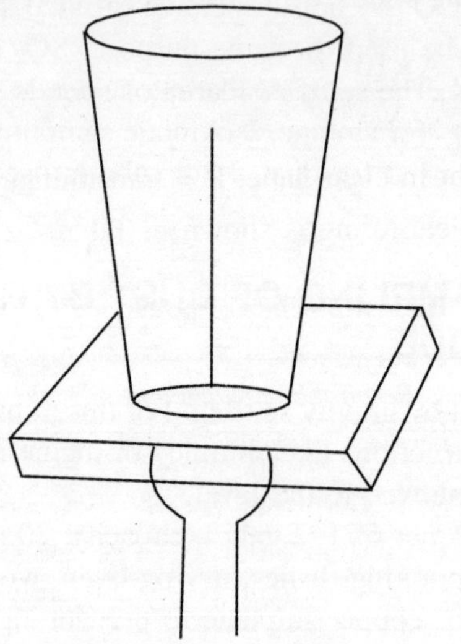

Fig. 4.8 Dust collector

Calculations

It collects microgram of suspended particulate matter within 24 hrs. The results are used for calculation as follows:

The sample air flow (measured by rotameter)

Clean Filter 56 cf/mt, Filter after exposure 52 cf/mt, Average = 54 cf/mt (56 – 52 = 54 mean)

Therefore $\frac{54\,\text{ft}^2}{\text{mt}} \times \frac{1440\,\text{mts}}{24\text{ hrs}} = \frac{76,760\,\text{ft}^2}{24\text{ hrs}}$ and $\frac{76,760\text{ hrs}}{24\text{ hrs}} \times \frac{0.0283\,\text{m}^3}{\text{ft}^3} \times \frac{2172\,\text{m}^3}{24\text{ hrs}}$

Weight of particulate matter

Weight before exposure = 3.417, weight after exposure = 3.925 difference 0.508 gms

Hence SPM concentration is

$$\frac{508 \times 10^3\,\mu\text{g}}{2172\,\text{m}^3} = \frac{234\mu\text{g}}{\text{m}^3} \text{ per 24 hr sampling at site.}$$

Dust Sampling by Tape Sampler

The greatest advantage of tape sampler is time required is much smaller ranging from 5 mts to 3 hrs. On drawing ambient air into sampler, particles are deposited on tape at fixed interval. The transmission of light through black spots developed by dust fallout is measured in photometer.

$$\text{Absorbance} = \log\frac{1}{T} = \log\left(\frac{P_0}{P}\right) = A$$

Calculations

Data is expressed as COH (coefficient of optical receptor 1000 ft).

COH is quantity of light scattering solids with absorbance of 0.01. We calculate

$$\frac{\text{COH}}{1000\,\text{ft}} = \log\left(\frac{P_0}{P}\right) \times \frac{A \times 10^3}{V} \quad \text{if A = area of spot, V = Sample volume}$$

P_0 = intensity of light in clean tape, P = transmittance after exposure

where COH is soil haze index. A typical rating is shown in Table 4.2.

Table 4.2 Kinds of pollution

S. No.	*Rate*	*Respirable contaminant*	*Soil haze index of pollution*
1.	0–1.0	80	Light pollution
2.	1.1–2.0	160	Medium pollution
3.	2.1–3.0	240	Heavy pollution
4.	3.0 >	240	Very heavy pollution

One can use unit of MURC (measure of undesirable respirable contaminant)

MURC = 70 $(\text{COH})^{0.7}$ *i.e.*, 0.5 COH = 43 MURC. When COH per 1000 ft spot of 1″ diameter with 50% transmittance after 2 hrs. of sampling average flow of 0.2 cfm

$$A = \frac{\pi d^2}{4} = \frac{\pi \times 1^2}{4} = \frac{1\,\text{ft}^2}{144\,\text{m}^2} = 5.45 \times 10^{-3}\,\text{ft}^2, V = \frac{0.2\,\text{ft}}{\text{mt}} \times 120\,\text{mts} = 24\,\text{ft}^2$$

$$\log\left(\frac{100}{50}\right) = \log 2 = 0.301. \text{ Substituting the values}$$

$$\frac{\text{COH}}{1000\,\text{ft}} = 0.301 \times \frac{5.45 \times 10^{-3} \times 10^5}{24} = 8.3. \text{ Also } \frac{\mu g}{m^3} = 240(\text{COH})^{0.01}$$

Such kind of calculations can be carried out during sampling of air pollutants.

ADDITIONAL READINGS

4.1 F.A. Patty, *Industrial Hygiene and Toxicology*, 2nd Ed. p. 97, Wiley-Interscience (1958).

4.2 P. Drinker and T. Hatch, *Industrial Dust*, 2nd Ed. p.152, McGraw-Hill and Co. (1954).

4.3 A.C. Stern, *Air Pollution*, 3rd Ed. p.4, 145, Elsevier Publishing House, Indian Edition (2006).
4.4 J.O. Ledbetter, *Air Pollution Analysis,* Part A, p.137, Marcell Dekker Co. (1972).
4.5 S.S. Butcher and R.J. Charlson, *Introduction to Air Chemistry*, Academic Press (1972).
4.6 R. Perry and R.J. Young, *Handbook of Air Pollution Analysis*, Chapmen and Hall, John Wiley (1977), p.84.
4.7 R.O. Warner, *Analysis of Air Pollutants*, John Wiley, (1976).
4.8 H.C. Perkins, *Air Pollution*, McGraw-Hill and Co. (1974).

CHAPTER 5

Introduction to Air Pollution

5.1 INTRODUCTION

The study of air pollution is of recent origin. In 1952, the fog in London promoted such studies in UK. In Los Angeles it further aggravated the need of investigation in USA. The Bhopal tragedy involving the leakage of killing gas like methyl isocyanate further accelerated the need for such investigations. With growing population which in turn accelerated the growth of traffic in cities, led to the discovery of pollution due to auto-exhaust emission. The depletion of ozone layer in the atmosphere caused serious problem of global warming. The indiscriminate cuttings of trees leading to deforestation of land promoted the discovery of green house effect which involved the release of heat by surplus accumulation of CO_2 in atmosphere, which in turn leads to warmer climate. All these episodes proved that air pollution is as important as water pollution.

The awareness led to systematic investigations of the structure of atmosphere, understanding presence of certain pollutants at particular level in atmosphere. The dispersion of air pollutants led to the exploration of meteorological parameters influencing air travel involving wind direction, wind velocity and phenomena of inversion. The accumulation of the foul gases in atmosphere promoted characterisation and determination of the air polluting gases which largely included oxides of sulphur, nitrogen and carbon. The suspended particulate matter added to growing problem of adverse effect on the health. Newer methods were developed for their characterisation which ultimately lead to the development of continuous monitoring instruments.

Therefore an endeavour is made in this chapter to identify the source of pollution, understand what green house effect is, what are methods for the sampling of aerosols as well as gaseous pollutants? Although some attempts were made to review the methods of analysis of SO_2, NO-NO_x, CO-CO_2, O_3 and NH_3, the detailed experimental details outlining its various steps in the analysis are presented in the separate chapter. This, in fact, will encourage the researcher to perform these experiments, gain confidence and get reliable information. The various stages involving preparation of reagents, their standardisation, and different calculations are provided in greater details in next two chapters. This chapter opens a vista of air pollution with an overall view.

5.2 STRUCTURE OF THE ATMOSPHERE

When we consider the abundance of elements in solar atmosphere, important components are metals as well as non-metals. It would be seen that (Table 5.1) carbon, nitrogen, oxygen, hydrogen are present in

Table 5.1 Abundance of elements in the atmosphere

S. No.	*Element*	*Atomic number*	*Atomic mass*	*Abundance ($\times 10^6$) atoms*
1	H	1	1.00	0.05×10^8
2	C	6	12.01	100
3	N	7	14.00	210
4	O	8	16.00	0.028×16^6
5	Mg	12	24.31	170
6	S	16	32.06	43
7	Ca	20	40.08	8.7
8	V	23	50.94	0.059
9	Cr	24	52.00	2.0
10	Mn	25	54.93	1.5
11	Fe	26	55.84	270
12	Co	27	58.93	0.55
13	Ni	28	58.71	4.7
14	Cu	29	63.54	0.87
15	Zn	30	65.37	0.31

various concentrations, e.g. carbon is present in the range of $10{,}000 \times 10^4$ atoms in the solar atmosphere, while nitrogen is $21{,}000 \times 10^4$, oxygen 2.8×10^{10} and hydrogen as 5.1×10^{12} atoms. On sun we have more of hydrogen and helium (1×10^{12}). Out of 104 elements in the Periodic classification, approximately 80 are metals and around 24 are non-metals. Out of these 24 non-metals, only 10 are of real concern to us, from the view point of atmospheric pollution. Table 5.2 lists the approximate composition of the atmosphere. This indicates that on an average there is 78% nitrogen, 21% oxygen and 1% of other gases. The major constituents among these other gases are CO_2 and water vapour. Now by anthropogenic activity if this composition changes and if there is undue generation of toxic gases like SO_2, H_2S, CO, CO_2, NO–NO_x, O_3, suspended particulate matter or organic compounds, we say the

Table 5.2 Important constituents of the earth's atmosphere

Gas	*Composition*	
	Vol. (ppm) ($\times 10^2$)	*Wt. (ppm) ($\times 10^3$)*
N_2	7809	755.0
O_2	2095	251.0
CO_2	3.0	0.466
CH_4	0.015	0.0094
N_2O	0.005	0.008
H_2	0.005	0.00035
O_3	0.004	0.007

atmosphere is polluted. We have already seen in the first chapter that level of SO_2 has risen from 40 to 172 μg/m^3 or NO–NO_x from 35–62 μg/m^3 and dust and particulate matter has gone up to 7000 μg/m^3 in last few years. These gases are all injurious to health. Table 5.3 describes the highest tolerance limit of industrial gases in general and tolerance limit for very short exposures. It will be seen that gases like phosgene ($COCl_2$), chlorine and methyl isocyanate are tolerated in very low concentrations. Further, they do not show much change in tolerance even on exposure for very short intervals. Table 5.4 gives the threshold limiting values of gases pollutant in the atmosphere. This table shows that only CO_2 is anthropogenic and not in the real sense a pollutant and it is tolerated in higher ratios also, while SO_2 (5), NO_x (5), CO (5O), NO (25), Cl_2 (1) and O_3 (0.1) are tolerated in very small concentrations.

Table 5.3 Tolerance limit of industrial gases

Gaseous pollutant	*Threshold limit ppm*	*Short exposure ppm*
NH_3	2.5	3.5
Cl_2	1.0	3.0
CO	5.0	300
SO_2	5.0	5.0
$COCl_2$	0.1	0.1
HCN	10	12
H_2S	10	15.0
CH_3CN	0.02	0.02

Table 5.4 Threshold limiting values of gaseous pollutants

Gas	*TLV, ppm*
CO	50
CO_2	5000
SO_2	5
H_2S	10
NO	25
NO_x	5
Cl_2	1
O_3	0.1
LPG/CNG	1000

It will be interesting to consider the structure of the earth's atmosphere (Fig. 5.1). If we consider altitude in km, the area up to 8–10 km is called troposphere; 10–50 km is called as stratosphere; 50–150 km is termed as mesosphere and 150–400 km is called as thermosphere. The lowest temperature in troposphere is about –60°C; in stratosphere –88°C; in mesosphere –100°C and in the thermosphere it begins to rise. The highest temperature in troposphere is 20°C, stratosphere 80°C, mesosphere 100°C and thermosphere greater than >500°C. An aircraft flies up to a height of 20 km and experiences a minimum temperature of –60°C. The highest peak of the Himalayas, viz. Gaurishankar is at a height of 6-8 km, with the approximate minimum temperature of –40°C. In the upper atmosphere, we have atomic oxygen and nitrogen. It is an atomic layer, while the middle air contains diffusion of oxygen and nitrogen and is called as homopause or diffusion layer, while the ground bottom layer contains a mixture of oxygen and nitrogen. One significant feature of temperature change is that as if we go beyond 10 km, temperature decreases continuously, then a further ascend up to 50 km shows a continuous rise in temperature up to 90 km temperature again falls and beyond 90 km till more than 800 km, temperature continuously rises. This drop in temperature and pressure with height in troposphere causes inversion and change in velocity and direction of wind (Fig. 5.1). The pressure drop is from 100 to 10 kpa or to 0.0001 kpa.

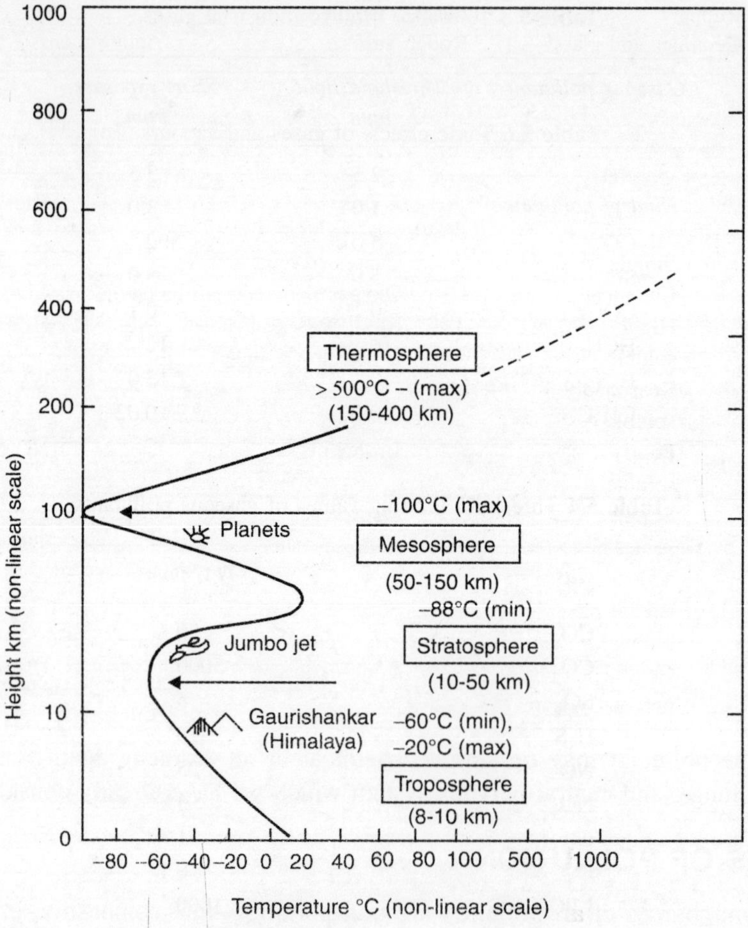

Fig. 5.1 Temperature zones of atmosphere

5.3 EFFECTS OF AIR POLLUTANTS ON MAN AND MATERIALS

It will be observed from Table 5.5 that there is significant effect of gaseous pollutants like SO_2, Cl_2, NO–NO_x, O_3 on materials like metals, paints, pigments, leather, paper, textiles, dyes, rubber, glass and ceramics. Table 5.6 describes the physiological action of toxic gases on human beings. This also includes toxic vapours. The principal toxic effects caused are asphyxiation, irritation, and anaesthetic

Table 5.5 Effect of pollution on materials

Material affected	*Effect*	*Pollutant*
Metals	Loss of metallic luster	SO_2
Pigment	Bleaching	SO_2, Cl_2
Paint	Discolouration	SO_2, H_2S, SPM
Leather	Bad gloss	SO_2
Paper	Brittle	SO_2, CO_2
Textiles	Decreased strength	SO_2
Dyes	Bleaching	NO_2, Cl_2
Rubber	Cracks, fatigue	O_3
Ceramics and glass	Rough surface	Acid/Vapours

Table 5.6 Toxic effects of gases and vapours

Effects	*Kind of pollutant*	*Action*	*Pollutant*
Asphyxiants	Simple	Causes oxygen deficiency	N_2, CH_4, CO, CO_2
	Chemical	Oxygen deficiency in blood	CO
Irritants	Highly soluble	Nasal passage irritated Irritate lower lungs	H_2SO_4, HCl, NH_3
	Reasonably soluble		SO_2, Br_2, Cl_2
	Poorly soluble	Bronchitis	NO-NO-$_x$
Anaesthetics	Simple	Depresent	Petrol
	Anaesthetic action	Exposure short	$CH_3 CH_2 OH$ $C_2 H_5 OC_2 H_5$
	Action on blood	Short exposure	$C_6 H_6$
	Central nervous system	Long exposure	$CH_3 OH$, CS_2

manifestations. The pollutant may be simple, chemical or an aesthetic agent. This action includes irritation of nose, lungs, and central nervous system which we have already considered.

5.4 SOURCES OF POLLUTION

The most important sources of air pollution are transportation, fuel combustion in stationary source, industrial process, solid waste disposal and other miscellaneous sources. It is noteworthy to see in histograms Fig. 5.2 that maximum CO is emitted by transportation, while maximum SO_2 is liberated due to fuel combustion from stationary sources. The auto-exhaust emission generates 10–8 million tons per year of oxides of nitrogen. Highest amount of hydrocarbons are liberated by transportation industry, while suspended particulate matter is produced by miscellaneous sources and also by fuel combustion. Amongst these pollutants SO_2 and NO–NO_x are major culprits. Figure 5.3 outlines the major sources of SO_2 emissions. It will be seen that coal combustion generates 41% pollutants, while petrol consumption liberates 16%, smelting gives 5%, refinery product gives 7% and other sources gives 5% of pollutants.

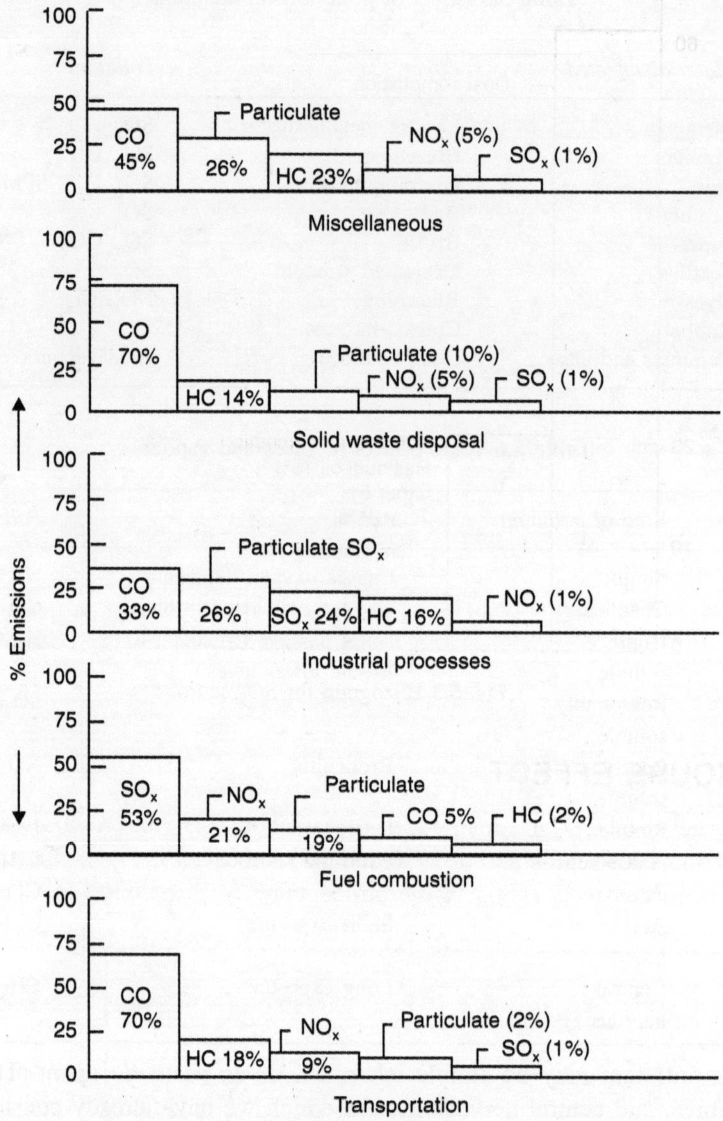

Fig. 5.2 Sources of atmospheric pollutants in histograms

Thus the major source of pollution is coal burning. The next serious emission of gases is NO–NO_x. Several surveys of industrial towns showed that auto-exhaust emission is responsible for the generation of NO–NO_x. Although NO-NO_x account for NO, NO_2, N_2O_3, N_2O_4 and N_2O_5, most harmful gases are NO–NO_2. By atmospheric lightening and photochemical reaction they form smog, NO gets converted into NO_2 and causes severe health problems. Amongst cities, Kolkata is notorious, for it has 58 $\mu g/m^3$ of NO–NO_x in its atmosphere. As opposed to this Delhi and Kanpur, in spite of being dusty towns have only 30–35 $\mu g/m^3$ of NO–NO_x. Next comes CO and CO_2. As stated earlier, CO_2 is not a pollutant in the real sense. It is an anthropogenic material in the atmosphere. The only consideration is, if its level in air goes up, it causes serious health problems. Every year the level of CO_2 in air is gradually rising.

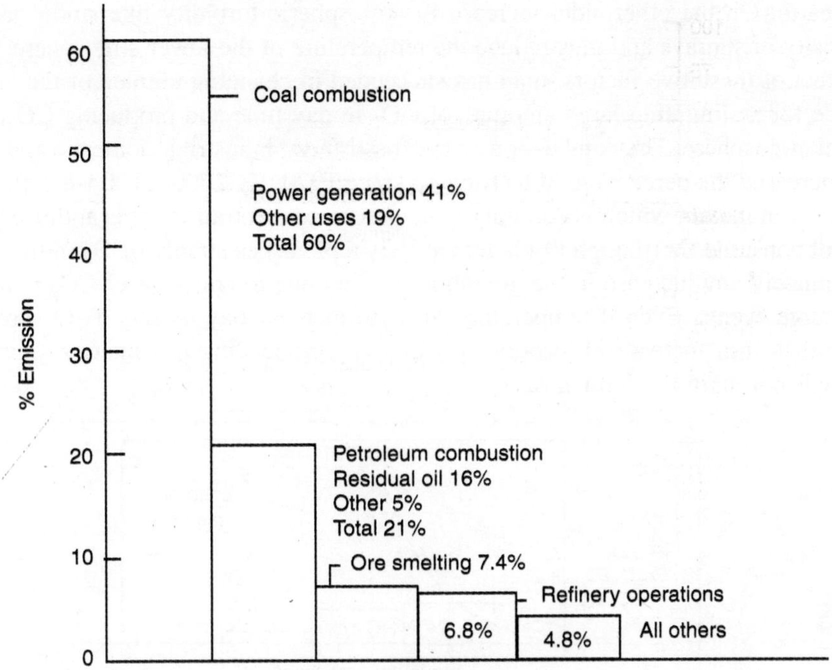

Fig. 5.3 Histogram for SO_2 source

5.5 GREENHOUSE EFFECT

If we consider the level of CO_2 in air in 1959, it was about 316 ppm, it rose to 320 ppm in 1962, it increased up to 324 in 1969 and is 350 in 1990 and 500 in 2000 (Fig. 5.4). One alarming feature is that the level of CO_2 is successively rising in the atmosphere. Such increase in CO_2 would warm up the

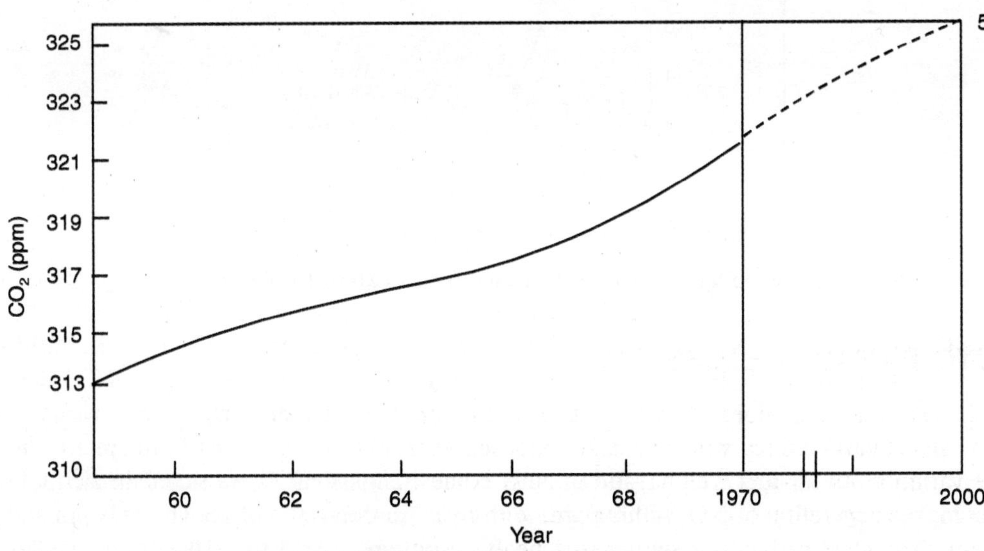

Fig. 5.4 Variation in CO_2 level in air

surface of the earth. On the other side, increase in atmospheric turbidity like smog or pollution can reduce the intensity of sunrays and thus reduce the temperature of the lower atmosphere. However, due to balancing effect of these two factors, man has succeeded in changing climate of the earth. Trees are important source for assimilating large amounts of CO_2 in day time and producing CO_2, in night time to biosphere and atmosphere. The combustion of the fossil fuels, industrial, domestic and transportation activities, has increased the percentage of CO_2 by 18 since 1970. By 2000–2100 we expect to have 380 to 495 ppm of CO_2 in the air, which is certainly a dangerous concentration. This anthropogenic production of CO_2 shall consume O_2 (though O-photosynthesis replenishes supply of O_2 in the atmosphere in day-time). Fortunately any increase in the greenhouse effect due to emission of CO_2 is far outweighed by more long-range events. Even if temperature of earth increase day by day, H_2O vapour in air will also rise and will in turn increase absorption of sun's energy, leading to runaway greenhouse effect, which in turn will not harm the human race.

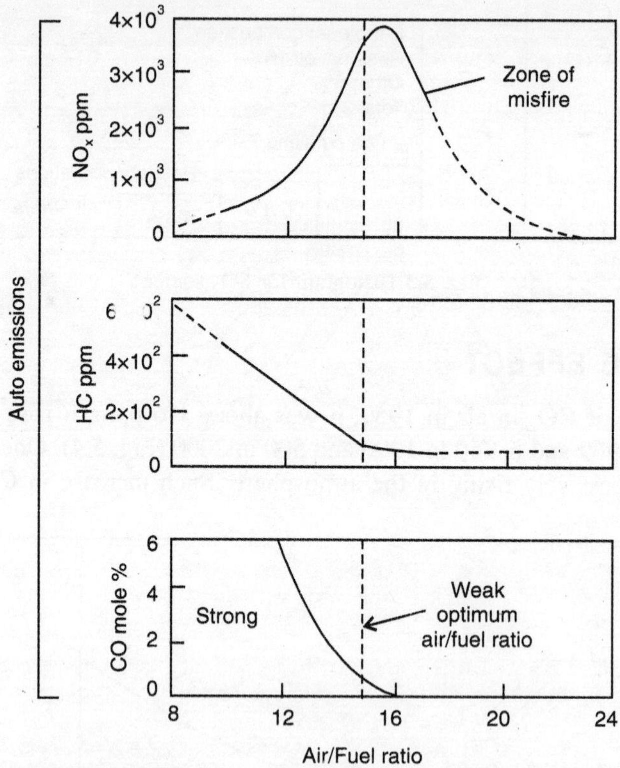

Fig. 5.5 Influence of fuel-air ratio on auto-exhaust emission

5.6 AUTO EXHAUST EMISSIONS

In auto-exhaust emissions along with CO_2, CO is also emitted, which is by far the worst pollutant emitted. As stated earlier, three wheelers, cars, generate enormous amounts of CO in the air. Figure 5.5 shows the variation of air-fuel weight ratio in auto exhaust emissions. If we maintain air/fuel ratio of 15, we get higher generation of NO_x (4000 ppm) with hydrocarbon (HC) of about 100 ppm and CO of about 2 ppm. The ideal carburettor setting is done at a ratio up to 1:13 to 1:16 ratio of fuel to air.

If we study the working of a four stroke engine with spark ignition (SI) Fig. 5.6. The position (I) represents intake stroke with inlet open and exit closed, when air and hydrocarbons are sucked in. Now (II) shows compression stroke with inlet and exit both closed when the piston moves up with compressing air-fuel in the ratio of 7:12. When spark ignition occurs in position (III), it represents power or work stroke when both inlet and exit are closed and combustion occurs. At this point temperature rises and leads to expanding combustion products push up the piston with exit open. Finally in position (IV), exhaust stroke exit is open, inlet is closed and exhaust gases go out. As a rule carburettor is located near the inlet control the mixture of air and fuel (15:1), which is sprayed by intake stroke (Fig. 5.6). Ideal fuel is iso-octane (C_8H_{17}, *i.e.* 2, 2, 4 trimethyl pentane). The combustion occurs as per the following reaction: $C_8H_{17} + 12\frac{1}{2} O_2 + (N_2) \rightarrow 8CO_2 + 9H_2O + (N_2)$. The consumption of fuel proceeds as per the following reaction: from (*a*) through (*b*) and finally (*c*) reaction.

(*a*) $C_8H_{18} \rightarrow C_{17}H_{15} + CH_3^+$ (cracking of hydrocarbons)

(*b*) $CH_3 + C_8H_{18} \rightarrow CH_4 + C_8H_{17}$

(*c*) $C_8H_{17} \rightarrow CH_3 + C_7H_{14}$

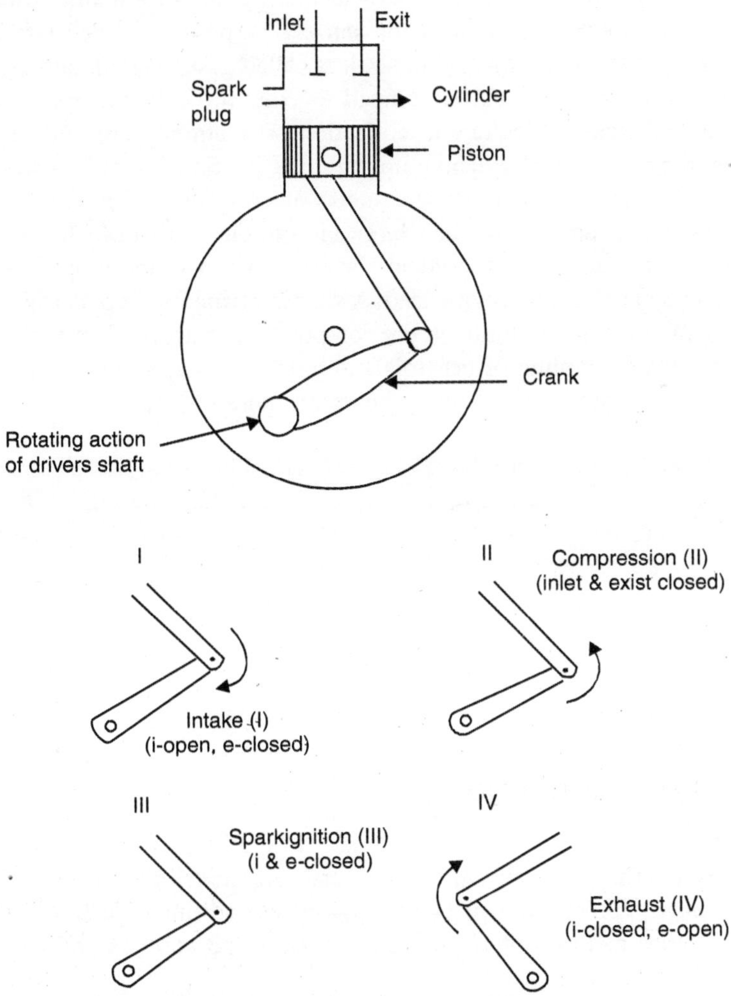

Fig. 5.6 Working of four stroke engine (i-inlet, e-exit)

A peroxide radical is formed along with peroxide chain. The final products in exhaust gases are H_2, O_2, formaldehyde, higher ketones, CO, CO_2 and water vapour. This may lead to octane dilemma. A clean air act (1970) stipulated definite ratios of HC, CO, NO_2 and TPS in air. The typical level in last 15 years are as presented in Table 5.7.

Thus, we can say that an ideal fuel-air ratio is 1:13. The best way is to add a catalyst to reduce emission of toxic gases. Pb $(C_2H_5)_4$ (ICA) is an antiknock, which curbs deposition of soot. The hydrocarbons which are emitted by petroleum combustion are olefins, parafins, aromatics, with carbon chain ranging from C_1 to C_{35} and residue has carbon chain ranging from C_{35} to C_{90}. The gasoline supply to engine can be enhanced by cracking, polymerisation, alkylation, or reforming high knocking agents to low knocking products. The common and most efficient catalyst used in the USA for such purposes is AK 33 X, which is really methylcyclopentadieneyl manganese carbonyl which reduces deposition of soot in the engine.

5.7 CLASSIFICATION OF AIR POLLUTANTS

Air pollution is due to the presence of substances in adequate quantities in air to cause human response, whether physiological, economical or in aesthetic sense. The pollutants are classified on the basis of their physical state. If a pollutant is dispersed in air, it is called gaseous pollutant, but a highly polymerised aggregate of liquids or solid substances lumped together are called aerosols. They contain solids and liquids. There are inherent difficulties in collection. The atmosphere with its load of pollutants constitute highly inhomogeneous and dynamic mixture, not in equilibrium or a steady state. The coexistence is possible due to dilution. An efficient mode of collection of pollutants is essential. If the concentration of pollutant in air is low, then large air samples will have to be collected to permit concentration of sufficient material for analytical purposes. All the interfering substances are assumed to be present before actual analysis is commenced. A simple syringe or hypodermic syringe with needle may cause severe fluctuations or variations in the composition to ascertain it with any degree of accuracy. Hence the methods of sampling of aerosols and gases pollutants are of crucial importance. Due to inhomogeneous nature of aerosols, sampling becomes more complex.

Table 5.7 Octane dilemma (%) and pollutants in air in various years

Pollutant	*1905*	*1970*	*1976*	*1980*	*1990*	*2000*
HC	11	2.2	0.5	0.25	0.06	0.01
CO	80	23	11	4.7	2.1	1.5
NO–NO_x	4	1.5	0.9	0.4	0.2	0.1
TPS	0.3	0.2	0.1	0.03	0.01	0.002

5.8 SAMPLING METHODS FOR AEROSOLS

We have already seen in last chapter different methods of sampling. Here we consider special methods for sampling of aerosols. They are inertial, gravitational, gradients, diffusional, sieving and filtration. They all need sampling equipment like air pump, flowmeter and finally analyser. A very brief discussion is taken to summarise methods of sampling of aerosols and gaseous pollutant.

(a) Inertial

A large group of techniques for separation of particles from gases may be termed as inertial. The inertia of a particle is greater than that of a gas molecule. If the direction of a gas stream is changed suddenly, due to collision the larger particulates gets deposited on the obstructing surface. With impingers, some sort of action occurs under liquid surface, as the liquid washes off the surface to provide more chance for air collection. The only limitation is, during bouncing action such particles are shattered and decomposed there by loosing their chemical identity.

(b) Gravitational

The particles are induced to leave with air as they travel, by induced gravity of centrifugal field. This principle is employed in sedimentation chambers. The centrifugal collections are used for the purpose of sampling. The larger particles settle quickly, provided there is no reaction there after collection.

(c) Gradient

The gravitational methods involve motion of particles under the influence of gravity. In gradient sampling, advantage is taken of some physical characteristics to effect sampling. Such a difference may originate due to a difference in properties of aerosols like electrical mobility, thermal character, diffusion, differences of light scattering capacity of aerosols. Electrophoresis is the basis of the electrostatic precipitator (ESP), which is useful for air pollution control and air-borne particulates. The sampling process is rapid and give high sampling rate. Due to generation of static electricity, the light particles are attracted by the precipitator, separating them from heavier components. The only demerit of this process is the generation of ozone, due to silent electric discharge. It further causes complications and collecting substrate poses a big problem, if it has no electrical conductivity. 'Thermophoresis' is the movement of particles under the influence of thermal gradient. A thermal precipitator draws sample on a heated surface and separates from those adhering to a cold surface. Thus particle collection on cold surface is beneficial in high concentration of aerosol particulates. The diffusophoresis is motion of particles in vapour concentration gradient, while photophoresis is migration of aerosols under the influence of intense radiation or light. However, both the forces are too small to provide rapid sampling. They are effective only for sampling photosensitive pollutants.

(d) Diffusion Method

This technique is very effective for sampling pollutant in form of fine particles. This is a statistical process, wherein collecting surface and small particles are brought together by the phenomenon of random Brownian movements of the suspended particles in the process of collision.

(e) Sieving

The largest air-born sample particulates may be collected by sieving with the help of fine mesh. This method is useful for filtering out insects from aerosols.

(f) Filtration

This is the most widely used single technique of removing particles from air. The membrane filters are used, which have small pore size. For dust and particulate matter, sampling, is the most efficient sampling technique.

5.9 SAMPLING OF GASEOUS POLLUTANTS

In comparison with sampling of aerosols which are heterogeneous in nature, the sampling of gaseous pollutants is somewhat simpler, because of the fact that by and large they are homogeneous. Unfortunately, both for aerosols as well as gaseous pollutants the task is difficult, because of sudden variation in wind velocity, wind direction. For such considerations we will have to account for several meteorological parameters. Now we would consider important methods for sampling of gaseous pollutants. The sampling methods for aerosols are classified as those involving storage or a grab sampling, condensation, adsorption and absorption.

(a) Storage for Grab Sampling

A portion of ambient air is closed in a container and stored. Mere isolation without concentration is possible by this technique. Grab sampling is also another name for this method. The storage at atmospheric pressure is made in rigid containers. Glass or stainless steel is used as container material. In a few instances, mylar—a special kind of material is used, as it is elastic and non-odorant. The nature of container selected will be decided on the basis of quality and kind of the air pollutant to be collected.

(b) Condensation

The low temperature condensation called cryo-sampling is an excellent method for air sampling at low temperature. The species condenses and gets concentrated. The cooling agent used consists of liquid hydrogen or even of liquid oxygen. This sampling technique provides an additional advantage of concentration.

(c) Adsorption

The gas is adsorbed on solid material at reduced temperature and then evacuated or flushed out with an inert gas like N_2 or Ar at elevated temperature. The material used for adsorption consist of porous material like silica, alumina, kieselguhr, diatomaceous earth or activated carbon. Large amount of gas can be easily sampled by this technique. Only limitation of this method is, polar substance cannot be easily desorbed from solid substrate. The process of adsorption is a surface phenomenon, which is used for purification of air for experimental purposes, like sample dilution.

(d) Absorption

It is popular method for sampling of gaseous pollutants. The reactive pollutants are absorbed in aqueous solution, e.g. SO_2 is absorbed in water or H_2O_2; ozone is absorbed in a solution of potassium iodide; the absorption of NO_2 is done in NaOH, as its absorption in water is very slow; CO can be absorbed in a solution of ammoniacal cuprous chloride; O_2 is absorbed in the solution of alkaline pyrogallol; CO_2 is absorbed in the solution of alcoholic KOH. In fact, this forms the basis of Orsat gas analyser. Thus by the process of absorption, a component of gas phase is transferred into a bulk of solid or liquid phase. The transfer process involves a chemical interaction and fixing of a particular species in the liquid phase. The ventilation, spraying, bubbling ensures intimate contact between gas and liquid or solid absorption; such sampling can also be classified as active and passive absorption sampling. The use of ion-exchange resin in packed bed can also be used as an excellent sampling medium. However, it is essential to maintain the sample in an extremely dry state so as to get successful, reliable and reproducible results of the final analysis.

5.10 ANALYSIS OF AEROSOLS

The aerosols are analysed to ascertain the composition of the entire sample in terms of its various constituents. This will enable us to obtain information of individual composition of the particles. The methods employed are bulk methods. The total content of atmosphere is expressed in terms of $\mu g/m^3$

(or ppm) for large samples, which are collected by electrostatic precipitator (ESP) or by filtration. The sample is then dissolved and analysed for its metal pollutant contents by AAS, ICP-AES, or spectrophotometry. Gravimetric or volumetric methods are useless for such purposes. Ring-oven technique is ideal for such a job. The single particle method is composed of three parts, viz. morphological examination, micrurgy test and microspot investigation. In morphological examination, an attempt is made to recognise a chemical species by mere observation. In micrurgy test, the sample identification or characterisation is done under light microscope with the use of precise micromanipulator. Finally in microspot investigation, a specific reagent is used for the identification of a material, e.g. nickel with dimethylglyoxime. The identification of radioactive material is done by radioactography.

5.11 ANALYSIS OF GASEOUS POLLUTANTS

The method for the characterisation and determination of atmospheric trace gases are more or less individual rather than collective. The use of different sampling methods for gases depends upon the availability and diversity of analytical techniques used for determination. The non-selective or the general method involve condensation of samples and analysis by mass spectrometry. Most reactive gases cannot be analysed by gas chromatography (GLC), as they are lost during collection. Non-selective methods are useful for less reactive gases, pollutants and closely related homologous series of hydrocarbons and aldehydes. GLC if used which involves use of FID (flame ionisation detector) or ECD (*i.e.* electron capture detector). In the specific analyser, polar traces of gases are absorbed in solution. No method of analysis is perfect and every method is subject of controversy. It is hoped that newer methods will be invented in course of time.

An endeavour is made to consider the basic principles in laboratory methods for the analysis of SO_2, H_2S, NH_3, NO-NO_x, CO, CO_2, O_3 and organic vapours. Such methods largely consist of spectrophotometric methods of analysis. In addition volumetric technique, conductometric analysis, infrared method, fluorescence measurement technique are also used. The methods for the continuous analysis or monitoring are described in subsequent chapters in great details.

5.12 SULPHUR DIOXIDE ANALYSIS

Its details of experimental process for analysis of SO_2 are given in chapter 6 in 6.1 to 6.4. These methods cover direct SO_2 analysis by hydrogen peroxide and lead candle methods.

(a) Acid Titration

Its important properties are acidity, reducing characteristic and precipitation method for gravimetry. There are also three spectrophotometric methods. In such method SO_2 is collected over H_2O_2 or H_2O and corresponding change in acidity is measured by titration or by measuring a change in conductivity, but the latter method is more reliable. Only drawback of this method is that, HF, HCI if present interfere but NO_2 does not interfere due to low solubility in water. NH_3 causes negative error.

(b) Reduction Method

For details of experimental procedure refer to chapter 6. In this method use of reducing property of SO_2 is made. The SO_2 is allowed to react with KI and the liberated iodine is titrated against standard solution of $Na_2S_2O_3$. It is possible to collect SO_2 over NaOH and continue with redox titration. The reduction method suffers from positive interference from other redox reagents, e.g. H_2S. Organic matter do interfere in such a determination, O_3 also interferes but NO_2 does not.

(c) Lead Candle Method

This is one of the classical method which was used for quite some time before photometric methods. A candle made of PbO_2 is exposed to atmosphere of SO_2 to undergo the following reaction: $PbO_2 + SO_2 \rightleftarrows PbSO_4$. The change in weight gives an idea about SO_2 in the air. The only limitation of this method is that dust and particulate matter also get deposited on lead peroxide to give erroneous results.

The best methods for laboratory analysis of sulphur dioxide is spectrophotometric methods. Such methods are summarised in Table 5.8.

Table 5.8 Determination of sulphur dioxide

(*a*)	SO_2 in H_2O_2, barium chloroanilate → Red complex → (λ_{max} = 530 nm)
(*b*)	$FeCl_3$, NH_4Cl, HCl, 1, 10—phenanthroline → coloured complex (λ_{max} = 510 nm)
(*c*)	PRA method
	Absorbing solution: NaCl, $HgCl_2$
	Reagent: PRA (para rosaniline) and formaldehyde (lucobase)
	λ_{max}: 560 nm; $\varepsilon = 3.0 \times 10^4$
	Colour development 20 min
	Working range: 1-10 μg
	Interference are less

(d) Spectrophotometric Determination of SO_2

Three methods are most common. The first method is used for direct analysis of SO_2 with absorption solution containing H_2O_2 and barium chloroanilate to give a red coloured complex with absorption maxima at 530 nm. The method is good for flue gas analysis, but lacks sensitivity required for ambient air analysis. The second method is based on absorption of SO_2 into a mixture of ferric chloride and ammonium chloride, $HClO_4$ and 1–10 phenanthroline. A colour formed is spectrophotometrically measured at 510 nm. The development of the colour is rapid, but the fading of colour is equally fast. H_2S in addition shows strong interference. The third method is the most widely used one known as PRA method. In this method, sodium chloride and mercury chloride are reacted with p-rosaniline (leucobase)

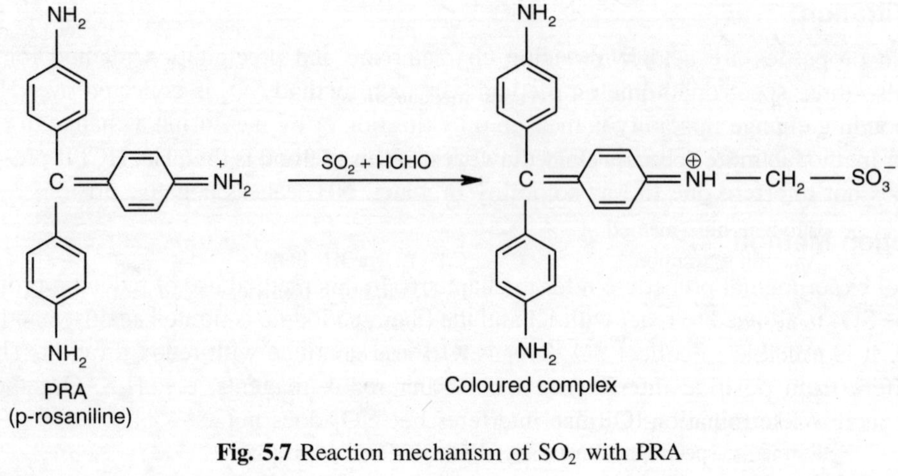

Fig. 5.7 Reaction mechanism of SO_2 with PRA

in formaldehyde, which gives intense red-violet colour of p-rosaniline methylsulphonic acid, showing strong absorbance at 560 nm. The mechanism of reaction is described in Fig. 5.7. At pH > 1.6, colour can be measured at 548 nm, which at pH < 1.2 is measured at 575 nm. The sensitivity of the method is in the range 1–10 ppm and the colour is stable.

5.13 H_2S ANALYSIS

The detailed experiment covering analysis of H_2S is described in chapter 6—section 6.4. In the absence of SO_2 it is determined by reduction technique. SO_2 if present can be removed by glass fiber filter impregnated with K_2CO_3. H_2S forms coloured precipitate with second group (of qualitative analysis) metals, e.g. Pb acetate turns black. In iodometric titration H_2S solution reacts with KI and liberated iodine is titrated with standard solution of thiosulphate.

(a) Spectrophotometric Determination of H_2S

There is only one reliable method called ethylene blue method (Table 5.9). In this method the absorbing solution used, contains $CdSO_4$ and NaOH. The dye ethylene blue so formed by interaction of $FeCl_3$ and N–N′–diethylphenylene diamine is measured at 670 nm. The Beer's law limit is 1–10 μg/ml with molar absorptivity $\varepsilon = 3.5 \times 10^4$. The reaction mechanism is depicted in Fig. 5.8. Gases like SO_2, O_3, NH_3 do not interfere, but sulphites, thiosulphates show strong interference, which can be eliminated with $FeCl_3$. The method is more sensitive if the complex is extracted in chloroform. For colorimetry, Cd, Pb or As are most useful for formation of precipitate or colour of sulphides.

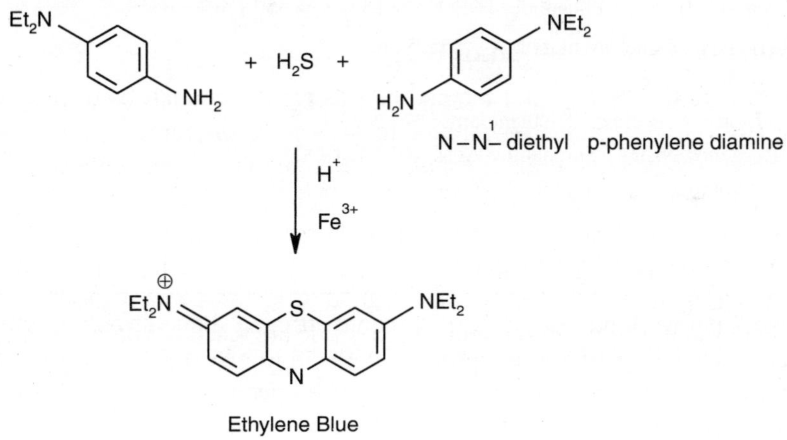

Fig. 5.8 Reaction mechanism of H_2S

Table 5.9 Spectrophotometric determination of hydrogen sulphide

Ethylene blue method		
Absorbing solution	:	$CdSO_4$, NaOH, $FeCl_3$ N–N′ diethyl p-phenylene diamine
Reagents	:	Ethylene blue
Colour	:	Instantaneous
λ_{max}	:	670 nm, $\varepsilon = 3.5 \times 10^4$
Working range 1–10 μg/ml		
Extractive spectrophotometry in $CHCl_3$ has better sensitivity		

5.14 NO-NO_x ANALYSIS

The details of method for analysis is provided in Chapter 6, section 6.7 to 6.10. All possible oxides of nitrogen exist in polluted atmosphere. NO–NO_x are most prevalent. They are determined by gas chromatography. NO converts to NO_2 to give rise to photochemical smog. The most useful method is Saltzman method, using a spectrophotometer.

(a) Spectrophotometric Determination of NO–NO_x

There are several methods of analysis as shown in Table 5.10. In phenoldisulphonic acid method, reaction of nitrogen oxide in H_2O_2 and H_2SO_4 is used to develop colour measurable at 400 nm. The reactions encountered are:

e.g.

$$2NO + H_2O_2 \rightarrow 2HNO_3$$

$$2NO + 3H_2O_2 \rightarrow 2HNO_3 + 2H_2O$$

$$HNO_3 + HO - C_6H_3\ (SO_3H)_2 \rightarrow HOC_6\ H_2\ (NO_2)\ (SO_3H)_2 + H_2O$$

Table 5.10 Spectrophotometric determination of oxides of nitrogen

(*a*) Phenoldisulphonic Acid (λ = 400 nm) useful up to 25 μg Organic nitrogen interfere

(*b*) Xylenol, $KMnO_4$, Prophylene glycol → Yellow complex λ_{max} 435 nm
Working range 2–500 μg/ml.

(*c*) $NaAsO_3$ (H_2O_2, sulphamide, H_3PO_4, NEDA) λ = 550 nm

(*d*) Triethanolamine, n-butanol, λ_{max} 550 nm

(*e*) Saltzman Method
Absorbing medium: Triethanolamine in NaOH
Reagents: NEDA, Sulphanilic Acid and CH_3COOH
λ_{max}: 550 nm, $\varepsilon = 4.0 \times 10^4$
Colour development: 15 minutes (stability: 24 hours)
Working range: 0.01–1.5 μg/ml

The method does not work beyond 25 ppm. Organic nitrogen shows strong interference. In xylenol method, nitrite ion reacts with $KMnO_4$ and propylene glycol to give a yellow complex with absorption maxima at 435 nm with limit of analysis as 2–5 μg/ml. All operations can be performed in single step. In the third method, sodium arsenite gas is absorbed on H_2O_2, in presence of sulphamide; H_2SO_4 and N–(napthyl) ethylene diamine hydrochloride (NEDA) a red coloured complex is measured at 550 nm. SO_2 does not interfere, the only limitation is that $NaAsO_3$ is toxic in nature. In triethanolamine method, the sample is absorbed in an aqueous solution of triethanolamine in butanol. Colour development is effected by diazotization reaction using NEDA to produce colour with maximum absorbance at 550 nm. The range of analysis is 0.005–0.5 ppm. In modified method, an absorbing solution in NaOH is treated with H_2O_2, sulphanilamide, H_3PO_4, NEDA to form a coloured complex with the limit of analysis as 0.04–1.5 ppm. However Saltzman's method is an excellent method for the analysis of NO–NO_x. The principal reactions involved in such analysis are depicted in Fig. 5.9. In this method absorbing solution is NaOH, trienthanolamine sulphanilic acid, glacial acetic acid and NEDA. Colour is developed instantaneously due to process of dye formation. It can be measured at 550 nm. The molar absorptivity is 4×10^4 with sensitivity of 0.01–1.5 μg/ml. The principal interferences are due to O_3 and high concentration of SO_2.

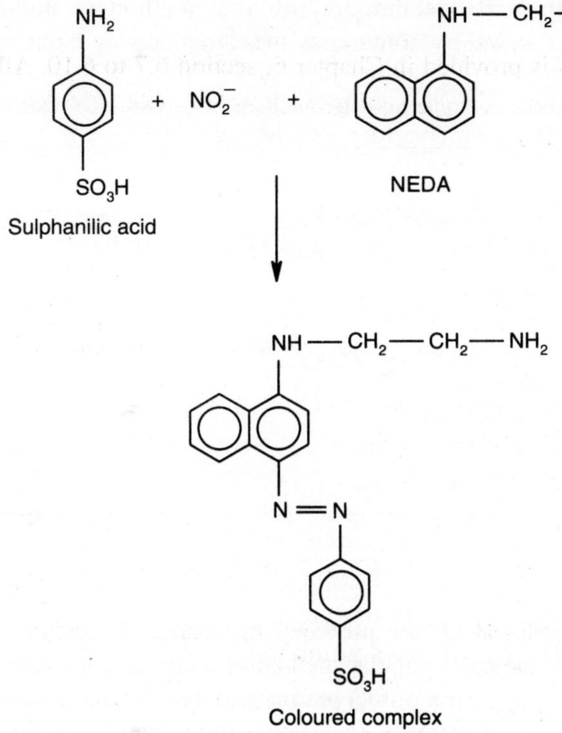

Fig. 5.9 Reaction mechanism of NO–NO_x

5.15 CO–CO_2 ANALYSIS

The experimental details for the analysis are provided in Chapter 6, section 6.13 to 6.16. The analysis of CO_2 is simple. It is bubbled through a solution of NaOH or KOH to form carbonates. The unreacted alkali is titrated against a standard solution of some acid. The measurement of CO_2 is the index of human and plant activity, as a significant portion of CO_2 in the air is anthropogenic. For CO analysis detection methods are many as shown below:

$$I_2O_5 + CO \rightarrow I_2 + CO_2 \text{ (}I_2\text{ vapour visual colorimetry)}$$

$$HgO + CO \rightarrow CO_2 + Hg \text{ (by AAS-cold vapour Hg analysis)}$$

$$PdCl_2 + CO \rightarrow COCl_2 + Pd \text{ (Pd by reflectance spectroscopy)}$$

Since CO is odourless, colourless, and tasteless gas, such indicator tubes are very useful in mines and deep sea operations.

(a) Spectrometric Method for Analysis of Carbon Monoxide

The indirect method for analysis of CO are already summarised in Table 5.11. In direct method CO reduces yellow silico molybdate to form molybdenum blue. The colour change from yellow-green to green and then to a permanent blue is gradual. Silica impregnated with ammonium molybdate in H_2SO_4 is reacted with CO to give blue solution with absorption maxima at 660 nm (Table 5.11). The method is sensitive. Only organic compounds, H_2S, moisture can reduce molybdenum and give wrong results.

This method is comparable with NDIR technique. But that method of non-dispersive infrared spectroscopic measurement is best suited by continuous measurements or monitoring.

Table 5.11 Spectrophotometric determination of carbon monoxide

Indirect method:

(*a*) Mercuric Oxide—reflectance spectroscopy ($HgCl_2$, CO) of Hg, H_2O, olefins, interfere

(*b*) Palladium Chloride—reflectance spectroscopy ($PdCl_2$, CO) metallic Pd

Direct method:

(*c*) SiO_2, $(NH_4)_2$ MoO_4, CO $\rightarrow$ molybdenum blue (λ_{max} = 660)
Absorbing medium: Silica gel impregnated by Ammonium Molybdate
λ_{max}: 660 nm
Colour development: Instantaneous
Working range: 1–10 μg
Interference: Organics, H_2S and Moisture

5.16 OZONE ANALYSIS

The experimental details for analysis of O_3 are provided in Chapter 7, section 7.1. The volumetric method involving redox reaction is the most popular method of analysis. The reaction involved is O_3 + 2KI + H_2O $\rightarrow$ O_2 + 2KOH + I_2 (at > 0.5 ppm of O_3) but interfering substance should be absent. 2% KI is buffered at pH 7. Final iodine concentration is measured volumetrically. It can also be determined spectrophotometrically at 352 nm. An alkaline solution of KI can also be used. The solution can be added to a mixture of H_3PO_4 and sulphuric acid. The method is useful for field analysis of O_3. Such titrations can be carried out by amperometry or coulometry. A fluorimetric method involves interaction of O_3 with fluorescent acridine to give fluorescence. The reliable techniques are spectral methods.

(a) Spectrophotometric Determination of Ozone

Various methods for such analysis are summarised in Table 5.12. The phenolphthalein method is based on the development of pink colour with reduced form of dye. It is oxidised by H_2O_2 with $CuSO_4$ as catalyst to give a pink colour with the maximum absorbance at 530 nm. SO_2 interferes in this method. The reaction of O_3 with 1,2–di(4-pyridyl) ethylene in acetic acid (glacial) gives a yellow coloured complex measurable at 442 nm. The aldehyde forms coloured compounds with 3–methyl–2–benzothiazolone hydrazine hydrochloride. The intense yellow coloured complex is measured at 442 nm. In N-tetramethyldiaminodiphenyl methane in dilute acetic acid gives a violet colour, measurable at 565 nm. When NO–NO_x are present, green colour turns to yellow (λ_{max} = 565 nm). In diphenylamine sulphonate method, $HClO_4$, sodium diphenylamine gives turquise-red colour (λ_{max} = 593 nm). While in 4–4′ dimethoxystilbene method, blue colour (λ_{max} = 610 nm) is obtained. NO_2, SO_2 do not interfere. Finally, reaction of phosphate buffered KI solution can be used at λ = 352 nm. The interference is due to SO_2, NO_2 and peroxyacetyl nitrate H_2S.

5.17 AMMONIA ANALYSIS

In Chapter 6 in section 6.10 provide experimental details for the analysis of ammonia. The air sampling methods for determination involves the absorption of the gas into dilute H_2SO_4. Various methods for

Table 5.12 Spectrophotometric determination of ozone

(*a*) Phenolphthalein, H_2O_2, $CuSO_4$ → pink colour
(λ_{max}: 530 nm)

(*b*) 1, 2–di (4–pyridyl) ethylene → (λ_{max} = 442 nm)
Yellow colour

(*c*) N-tetramethyl diaminophenyl methane → Blue-violet colour
λ_{max} 565 nm

(*d*) Diphenyl amine sulphonate
Sodium diphenyl amine, $HClO_4$ → λ_{max} = 593 nm
Turquise red-violet colour

(*e*) 4–4′ dimethyl stilbene → Blue (λ_{max}: 610 nm)

(*f*) Iodide Method
Absorbing solution: Phosphate buffered KI solution
Reaction: $O_3 + 3KI + H_2O \rightarrow KI_3 + 2KOH + O_2$ λ_{max}: 353 nm
Interference: SO_2, NO_2, peroxy acetyle nitrate and H_2S

analysis are summarised in Table 5.13. In indophenol method, $(NH_4)_2SO_4$, phenol, sodium-nitropruside and NaOCl are reacted to give a blue complex, measurable spectrophotometrically at 630 nm. Only formaldehyde interfere. Sensitivity is 0.025–1 ppm. The nitrite method is based on reaction of $(NH_4)_2SO_4$ with HOCI and Br_2 gas to give coloured complex (λ_{max} = 550 nm). The direct determination of NH_3 in air can be achieved at 204 nm in a 100 mm quartz cell where $\varepsilon = 2.7 \times 10^3$ with sensitivity of 7 ppm. A direct one-step method for the analysis of ammonia involves its absorption in dioxone containing o–(benzene sulphonamido) p–benzoquinone. It gives a coloured complex (λ_{max} = 480 nm). Finally in the Nessler method which is the best method, the brown colour is measured at 450 nm. The important reaction is $2HgI_4^{2-} + NH_3 + 3OH^- \rightarrow NH_2Hg_2IO$ (brown) $+ 7I^- + 2H_2O$. This is a sensitive method; the method however shows interference of aliphatic amines and H_2S. The method is simple and as such suffers minimum interference.

Table 5.13 Spectrophotometric determination of ammonia

(*a*) Indophenol–$(NH_4)_2SO_4$, Phenol, sodium nitropruside NaOCI
Blue λ_{max} = 630 nm
Formaldehyde interfere (23)

(*b*) $(NH_4)_2SO_4$, HOCI, Br_2 → NO_2^- formed diazotization orange (λ_{max} = 550 nm)

(*c*) NH_3 in quartz cell at λ = 204 nm and $\varepsilon = 2 - 6 \times 10^3$

(*d*) 1, 4-Dioxane, Benzene sulphimido-p-benzoquinone → λ_{max} 480 nm

(*e*) Nessler's Method
Absorbing Solution: H_2SO_4
Reagents $HgCI_2$, KI in NaOH
λ_{max}: 450 nm
Working range 10-100 μg/ml
Interference: ammonium salts, aliphatic amines, H_2S

5.18 ORGANIC GASES AND VAPOUR ANALYSIS

The Chapters 15 and 16 provide details of experimental procedures for analysis of few organic compounds. The best method is GLC. Lighter organics are determined by adsorption chromatography (GSC), CH_4 is determined at ppb level by GLC. 3–4 benzopyrene and polycyclic compounds are determined by GLC. Formaldehyde is a component of photochemical smog and is determined by chromotropic acid in H_2SO_4. Olefins and acroleins are also analysed by gas chromatography. It is not always possible to get definite information by GLC, but hyphenated GC-MS offers useful information in characterisation of organic pollutants.

It will be noted that for rapid analysis of gaseous pollutants in our environment, the best method is spectrophotometry. The methods are simple, inexpensive, with availability of a plethora of organic ligands. In the next chapter we would consider other methods for measurement of air pollutants.

ADDITIONAL READINGS

5.1 Mayuri Gandhi and S.M. Khopkar, *Ind. J. Envon. Prot.* **9**, 343 (1989).
5.2 *Scope Environmental Pollutants: Selected Analytical Methods*, Butterworths, London (1975).
5.3 A.K. De, *Environmental Chemistry*, New Age International Publishers (P) Ltd., New Delhi (2006).
5.4 A.C. Stern, *Air Pollution*, Vol. 3, 3rd Ed., Academic Press, New York (1976).
5.5 A. Attri, T.P. Igielbki and B. Jaselskis, *Anal. Chem.*, **42**, 1282 (1970).
5.6 P.W. West and G.C. Gaceke, *Anal. Chem.*, **28**, 1816 (1956).
5.7 M.B. Jacobs, M.M. Braveman and S. Hochheiser, *Anal. Chem.* **29**, 1349 (1957).
5.8 S.M. Khopkar, *Res. Ind.*, **29**. 86 (1984).
5.9 B.E. Saltzmann, *Anal. Chem.*, **26**, 1949 (1954).
5.10 APHA, *Methods for Air Sampling and Analysis*, American Public Health Association, p. 337 (1972).
5.11 H. Yagoda and F.H. Goldmann, *J. Ind. Hyg. Toxicol.*, **25**, 440 (1943).
5.12 A.A. Christie, R.G. Lidzey and D.W.F. Radford, *Analyst,* **95**, 519 (1970).
5.13 D.A. Levaggi, W. Sin and M. Feldstein, *J. Air Pollut. Cont. Ass.*, **23**, 30 (1973).
5.14 EPA United States Environmental Protection Agency, *Fed. Regist.*, **36**, 8200 (1971).
5.15 R.C. Robbins, K.M. Borg and E. Robinson, *J. Air Pollut. Contr. Ass.*, **18**, 106 (1968).
5.16 A.J. Haagensmit and M.F. Brunnelle, *Int. J. Air Pollut.*, **1**, 31 (1958).
5.17 T.R. Hanser and D.W. Bradley, *Anal. Chem.* **38**, 1529 (1966).
5.18 C. Arnold and C. Mentzel, *Ber. Dent. Chem. Ges.*, **35**, 1324 (1902).
5.19 H.H. Bravo and R.J. Robinson, *Anal. Chem.*, **33**, 1115 (1961).
5.20 H.H. Bravo and J.P. Lodge, *Anal. Chem.* **36**, 671 (1964).
5.21 D.H. Byers and B.E. Saltzmann, *Am. Ind. Hyg.* **19**, 252 (1958).
5.22 HLS Intersociety Committee, *Health Lab. Sci.*, **10**, 115 (1973).
5.23 V.W. Truesdale, *Analyst*, **96**, 584 (1971).
5.24 F.A. Gunther, J.H. Borkley, M.J. Kolbenzen, R.C. Blinn and E.A. Staggs. *Anal. Chem.*, **28**, 1985 (1956).
5.25 D.N. Kramer and J.M. Sech., *Anal. Chem.* **44**, 395 (1972).
5.26 M. Buck and H. Stratmann, *Z. Anal. Chem.*, **213**, 241 (1965).

Chapter 6

Analysis of Major Air Pollutants

6.1 INTRODUCTION

In the earlier chapters we had considered the importance of the methods of sampling of air pollutant. Now we would consider the details of the experimental procedures for the laboratory analysis of major air pollutant. In the earlier chapters we had taken an overall view of the kind of the air pollutants encountered in the atmosphere with brief mention of principles used for its characterisation or determination.

However mere indication of principles of analysis may not satisfy the dedicated worker as he would be more interested in carrying out the analysis of the air pollutant in the laboratory or on sight. Such analysis consist of the various methods. Therefore, we would consider analysis of oxides of the sulphur, nitrogen, carbon and hydrocarbons and analogous compounds.

For the analysis of SO_2, one prefers to use photometric methods involving use of p-rosaniline, or H_2O_2, or lead peroxide while for the analysis of H_2S we use N-N′ dimethyl p-phenyldiamine dihydrochloride in colorimetry. For compounds of nitrogen specially for NO_2 we use Saltzman method involving diazotisation of N (1-napthyl) ethylene diamine (NEDA) as the chromogenic reagent in spectrophotometry. One can also use for analysis of NO_2 arsenite method in photometric analysis. Apart from Nessler's reagent one can also use phenol-nitropruside for the colorimetric analysis of ammonia. Iodine pentoxide absorption of CO is good method for analysis of CO but p. sulphamine benzoic acid with silver as catalyst is also effective for the colorimetry, however NDIR is most dependable method for analysis of CO.

In all these experiments attempts have been made to give small details of reagent concentration, pH of complexation, use of appropriate buffers, methods for their preparation and the chromogenic ligand, stabilisation of colour of complex, preparation of the standards as well as construction of a calibration curve for assessing the concentration of a specific air pollutant from given unknown sample. All details of the various calculations have been also provided for easy comprehension.

Now we would consider actual experimental details for the analysis of pollutants. As regards the compounds of sulphur we would consider SO_2, H_2S, SO_3^{2-}, SO_4^{2-} analysis while for the compounds of nitrogen we would consider the oxides NO-NO_x if $x = 1, 2, 3, 4, 5$ as well as ammonia and nitrates and nitrites. In the following section on carbon, we would discuss the methods for the analysis of CO, CO_2

from the atmosphere. While in the subsequent section on halogen, we would consider the analysis of halogen compounds and compounds of oxygen such as O_3, H_2O_2 and finally analysis of suspended particulate matter from the atmosphere.

6.2 ANALYSIS OF SULPHUR DIOXIDE

The main laboratory methods include acidimetry, colorimetry and coulometry or conductometric methods of analysis. One can also use the flame photometry, emission and absorption spectroscopy. Several methods were used but they all lack sensitivity (0.01–0.02 ppm) (Table 6.1). The chemical methods are first discussed.

Table 6.1 Analysis of sulphur dioxide

S. No.	*Method*	*Absorbing media*	*Comments*
1.	Titrimetry	H_2O_2/water	SO_2 in H_2O oxidised by H_2O_2 to give H_2SO_4
2.	Titrimetry	KI/I_2	excess iodine titrated with $Na_2S_2O_3$
3.	Titrimetry	NaOH	SO_2 reacts with OH^- in alkaline media
4.	Nephelometry	H_2O_2	SO_2 detected as $BaSO_4$ turbidimetry
5.	Colorimetry	NaCl, $HgCl_2$	Colour developed with PRA as ligand
6.	Colorimetry	H_2O_2-Bachloro anilate	Absorbance measured at 530 nm
7.	Spectrophotometry	$FeCl_3$, NH_4Cl, HCl	o-phenonthroline gives colour to Fe^{2+}

(a) Colorimetric Method (Experiment No. 6.1, Analysis of SO_2 from Air)

SO_2 is absorbed in solution of potassium tetrachloromercurate (TCM). With the addition of p-rosanilinemethyl sulphonic acid and formaldehyde produces intensely coloured sample at pH 1.6 in the presence of 3M H_3PO_4 ($\varepsilon = 47.7 \times 10^3$ at 548 nm). The interference of nitrogen oxides by using sulphonic acid eliminates nitrite and addition of EDTA complexes heavy metals from the atmosphere.

Reagents: (1) Absorbing reagents are 0.04 M TCM (potassium tetrachloromercurate). Dissolve 10.80 g $HgCl_2$ + 5.96 g KCl + 0.066 gm EDTA in one liter of water. Adjust the pH ~ 5.2 (2) Sulphamic acid. Dissolve 0.6 g of acid in 100 ml water. (3) Buffer solutions: 100 ml of 0.1 M CH_3COONa–CH_3COOH (pH 4.7). (4) H_3PO_4 3M-dilute it to 20.5 ml. Use H_3PO_4 (85%) to one liter of water (5) 0.2% pararosaniline-leucobase–Dissolve 0.2 g reagent in 100 ml HCl (1M) (6) Formaldehyde 0.2%. Dilute 5 ml of 40% formaldeyde to one liter with water (7). Standardise sulphite solution, dissolve 0.4 g Na_2SO_3 in 500 ml of hot water. Keep it in refrigerator as solution is unstable at room temperature.

Procedure: Add 10–20 ml of absorbing reagent from midget impinger with flow rate of 0.5–2.5 L/mt. The sample is shielded from direct sunlight. Prepare reagent blank using same procedure (in absence of SO_2 gas). To sample solution add 1 ml of 0.6% sulphamic acid + 2 ml formaldehyde (0.2%) and 5 ml of chromogenic ligand (*i.e.* PRA). Make up to 25 ml and measure the absorbance at 548 nm.

Calculations: The concentration of SO_2 from the sample is calculated as

$$SO_2 \text{ (ppm/v)} = \frac{(A - A_0) \times 0.382}{V}$$

If A, A_0 are absorbance of sample and blank then 0.382 is volume in µl of SO_2 of 760 mm pressure and 25°C while *V* is the sample volume. The calibration curve can be also plotted from standard SO_2 gas in water. H_2O_2 is another method for automatic colorimetry.

(b) Hydrogen Peroxide Method for SO_2: (Experiment No. 6.2, Analysis of SO_2 with H_2O_2)

SO_2 is bubbled in volume H_2O_2 when SO_2 is oxidised to H_2SO_4. The amount of sulphate is measured in order to know amount of SO_2 gas. A decrease in absorbance of Ba-thoronol complex is measured at 520 nm. Thoronol is 1-(orthoarsenophenyl azo) 2-napthol–3,6 disulphonic acid used as sodium salt. Limit of SO_4^{2-} detection is 0.1 μg/ml. The sampling unit shows how to collect SO_2 in H_2O_2 (Fig. 6.1).

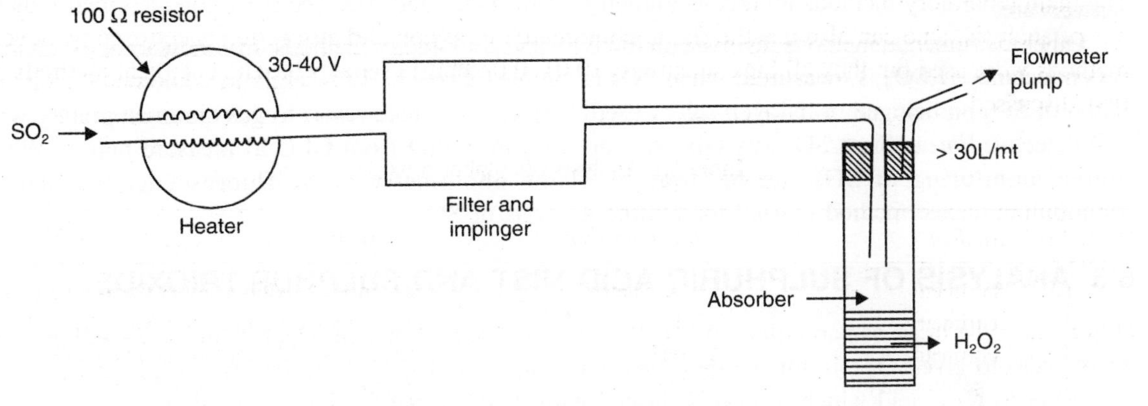

Fig. 6.1 Collection of SO_2 in H_2O_2

Reagents: (1) Absorbing solution (One vol H_2O_2): Dilute 10 ml of 100 vol H_2O_2 (30%) to one liter (2) $BaClO_4$ solution: weigh 0.90 gm perchlorate of barium and dissolve it in one liter of 0.1 M $HClO_4$ (8.6 ml $HClO_4$ 70% spgr) in distilled water; (3) Thoronol; weigh 0.20 gm and dissolve it in one liter of water, (4) K_2SO_4 solution. Add 0.183 gm K_2SO_4 to one liter of water. It contains 100 μg/ml of sulphate anion.

Procedure: Take 40 ml of aliquot of absorbing solution in the impinger (Fig. 6.1). Heat sampling unit with 100 ohm resistance at 30–40 V potential at flow rate of 1–30 L/mt. The standards are prepared by filling solution with unexposed reagent with requisite amount of K_2SO_4. A calibration plot is prepared and then calculation proceed as

Calculations: $SO_2\ (\text{ppm/v}) = \dfrac{\mu\text{g/ml of } SO_4^{2-} \times 0.255 \times V_a}{V}$ if V_a = vol. of absorbing solution

V = vol. of air, 0.255 is vol. of SO_2 corresponding to 1 μg/ml of SO_4^{2-}.

In above technique, $BaSO_4$ so formed can be analysed by turbidimetrically or barium chloroanilate by spectrophotometric methods.

(c) Lead Peroxide Method (Experiment No. 6.3)

SO_2 with lead dioxide gives insoluble sulphate. In 1934 Wilsdon and McConnel proposed this method. Only H_2S and H_2SO_4 interfere in quantitative analysis. A candle of PbO_2 with fiber glass is prepared. The preparation is carried out as follows: Take 7 gm of gum + 7 gm Gelman glass fiber filter + 112 gm PbO_2 in 700 ml of water. For emulsion preparation one can use dioxane. Each time 10 ml of mixtures is used. It is placed in dish and dried at 60°C. The plate is then exposed to SO_2 gas [Use 5% Na_2CO_3 to remove excess of insoluble $PbSO_4$ as $PbCO_3$]. The remaining PbO_2 is separated by filtration and SO_4^{2-} analysed by suitable method. The solution filtered and the filtrate is made to 100 ml. Then to

10 ml of aliquot 0.1 gm of sulphur powder is added and turbidimeter is used to measure turbidity at 450 nm.

Calculation: $\text{mg of } SO_4/100 \text{ ml} = 1029 \dfrac{w_1 - w_2}{A_T}$

wherein w_1 is wt of $BaSO_4$ exposed, w_2 = wt of blank, A = area of exposed surface, T = time (duration) of exposure.

Other instrumental methods involve conductivity and coulometry methods. In conductometry, change in conductance H_2SO_4 is measured, when SO_2 is bubbled through H_2O_2 while in coulometry reducing action of SO_2 on halogen or halogen salt is used. One can also take resort to gas chromatography with FID detector. For ambient SO_2 analysis one can use direct FID from GLC to measure pulses. While source monitoring of SO_2 can be done by flame photometry or by fluorescence method. A chemiluminescence method is used for routine analysis of SO_2.

6.3 ANALYSIS OF SULPHURIC ACID MIST AND SULPHUR TRIOXIDE

H_2SO_4 can exist in air as particulate. It is called mist. In the presence of water it combines with NO (or air oxygen) to give SO_4^{2-}. In this method involving analysis of acid mist, mist is passed through filter and water to form acid which in turn is titrated with 0.002 N NaOH. The calculations is

$$\text{ppm } H_2SO_4 = \frac{(\text{ml of NaOH}) \times N \times 22.4 \times 10^2}{2 \times 28.3 \times V}$$

where V = vol. of air, in cubic feet.

In Europe H_2O_2 method is preferred while in USA, the colorimetric PRA method is most popular. Very few users use coulometric technique. However the analysis of SO_3 is carried out by instrumental technique involving barium chloroanilate which can be measured by spectrophotometer. In GLC SO_3 when passed on $H_2C_2O_4$ gives CO, CO_2 where gas chromatography is used to analyse CO using special detector.

6.4 ANALYSIS OF H_2S: Experiment No. 6.4

The ISC (*i.e.*, Intersociety committee for manual methods of air sampling and analysis) has recommended methylene blue method with code No. ISC-42402-01-70 T. H_2S is collected on alkaline $Cd(OH)_2$ solution to give CdS, which in turn is measured by colorimetry with methylene blue as chromogenic ligand. A midget impinger is used to store absorbent. Pump with capacity of 2 L/mt is used.

Reagents: (*i*) Amine: H_2SO_4 (stock solution) in 30 ml water and 50 ml con. H_2SO_4 After cooling add 12 gms N-N dimethyl-p-phenylene diamine dihydro chloride in solution. (*ii*) Amine-H_2SO_4 (working, solution) Dilute 25 ml of stock solution to one Liter with 1:1 H_2SO_4 (*iii*) $FeCl_3$ solution dissolve 100 gm $FeCl_3$ in 100 ml H_2O. (*iv*) Ammonium phosphate: dissolve 400 gm of $(NH_4)_2$ HPO_4 in one litre water. (*v*) Absorbing solution dissolve 4.3 gm $CdSO_4$ and 0.3 gm NaOH in water to dilute it to one liter (*vi*) standard solution: A stock solution contains 400 $\mu g S^{-2}$. It is obtained by dissolving it in one litre 0.1 N NaOH (by gases H_2S) or Na_2S (3g in distilled water). Then standardise it with I_2 and $Na_2S_2O_3$ solution. Dilute standardised solution.

Procedure: Aspirate air sample to 10 ml absorbing solution in midget impinger at 1500 ml/mt for 2 hrs. Add 5 ml ethanol to prevent foaming. For analysis add 1.5 ml of amine solution to absorbing solution, add 1 drop $FeCl_3$ and transfer to 25 ml flask. Add 1 drop ammonium phosphate solution to discharge yellow solution. Make it to know volume. Keep it for 30 min. to develop colour and then measure the absorbance spectrophotometrically at 670 nm against reagent blank. Use standard solution to make calibration curve which is prepared as follows. To 10 ml of absorbing solution, add 1.5 ml amine solution + 1 drop $FeCl_3$. Developed colour is measured at 670 nm against sulphite free blank. In each analysis it is preferable to use fresh solution, Gas chromatography can be also used for measurement.

Calculation: The amount of H_2S is computed by following procedure

$$H_2S\ \mu g/ml = \frac{\mu_g H_2S \text{ from standard curves} \times 10^4}{v_s} \text{ wherein } v_s = \text{Volume of sample}$$

Only strong reducing agents interfere. NO_2 also shows interference. Sulphides other than H_2S interfere unless one adds Stractan-10 to eliminate interferences.

6.5 ANALYSIS OF MERCAPTANS AND ANALOGOUS COMPOUNDS: Experiment No. 6.5

They are measured spectrophotometrically with N–N dimethyl p-phenyldiamine (DMPDA) and $FeCl_3$.

Reagents: (*i*) Amine HCl solution—dissolve 5 gm DMPDA in one liter conc. HCl. (*ii*) Reissner Solution—dissolve 67.6 gm. $FeCl_3$ in 500 ml H_2O_2 and add 500 ml of solution containing 72 ml of conc. HNO_3 (*iii*) Mix 3 vol. of DMPDA with 1 vol. of Reissner solution. (*iv*) Absorbing solution—dissolve 50 gm mercuric acetate in 400 ml water + 25 ml glacial acetic acid (*v*) Allow it to stand the solution. Weigh 156.6 mg of lead mercaptide in 100 ml absorbing solution.

Procedure: Aspirate sample in 15 ml absorbing solution in a midget impinger at 1.5 L/mt. Add 2 ml of chromogenic ligand. The colour is developed in 30 minutes which is measured at 500 nm.

Calculation: $(\text{ppb})\ CH_3SH = \frac{A \times 0.51 \times B}{V}$ If A = absorbance, 0.510 is vol. of 1μg CH_3SH, B = calibration factor, V = vol. of sampled air.

6.6 ANALYSIS OF SULPHATE (SO_4^{2-}): Experiment No. 6.6

It is determined gravimetrically as insoluble $BaSO_4$. This insoluble sulphate can also be measured by making use of turbidimeter or nephelometer.

Reagents: (*i*) HCl (conc) (*ii*) Glycerol-alcohol solution (1:2 by volume) (*iii*) standard SO_4^{2-} solution, dissolve 0.1479 gm Na_2SO_4 in one liter of water. It is 100 μg/ml) (*iv*) $BaCl_2$ solution.

Procedure: Draw air through filter at rate of 10L/mt for 24 hours. Extract precipitate from filter paper and adjust volume to 25 ml. Use standard solutions of SO_4^{2-} from stock solution in 25 ml volumetric flask. To the sample as well as to blank add 10 N HCl (1 ml) 4 ml of glycerol-alcohol mixture. Add known amount of $BaCl_2$ crystals. Allow it to stand for 30 minutes. Decant turbid solution in cuvette and measure its absorbance at 500 nm against reagent blank by turbidimeter.

Calculations: Prepare a Beer's law plot for varying amount of SO_4^{2-}. Ascertain concentration of SO_4^{2-} from standard calibration plot. The atmospheric concentration of SO_4^{2-} is given by following equation

$$\text{Amount } SO_4^{2-} = \frac{(x-y) \times 25}{\text{Vol. of air sample } (m^3)} \mu g/m^3$$

where in this equation x = sulphate concentration of sample while (y) = filter blank solution reading.

6.7 ANALYSIS OF COMPOUNDS OF NITROGEN

Amongst the compounds of nitrogen in atmosphere, we have NO, NO_2, N_2O_3, N_2O_5, etc. (Refer Table 6.2). They are collectively designated as NO–NO_x if $x = 1, 2, 3, 4, 5$. In addition to these oxides we have ammonia (NH_3) and Peroxynitrate (PAN) which are also encountered in polluted atmosphere. The chemiluminescence is most dependable method for the analysis of oxides of nitrogen. Since NO gets oxidised to NO_2 there are several methods for the analysis of NO_2. The principal one amongst them in Saltzman diazotisation method. One can also use here coulometric method of analysis. The azo dye method was first suggested by Griess-Isolvay using NEDA (napthyl ethylenediamine) to develop colour in the absorbing solution. NO_2 is dissolved in ethanolamine solution. In coulometry action of NO_2 on KI solution is used for the purpose of estimation.

Table 6.2 Oxides of nitrogen

S. No.	*Oxidation state*	*Formula*	*Name*	*Kind of species*
1.	+1	N_2O	Dinitrogen oxide	Colourless
2.	+2	NO	Nitric oxide	Colourless
3.	+3	N_2O_3	Dinitrogen trioxide	Deep blue liquid
4.	+4	$2NO_2 \rightleftharpoons N_2O_4$	Nitrogen dioxide	Red brown gas
5.	+5	N_2O_5	Dinitrogen pentaoxide	White volatile solid

Analysis of NO_2 (Saltzman method for analysis of NO_2) (Experiment No. 6.7): About 0.005–5 ppm of NO_2 can be analysed. The strong oxidising or reducing agents interfere. The apparatus is shown in Fig. 6.2. A specially designed bubbler is used to collect NO_2 gas. Air flow is 0.4L/mt. The sample is drawn for 30 mts. Rotameter is used to know the rate of collection.

Reagents: Water should be nitrite free. $Ba(OH)_2$ and $KMnO_4$ is used to get rid of nitrite in water. During analysis it is better to use brown glass-ware in order to prevent photochemical reactions (*i*) NEDA *i.e.*, N (1-napthyl) ethylene diamine 0.1%, dissolve 0.1 g of ligand in 100 ml of water.

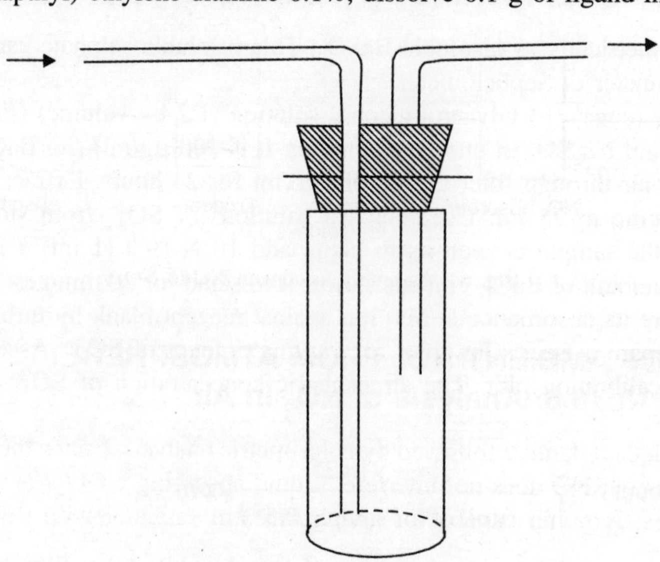

Fig. 6.2 Bubbler for NO_2 gas

(ii) Absorbing solution: Dissolve 5 gm sulphanilic acid in one liter water, containing 150 ml acetic acid. To 20 ml of mixture add 20 ml of reagent and dilute to liter. *(iii)* Standard $NaNO_2$ solution (0.0203 g/L). 1 ml of this working solution gives colour equivalent to 10 μg/ml of NO_2 in atmosphere (Refer to Table 6.3 for comparison).

Table 6.3 Spectrophotometric determination of oxides of nitrogen

Spectral method	*λ max (nm)*	*Range of analysis (μg/ml)*	*Remarks*
Phenol disulphonic acid	400	1–25	Organic N interferes
Xylenol orange	435	2–500	$KMnO_4$ acts as oxidant
Sodium arsenite	550	–	H_2O_2 sulphanolic acid
Triethnanol amine	550	–	n-butanol solvent
Saltzman method	550	0.01–1.5	NEDA, sulphanolic acid are used

Procedure: Take an aliquot of 10 ml of absorbing solution and assemble sampling unit (Fig. 6.2). Draw air sample through a 0.4 L/mt to develop colour. Measure absorbance of diazotised orange coloured complex at 550 nm on spectrophotometer against blank. Prepare a calibration curve with known concentration of $NaNO_2$ solution. Measure slope of line in graph.

Calculation: NO_2 (ppm) = absorbance × K/V if V = vol. of sample, K = constant

6.8 ANALYSIS OF NO

In 1972 it was proposed as ISC method as 42001-01-7T are made by Inter Society Committee for manual methods for air sampling and analysis. NO is oxidised to NO_2 by chromic oxide. NO_2 is then determined by Saltzman method. The apparatus used is shown in Fig. 6.3.

A 10 ml Saltzmann reagent is put in absorber and air is drawn through sample by a pump at rate of 0.4 L/mt. Note the volume of air passed and measure the absorbance at 550 nm. The calibration may be carried out using standard solution of sodium nitrite.

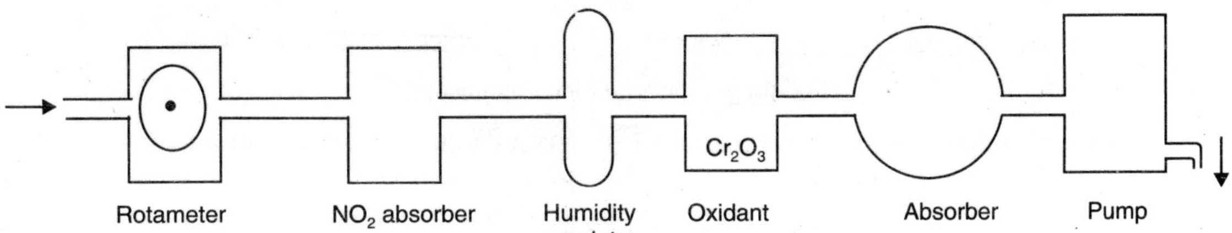

Fig. 6.3 Assembly to convert NO to NO_2

6.9 ANALYSIS OF AMBIENT NO_2 FROM ATMOSPHERE: Experiment No. 6.8: Analysis of NO_2 in Air

NO_2 is absorbed in triethanolamine followed by colorimetric analysis. Factor for converting $NO_2 \rightarrow NO$ is 0.72. SO_2 interferes but NO does not interfere. Liquid absorbing bubbler is used. Solid adsorbent is packed in glass tubes. A pump capable of giving 0.5 atm vacuum, with flowmeter is used during analysis.

Reagent: (*i*) Liquid absorber-dissolve 15 g triethanolamine in 500 ml water + 3 ml n-butanol and then made up to one liter (0.1 N) (*ii*) Solid absorber 25 gm ethanolamine + 4g glycerol + 50 ml acetone + water (*iii*) H_2O_2 dilute 0.2 ml 100% vol. of H_2O_2 to 250 ml. (*iv*) Sulphanilamide: dissolve 10 gm 400 ml water + 25 ml H_3PO_4 (conc). All made up to 500 ml (*v*) NEDA-dissolve 0.5 g in 500 ml of water (*vi*) The standard nitrite solution dissolve 0.15 g $NaNO_2$ in one liter water.

Procedure: Adjust flow rate to 150-200 ml/mt. Then sample the air for 24 hours. In liquid absorber add 50 ml water. Take 10 ml aliquot and add reagent. Then add 1 ml of H_2O_2 + 10 ml of sulphonamide solution + 1.4 ml NEDA. Mix thoroughly and after 10 min measure the absorbance at 550 nm. Then transfer solid absorber in 50 ml test tube. With 10 ml water make it up to known volume. Measure absorbance of coloured sample at 550 nm.

Calculations: $NO_2\,(\text{ppm}) = \dfrac{\mu g\ NO_2^- \times 5}{1.88 \times 0.85 \times V}$ if V = vol. of air passed through while 1.88 is factor for converting NO_2 (gm) → NO_2 (ml) (*i.e.* weight is converted to volume).

6.10 ANALYSIS OF NO_2 BY ARSENITE METHOD: Experiment 6.9

Saltzman method has major limitations of being not useful for field analysis or continuous monitoring Jacob-Hochheiser developed Arsenite method. The method eliminates delay in colorimetric analysis and sampling for 24 hours. NO however interferes in analysis and has low collections efficiency. NO_2 is bubbled to NO_2^- and is used for diazotisation reaction. The apparatus used is shown in Fig. 6.4. It permits analysis of 5-750 μg/m³ of NO_2 from air. 40 μg NO_2 gives 0.020 absorption.

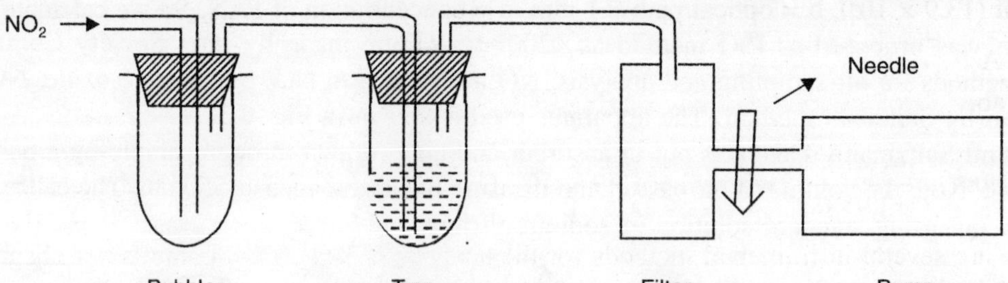

Fig. 6.4 Sampling of NO_2 (continuous)

Reagents: (1) Sulphonamide 20gm dissolved in 750 ml water (2) H_3PO_4 50 ml (85%) is mixed sulphonamide. Keep it in freeze.

(3) Absorbing solutions: 4 g NaOH + 1 gm of Na_2AsO_3 dissolved it in one liter water. (4) Saltzman reagent dissolve 0.5 g NEDA in 500 ml water and refrigerate it (5) Standard nitrite (NO_2^-) solution dissolve 1.5 gm of $NaNO_2$ in one liter water. It will have 1000 μg/ml of nitrite. Use diluted solutions if necessary. (6) H_2O_2—dil 0.2 ml to 250 ml with water.

The conversion factor NO_2^- to NO_2 is $= \dfrac{1.5 \times 100}{\%\ Na\ NO_2}$

Procedure: A 50 ml of absorbing solution is placed in absorber tube. Flow of air is started and sampling continued for 24 hours. After adding water absorbing solution is taken, 1 ml of H_2O_2 is added (to oxidise $SO_2 \rightarrow SO_4^{-2}$) + 10 ml sulphonic acid and 1.5 ml (NEDA, Saltzman reagent). Allow 10 ml to settle and develop a colour. Measure the absorbance of coloured solution at 510 nm. Carry out calibration and plot the Beer's law plot using standard solutions of known concentration.

Calculations: If V-vol of sampled air. The ambient level of NO_2 is determined as

$$\mu g\ NO_2^- /m^3 = \frac{\mu g\ NO_2^-/ml \times 50}{0.85V}$$

where 50 is vol. of absorbing solution, 0.85 is collection efficiency, V = air volume sample of NO_2 in parts per volume it is calculated from following formula

$$\mu g \frac{NO_2}{m^3} \times 5.32 \times 10^{-4} = \text{ppm } NO_2 \text{ (in volume)}$$

6.11 PEROXY ACETYL NITRATE (PAN) ANALYSIS

PAN is formed by oxidation of hydrocarbons. The compound is based on reaction shown as per equation.

$$R-\underset{\underset{O}{\|}}{C}-OO^+ + :NO_2 \rightarrow R-\underset{\underset{O}{\|}}{C}-OONO_2$$

Gas chromatography is best method for the analysis of PAN.

The retention time is 30 seconds with sensitivity of analysis as 1 ppb. The reagent includes pure nitrogen gas. A calibration curve is prepared with known concentration of PAN samples. 50 μl of ethyl nitrite vapour is injected into chamber. PAN can be standardised with IR spectrometer. Using Beer's Lamberts law we get PAN concentration as A = abc if A = absorbance measured, a = molar absorption coefficient (13.9 × 10^4), b = optical path 0.1 mm, c = concentration of PAN. So we calculate it as A = abc, $c = \frac{A}{ab}$

$$c(\text{ppm}) = \frac{A \times 10^{-4}}{13.9 \times 0.1}$$ where a = 13.9 × 10^4, b = 0.1, A = absorbance, c = concentration.

There are several instrumental methods for the analysis of NO_2. They comprise of chemiluminescence which is universally used in continuous monitoring.

6.12 ANALYSIS OF AMMONIA FROM AIR: Experiment No. 6.10: Analysis of Ammonia

The common method of the analysis is acidimeteric titration, but acidic and basic components interfere. Two colorimetric methods involving use of indephenol or Nessler reagent are quite common, but H_2S, formic acid interfere. Ion selective electrodes can be also used for rapid analysis of ammonia from air or water.

(a) Indephenol Blue Colorimetric Method for Ammonia

NH_3 is collected over H_2SO_4 in the standard impinger. NH_3 converts to NH_4^+ which is then analysed colorimetrically. Here SO_2, O_3, NO_3 fortunately do not interfere. 0.02 μg/ml of NH_3 can be analysed by this method.

Reagents: (1) Absorbing solution 0.0025 M H_2SO_4. Prepare with 1 M H_2SO_4 by mixing 28 ml conc. H_2SO_4 in 500 ml water made finally to one litre (2) Phenol nitropruside: dissolve 5 g phenol + 25 mg sodium pentacyano nitrosyl ferrate. Na_4 $(FeCN)_3$ NO $2H_2O]$ in 500 ml water. (3) Alkaline hypochlorite dissolve 5 g NaOH in 500 ml water (4) Standard $(NH_4)_2SO_4$ dissolve in 0.370 gm salt in one litre of water, (5) Dilute $(NH_4)_2SO_4$ solution during analysis.

Procedure: Take 50 ml absorbing solution. Aspirate air for 30 mt at rate of 30 L/mt. Note volume of air sampled at 760 mm pressure and 25°C temperature. Transfer 5 ml of absorbing solution and add 10 ml phenol-nitropruside reagent. Then add 10 ml alkaline hypochlorite. Prepare blank similarly in absence of ammonia. Allow it to stand for 30 mt. Measure the absorbance of coloured solution at 625 nm with blank using standard solution. Use Beer's law plot to find out unknown concentration of the ammonia in atmosphere.

Calculations: The amount of ammonia from sampled air is evaluated as

$$NH_3\,(\text{ppm}) = \mu\text{g/ml } NH_4^* \times 1.36 \times V_a / V_s$$

where V_a is final volume of absorbing solution, V_s = Volume of air sampled

1.36 is factor for conversion of μg ammonium ion to μl ammonia.

6.13 ANALYSIS OF NITRATE: Experiment No. 6.11: Analysis of Nitrate

It is formed in photochemical smog. NO_3^- is leached on filter paper and converted to NO_2^- by copper catalysed with reduction NH_2–$NHSO_4$ in alkaline media. Then NO_2^- is determined by Saltzman method.

Reagent: (1) Diazotisation reagent—dissolve 10 gm sulphonamides and 0.5 g NEDA in 10% v/v H_3PO_4 and made to one litre (2) $CuSO_4$—dissolve 2.5 g to one litre with H_2O (3) Hydrazine sulphate—dissolve 27 gm in one litre (4) NaOH—dissolve 12 gm in one litre water (5) NO_3^- standards—dissolve 0.163 g KNO_3 in one litre (1 ml = 100 μg/ml NO_3^-)

Procedure: Draw air through filter at rate of 10 L/mt for 24 hours. Extract sample from filter paper. Make it to known volume. Prepare calibration curve using standard potassium nitrate solution. To sample add 6 ml $CuSO_4$ + 6 ml NaOH + 4 ml hydrazine sulphate. Mix well and heat on water bath. Cool it and add 6 ml of diazotisation reagent. Measure coloured absorbance at 540 nm against reagent blank.

Calculations: The amount of NO_3^- is found as

$$NO_3\,(\text{ppm}) = \frac{(x - y) \times 10}{\text{Vol. of sampled air}}\ \mu\text{g/m}^3$$

where in x = μg/ml sample concentration y = μg/ml of blank solutions.

6.14 AMMONIUM ION: Experiment No. 6.12: Analysis of NH_4^+ ion

Ammonium ion from air can be analysed by colorimetric method. Air is drawn at rate 10L/mt for 24 hrs. Filter paper is extracted so also blank. To 5 ml of a washed solution add 10 ml sodium nitropruside reagent and proceed as described in procedure for the analysis of ammonia.

Calculation: It is similar to one described earlier as the ammonium from air is calculated by the following equation as:

$$NH_4^+ \text{ in air} = \frac{(x - y) \times 25}{V}\ \mu\text{g/m}^3 \text{ if } x = \mu\text{g/ml of sample solution}$$

$$y = \mu\text{g/ml of blank solution.}$$

6.15 ANALYSIS OF CARBON MONOXIDE

The main source is the incomplete combustion of fuels. It readily reacts with blood haemoglobin to produce carboxyhaemoglobin. Number of methods are proposed for its analysis in stream of blood involving colorimetry. Auto-exhaust emission generates substantial amount of CO. There are very good methods for the quantitative analysis of CO. They include absorption method, Iodine penta-oxide method, colorimetry with p-sulphaminobenzoic acid and NDIR method.

(a) Absorption Method

Cu_2Cl_2 is used to absorb it as $Cu_2Cl_2{\cdot}2CO{\cdot}4H_2O$ in acidic/alkaline media. The basis of orsat analysis involves absorption of CO in alkaline cuprous chloride (cosorbent is used which is a mixture of β-napthol + Cu_2Cl_2). As small as 10 ppm can be analysed. A contraction in volume due to sorption of CO in Cu_2Cl_2 gives idea in Orsal gas analyser.

(b) Analysis of CO_2 (Iodine Penta-oxide Method): Experiment 6.13

The reaction of absorption proceeds as activated carbon is used to remove the hydrocarbons, while chromic acid is used to absorb NO and O_3 on glass beads. CO generated is passed on to form CO_2 on I_2O_5. The calculations are:

$$I_2O_5 + 5CO \underset{150°C}{\rightleftharpoons} I_2 + 5CO_2 \uparrow$$

The liberated iodine can be determined by colorimetry or by volumetry by titration. The details of the process are shown in Fig. 6.5 and Fig. 6.6.

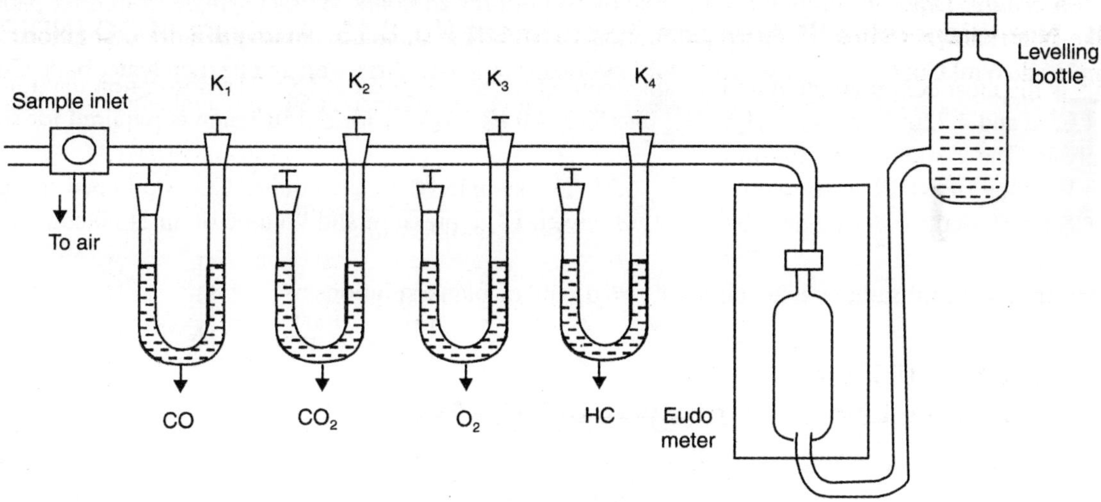

Fig. 6.5 Orsat gas analyser

(c) Manual Colorimetric method for CO: Experiment No. 6.14: Analysis of CO_2

Silver salt of p-sulphamino benzoic acid reacts with gas A colloidal. Ag is formed which can be measured spectrophotometrically at 425 nm.

Reagents: (1) 0.1 p-sulphaminobenzoic acid (2g) in 100 ml of 0.1 M NaOH (2) 0.1 M $AgNO_3$, (3) 1 M NaOH.

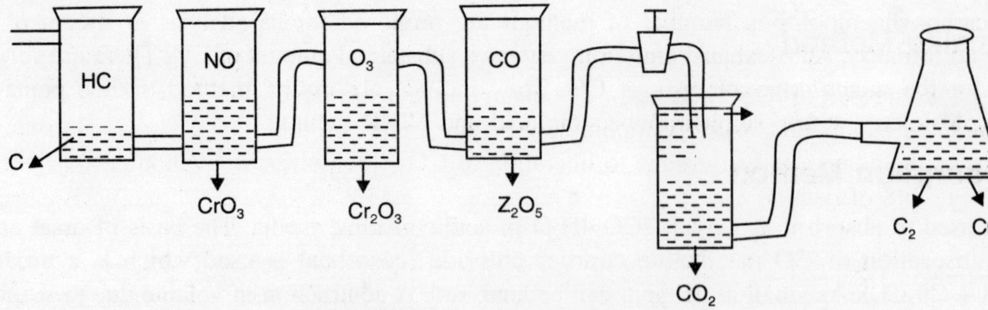

Fig. 6.6 Iodine pentoxide absorption

Procedure: To 20 ml of p-sulphaminobenzoic acid, add 20 ml of 0.1 M $AgNO_3$ then 10 ml of 1 M NaOH solution. It is absorbing solution. Take 10 ml absorbing solution and admit measured train of air containing CO in it. It generates suspended silver which is slightly coloured. It is measured at 425 nm.

Calculation: Prepare first standard Beer's law plot using pure CO

ppm of CO = μl of CO/Volume of sample air.

Other laboratory methods are less common and very rarely used e.g. The combustion of CO to CO_2 and then subsequent measurement of CO_2. HgO can be also reduced by CO to pure Hg and CO_2. Hopcalite method can be used to liberate CO_2 which is absorbed and washed. The Hopcalite catalyst mainly consists of (MnO_2 + CuO). As much as 20 ppm CO can be analysed.

(d) Non-dispersive IR Analysis: Experiment No. 6.15: Analysis of CO (NDIR)

This is the most common, rapid and reliable method for the analysis of CO. The absorption maxima is at 4.67 μ and 4.72μ. Cyanogen, H_2S, NO interfere. An IR model with 2–15μ range is required for such analysis. The working is as follows:

Working of NDIR is simple (see Fig. 6.7). The sample of air containing CO is pumped through $CaCl_2$ latter used for drying and then into gas cuvette. The pressure and volume of air is noted and the

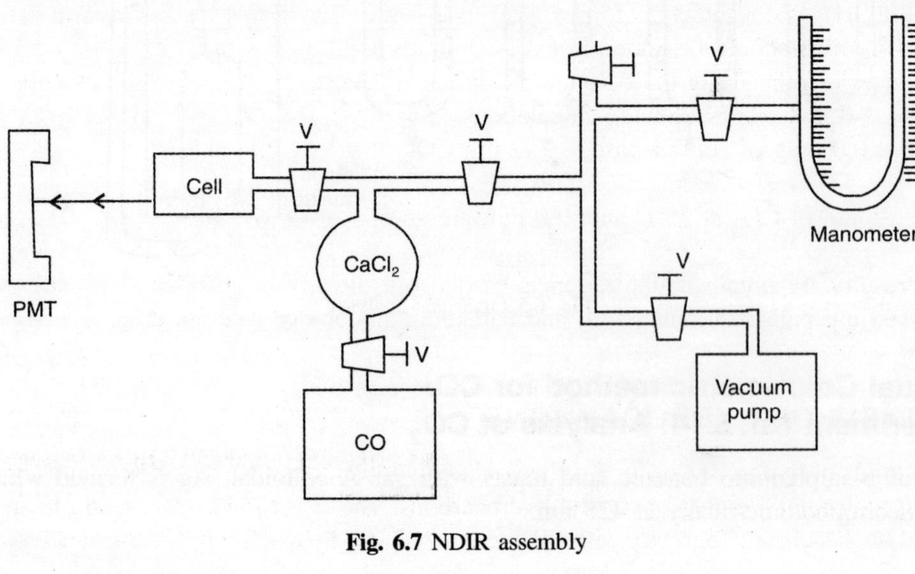

Fig. 6.7 NDIR assembly

(CO) in cuvette is measured spectrophotometrically at 4.67μ. The manometer measures air volume and pressure.

(e) Hopcalite Method

This catalyst is composed of ($CuO + MnO_2$). The change in temperature or change in volume of CO is measured but latter technique is more commonly used. Temperature used is 100°C. Gravimetric method is used to know the change in weight due to liberation of CO_2 when measured volume of air is introduced. (See Fig. 6.8 of details).

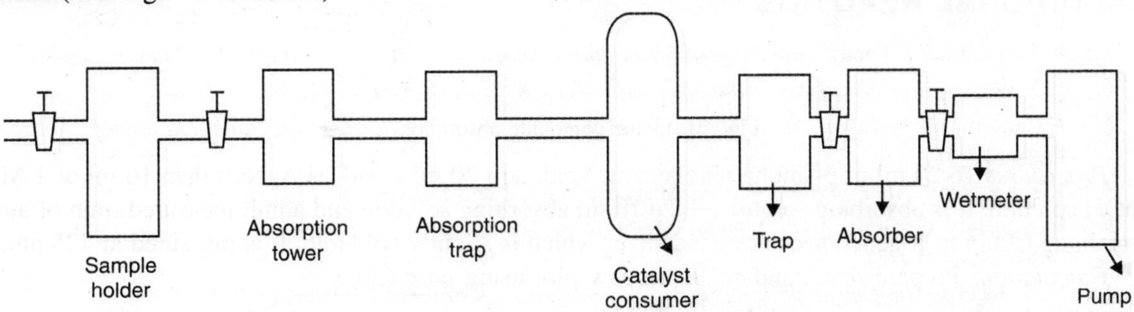

Fig. 6.8 Hypochlorite method for hydrocarbons

Calculation: CO ppm at 20°C $= \dfrac{517.6 \times \text{mg CO}}{\text{Vol. of air (V)}}$ Also CO ppm $= \dfrac{\text{mg CO} \times 555.54}{\text{Vol. of air}}$

Change in weight of catalyst = mg of CO as 1 mg CO at 25°C = 555.54 ppm CO.

6.16 ANALYSIS OF CARBON DIOXIDE: Experiment No. 6.18: Analysis of CO_2

On average level of CO_2 in the atmosphere varies from 300–350 ppm. This is safe limit for mankind.

Procedure: Air is passed to glass bubbler containing 20-50 ml of 0.01 $Ba(OH)_2$ (6%) and 0.2–0.4% butanol. The $Ba(OH)_2$ is consumed to form $BaCO_3$ and water. $Ba(OH)_2 + CO_2 \rightarrow BaCO_3 + H_2O$. The unused $Ba(OH)_2$ which should be added in excess initially is titrated with 0.1N oxalic acid with 1% phenolphthalein as an indicator or 0.5% thymolpthalein in alcohol or both can be used in a mixture. The amount of CO_2 is found from following calculations.

Calculation: 0.01 mg of acid or base = 0.22 mg CO_2

$$\text{ppm of } CO_2 \text{ at 25°C and 760 mm pressure} = \frac{\text{mg } CO_2 \times 873}{\text{Vol. of air in litres}}$$

The best results are obtained if titration is carried out directly in absorber. The indicator tubes containing potassium palladosulphite. I_2O_3 and Palladium silicomolybdate are used to absorb CO_2 in qualitative test.

6.17 ANALYSIS OF HYDROCARBONS

There are two kinds of hydrocarbons, one is the volatile hydrocarbons with low molecular weight while others are high molecular weight nonvolatile hydrocarbons. The lower molecular weight hydrocarbons are determined by gas chromatography with FID detector. C_1-C_5 hydrocarbons are thus analysed. While high molecular weight hydrocarbons if volatile are generally analysed by GC/MS. With capillary columns, they form C_6-C_{20} fractions. In addition to GLC one can use spectroscopic methods like Laser Raman Spectroscopy or high resolution infrared spectroscopy. The specific methods for the analysis of organic compounds are presented in subsequent chapters.

In conclusion, we can summarise that analysis of major air pollutants like oxides of sulphur, nitrogen and carbon, instrumental methods are quite useful. Thus amongst then spectrophotometric methods are by and large most popular for the analysis of air contaminants.

ADDITIONAL READINGS

6.1 R. Perry and R.J. Young, *Handbook of Air Pollution Analysis*, Chapman and Hall, John Wiley & Sons (1977).

6.2 P.O. Warner, *Analysis of Air Pollutants*, John Wiley & Sons, New York (1976).

6.3 A.C. Stern, *Air Pollution* Vol. 3. *Measuring Monitoring and Surveillance of Air Pollution*, 3rd Ed., Academic Press (2006).

6.4 J.O. Ledbetter, *Air Pollution,* Part A, Analysis, Marcell and Dekker, Inc. (1972).

6.5 Intersociety Committee for Manual Methods of Air Sampling and Analysis, American Public Health Association (1972).

6.6 M.B. Jacobs, *The Chemical Analysis of Air Pollution*, Interscience Publishers (1960).

6.7 R. Perry and D.H. Slater, *Chemistry and Pollution*, Ed. F.R. Benn, C.A.McAulffe, MacMillan Press, London (1975).

6.8 M.B. Jacobs, *The Analytical Chemistry of Industrial Poisons Hazards and Solvents*, Wiley-Interscience (1960).

6.9 B.F. Saltzman, *Anal. Chem.*, **26**, 1949 ibid (1954). **38**, 760 (1961), Anal. Chem. **37**, 1261 (1965).

6.10 S.M. Khopkar, *Environmental Pollution, Monitoring and Control*, New Age International Publishers (P) Ltd., New Delhi (2005).

CHAPTER 7

Analysis of Minor Air Pollutants

7.1 INTRODUCTION

We had considered so far the experimental details of the analysis of major air pollutants. The atmospheric pollutants included were oxides of sulphur, nitrogen and carbon. In fact the SO_2, NO-NO_x, CO_2 constitute the important components of the major pollutants in air. Apart from these we had considered ammonia and hydrocarbons which foul the atmosphere. In majority of these pollutants spectrophotometric methods of analysis involving the use of the chromogenic ligands were used popularly. However, in case of CO the non-dispersive infrared spectroscopy was predominantly used.

Apart from these compounds, several minor components of atmospheric pollutants foul the air. They consist of ozone, hydrogen peroxide, all kinds of halogens, specially chlorine and its compounds with gases like phosgene, freon, hydrocarbons and the suspended particulate matter (SPM). The last component is most harmful to the health as the damage caused to the respiratory track which depends upon the size of the particulate matter. Ozone does not occur freely in the atmosphere. In fact degradation of ozone layer in atmosphere leads to the global warming. The halogens are most common in occurrence specially chlorine gas and HCl vapour. They are primarily used for the disinfection of water. Fluorocarbons like Freon released from cryogenic industry are very harmful for the human health.

There are specialised methods for analysis of these pollutants. Thus O_3 is analysed by colorimetry or coulometric methods, Fluoride is best analysed by alizarine or SPADNS photometrically. The determination of dust and respirometric suspended particle matter is done by cascade impactor coupled with nephelometric methods or by conventional gravimetric analysis of suspended particulate matter.

In this chapter, every attempt is made to provide optimum experimental conditions such as ligand concentration, pH, methods for preparation of standards, and buffer solutions, needed for their quantitative analysis. This would certainly facilitate carrying out experiments in the laboratory.

So far we had considered the experimental methods for the analysis of important air pollutants. Some of them included SO_2, SO_3^{2-}, SO_4^{2-} as well as NO, NO_2^-, NH_3, PAN. Also there was a discussion on the analysis of CO, CO_2 and brief information on analysis of hydrocarbons. We propose to describe the experimental methods for the analysis of ozone, hydrogen peroxide and analogus products. First we would consider the analysis of halides like fluorine, chlorine, chlorinated and halogenated hydrocarbons. The chapter will be concluded with the detailed description on the methods for the analysis of suspended particulate matter on account of grab sampling. Cascade impactor and analogous technique shall be discussed once again though it has been thoroughly covered in the earlier chapters.

7.2 ANALYSIS OF OZONE: Experiment 7.1

This is the most important oxidant in air. It is a reactive gas. There are chemical as well as physical methods for its analysis. The average concentration in air is < 0.08 ppm. Other methods include the redox titration or colorimetry or coulometry. The simplest method involving iodine is colorimetry which involves liberation of iodine in following manner

$$O_3 + 3KI + H_2O \rightleftharpoons KI_3 + 2KOH + O_2 \uparrow$$

Reagents: (1) Absorbing solution (1% KI + 0.1M phosphate buffer). Dissolve 10 gm KI in 20 ml of water. Add buffer of pH 6.8 (which is obtained by dissolving 13.6 gm of KH_2PO_4 + 14.2 gm of Na_2HPO_4 in 500 ml water and made upto one litre (2) Iodine stock solution (0.05 N). Dissolve 16 gm KI + 3.18 gm resulbimed I_2 crystals dissolved in one litre of water. Keep it aside for 24 hrs. for stabilisation. Then standardise it against 0.025 M $Na_2S_2O_3$ and $K_2Cr_2O_7$ (3) For working solution (0.001 M). Dilute 5 ml of 0.05 stock solution to 100 ml. Protect it from sunlight (4) Standard solution of iodine (5) SO_2 absorber (6) Chromic oxide is used as scrubber during reaction.

Procedure: The assembly for ozone analysis is shown in Fig. 7.1. A 10 ml of absorbing solution is put in the midget impinger at the flow rate of 0.5–3 L/mt for 15 mts. Measure total volume of air passed.

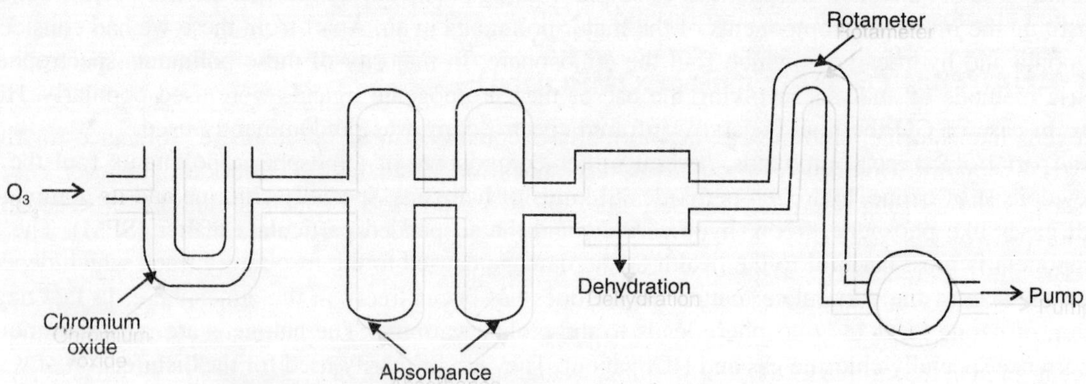

Fig. 7.1 Assembly for analysis of ozone

So also measure temperature and pressure. Transfer absorbing solution to 10 ml volumetric flask. Within 30 minutes measure the absorbance of coloured solution at 352 nm on spectrophotometer.

Then prepare a standard calibration curve from the stock solution of iodine.

Calculations: The corrected sample of volume of air at 760 mm and 25°C is

$$O_3 \text{ ppm} = \frac{\mu\text{l of } O_3 \text{ from calibration curve}}{\text{Vol. of sample of air passed}}$$

$$O_3\ \mu\text{g/m}^3 = \frac{\mu\text{g of } O_3 \text{from calibration curve}}{\text{m}^3}$$

(b) Phenolpthalein method for analysis: Phenolpthalein with $CuSO_4$ give rose coloured solution but H_2S, NO_2 interfere in such determination.

Reagents: (1) Phenolpthalein solution. Reflux mixture of 10 g NaOH + 5 g powder zinc + 4 gm phenolpthalein + 20 ml water. Dilute it to 50 ml. (2) $CuSO_4$ (0.01M) *i.e.*, 0.250 g dissolved in 100 ml water (3) Absorbing solution: To 100 ml flask add 10 ml of 0.01M $CuSO_4$ and prepare dilute solution as required during the course of analysis.

Procedure: Fill midget impinger with 10 ml of absorbing solution. Pass air at flow rate of 0.5 - 1 Lt/mt for 60 mts. Measure the absorbance of coloured solution. The standards can be obtained from H_2O_2. The solutions have 0.1 µg peroxide/ml and 0.1 ml Cu_2SO_4 and phenolpthalein solution per 10 ml volume. There are many colorimetric methods mainly involving bleaching action by ozone which results in decrease in magnitude of absorbance by the reaction. The O_3 with chromogenic ligand or the formation of ozonolysis products undergoes sensitive colour reaction.

The other methods include chemiluminescence, UV absorption, 1R absorption but chemiluminescence method with ethylene is most accurate and reproducible. In UV spectrometry O_3 absorption at 253.7 nm can be measured by using mercury resonance line.

7.3 ANALYSIS OF HYDROGEN PEROXIDE

The chemical method involves interaction of H_2O_2 with KI in presence of $(NH_4)_2MoO_4$ acting as the catalyst for rapid colour development: H_2O_2 on acid hydrolysis with Ti (IV) gives peroxysulphate complex which is measured colorimetrically at 407 nm ($\varepsilon = 2.4 \times 10^4$). In another method air containing H_2O_2 is bubbled through $Ti(SO_4)_2$ solution. This mixture with 8 quinolniol (0.1%) in chloroform gives coloured absorbance which can be measured at 450 nm ($\varepsilon = 3 \times 10^3$).

7.4 ANALYSIS OF HALOGEN COMPOUNDS

Halogens like fluorine, chlorine etc. are normally encountered in the atmosphere. Tolerance for fluorides is 2.5 mg/m^3. Colorimetry provides best methods for the analysis of fluorides. However, several substances show strong interference. Therefore only two prominent methods are used for the quantitative analysis of fluoride.

(a) The most commonly used method is to form complex of ZrO^{2+} with Alizarine solution which develops intense red coloured complex. This complex on interaction with soluble fluoride like NaF reduces the absorbance of the complex resulting decrease in absorbance proportional to fluoride concentration in the sample SPADNS (Sodium 2-p sulphophenyl) 1–8 di hydrocyanapthol-3-6 disulphate) was developed in Geological Survey of India Laboratories, Kolkata and it remains to be most reliable method where bleaching of Zr (SPADNS) complex occurs analogous to bleaching of alizarine.

The interfering ions are removed on anion/cation exchange resin in column packed.

(b) Analysis of Fluoride: Experiment 7.2

Titrimetric analysis consists of following operations:

Reagents: Alizarine solution, Fluoride, Thorium titrate (0.1 N), HCl (0.25 M), NaOH (0.25 M), chloroacetate buffer of pH 4.5–5.1. Pipet 10–15 ml of sample solution, add 2 drops of indicator (*i.e.* alizarine) add HCl till pink colour disappears then add 5 drops of buffer solution and titrate against Thorium nitrate till pink colour is obtained. A blank can be obtained in same manner by developing colour in the absence of fluoride ion. One can use spectrophotometric method with several advantages.

Procedure: Alizarine forms complex with Lanthanum salt at pH 4.3–5.2 to give blue coloured complex with fluoride anion. This is used for analysis of fluoride ion by the spectrophotometry. The interfering ions are eliminated by masking with the sequistering agents (Alizarine 0.000005M, 0.02M, $La(NO_3)_3$, buffer pH 5.2, acetone). To an aliquot of solution containing fluoride anion, add 10 ml of alizarine solution to form coloured complex and 2 ml of buffer pH 5.2 and finally add 10 ml of

$La(NO_3)_3$ solution. Add 25 ml acetone and dilute it to 100 ml. Allow it to stand for 90 minutes and ascertain that the colour is developed, it is measured spectrophotometrically at 618 nm against reagent blank. Find out the concentration of fluoride in unknown sample from caliberation plot of standard solution.

Alternative procedure: Same procedure as described above can be repeated with minor modification viz. use 2 ml buffer (pH 4.3). With no addition of acetone, the resulting coloured complex can be measured in UV region at 280 nm.

7.5 ANALYSIS OF CHLORINE FROM ATMOSPHERE

It is determined with o-toludine by spectrophotometer at 435 or 490 nm in the alkaline solution. The OTA method is most popular. Use of N–N′ dimethyl p-phenylene diamine or 3 3′ dimethyl napthadine instead of o-toludine can be used with some advantage. These methods are reliable provided chlorine is present at low concentration.

Methyl orange method: Methyl orange solution, chlorine solution.

Procedure: Air is sampled through 100 ml sample at rate of 1.2 L/mt. The resulting colour solution is then transferred to glass cuvette and absorbance is measured at 505 nm. From the calibration curve with standard solution, amount of the chlorine from unknown sample can be calculated. However OTA is best method. Here Toludine at pH ~ 1.3 gives yellow coloured complex. The sample of air is drawn through impinger containing NaOH. The hypochlorine so formed is treated with o-toludine to give yellow coloured complex with absorption maxima at 440–490 nm against reagent blank prepared similarly. The NaOH concentration used is 0.0125 M. After neutralising alkali with nitric acid o-toludine is added. The absorbance is measured at 490 nm. The mercury thiocyanate and iron (III) per chlorate methods are used for analysis of hydrochloric acid.

7.6 ANALYSIS OF HYDROCHLORIC ACID: Experiment 7.3

The following two methods are worth mentioning:

(*a*) *Mercury thiocyanate method:* Reagents are 2M HNO_3; absorbing solution, 0.01 M NaOH, 0.1M ferric ammonium indicator, saturated solution of $Hg(SCN)_2$ in methanol. The stock solution of chloride anion is required.

A known volume of air is passed through impinger containing 50 ml of 0.01 M NaOH at rate of 30L/mt. Take 20 ml of solution from it and add 2–3 drops 2M nitric acid + 2 ml of 0.1 M ferric alum sulphate + 2 ml mercury thiocyanate in methanol. The intensity of developed colour is measured at 460 nm.

(*b*) *Iron (III) perchlorate method:* Reagents are 0.1 M Iron (III) perchlorate in 5M perchloric acid and standard chloride (10 μg/ml) solution.

Procedure: First pass the air through bubbler containing sample solution for 10 mts at rate of 30 L/mt. Take coloured complex which has intense yellow colour and measure the absorbance at 350 nm. One must use 8.5 M $HClO_4$ to get reproducible results. Then calculate the concentration of chloride ion from calibration curve obtained by standard solution containing hydrochloride vapour.

7.7 ANALYSIS OF FLUOROCARBONS AND CHLORINATED HYDROCARBONS

CCl_2F_2 is used in refrigerator and air conditioner which directly interacts with atmospheric ozone to decompose it into injurious products. Gas chromatography with ECD can be used for its analysis but

GC-MS works far better for both $CFCl_3$ as well as CF_2Cl_2 cryogenic sampling is best. Fluorocarbon-11G-12 were analysed with ECD detector. These are most injurious as they react with O_3 to decompose it and promote global warming. The chlorinated hydrocarbons can be also analysed by same methods as used for the analysis of hydrocarbons with chlorine as described above.

7.8 ANALYSIS OF HALIDES: Experiment 7.4

(a) Ambient Chloride Analysis: Reagent

The 0.01 HN mercuric solution, (To 25 ml, H_2O add 0.25 M concentrated HNO_3 + 2.40 g mercuric nitrate–make it to one litre. (2) Standard solution of NaCl 0.025 N dissolve 1.46 gm NaCl in one litre. (3) Mixed indicators: To 100 ml methanol add 500 mg diphenyl carbazine + 500 mg bromophenyl blue (4) Alkali: dissolve 10g NaOH in one litre water) (4) Hydroquinone solution. Mix 1 gm of hydroquinone in 10 ml water and make it upto 100 ml.

Procedure: To an aliquot of solution, add 5 drops of mixed indicator + 2ml hydroquinone. Add HNO_3 dropwise till colour changes to yellow from original blue-violet colour. Then check pH 3.0–3.5 of solution. Take standard mercuric nitrate solution and titrate it till blue-violet colour is obtained with the analyte solution.

Calculation: $\text{ppm HCl} = \dfrac{VN_m T \times 6.24 \times 10^4}{A\,Pm}$ If sample volume(s), (S – B) = V in ml of B = Vol. req. to titrate.

N = Normality of Hg $(NO_3)_2$, M = Total vol. of sample, m = vol. of sample aliquot of solution.

T = Sampling temperature, P = atmosphere pressure, A = volume of sample air passed through.

(b) Method for analysis of fluoride: Experiment 7.5 (colorimetry)

Reagents: (1) SPADNS-dissolve 0.985 gm in 500 ml water, (2) Fluoride stock solution: dissolve NaF, 2.210 gm in one litre (3) Reference solution to 100 ml water add 10 ml SPADNS and maintain acidity, 1:1 with hydrochloric acid. (4) Zirconium solution: To 0.265 g $ZrOCl_2$ in 50 ml add 700 ml contact and make it to litre (5) Zr-SPADNS combine 50 ml SPADNS and 50 ml zirconium solution (1:1) reagent.

Procedure: To an aliquot of solution, add 5 ml Zr-SPADNS. Read absorbance at 570 nm. Prepare calibration curve containing standard fluoride ion. The colour is developed in 30 mts. Prepare calibration curve to find out unknown ion.

7.9 ANALYSIS OF PARTICULATE MATTER (SPM)

The source of these particulate matter is from stationary and mobile sources. The interaction of anthropogenic matter also generates enormous amount of suspended particulate matter. Let us consider special methods of analysis followed by physical and chemical examination as well as quantitative determination of SPM. They are classified as suspended materials and deposited products in the atmosphere. Their difference in size of two variety is 10 μm. They occur in plumes from chimney.

The consideration of meteorological parameters is of utmost importance while measuring particulate matter. As told earlier, we must consider the wind direction, wind velocity, temperature. Wind varies and a nemometrics are some of the instruments used for measuring of meteorological parameters. Let us consider sampling of suspended material, dust fall and classification of particulates.

Sampling of Suspended Material

(1) *Filter paper technique:* With the good suction pump particulates are sucked through filter paper. Note weight of paper before and after suction is taken to provide the difference. Since we know amount of suspended material, cellulose and glass fiber filters are commonly used. Whalman No. 1 and 4 are ideal filter papers for this task. Glass fibre filters are fragile, but can be used at the high temperature. Membrane filters are also specially used for microscopic work, but they have high resistance to air flows during suction Nucleopore filters are also good. They both have high efficiency. There are several methods for quantitative analysis of particulates.

(2) *Determination of particulate concentration:* Gravimetric method: This is most widely used and simplest technique of characterisation and of quantitative analysis. The change is weight before and after sampling is taken to find out amount of suspended material deposited on filter when high volume sampler is used. The size of inlet orifice must be known to ascertain size of particulate matter. On an automatic devise, Drechsel bottles and filter papers are used. Tape sampling with filter paper spool is most dependable for continuous monitoring of SPM, the other method is with β-radiation monitor. Especially it is good for monitoring of smoke. When working on fields, pumps operating on patterns can be used. In the mass monitoring air is drawn through chamber containing electrostatic precipitor which precipitate particles from air on crystal. Crystal must be cleaned from time to time to procure reproductive results. Figure 7.2 shows the typical cascade impactor. It separates particles into fractions on basis of their aerodynamic impaction properties. Anderson Cascade impactor is best. There are 5 stages in membrane filter. Each stage has accurately drilled holes (Nos. 360–400 per each stage, *a*, *b*, *c*, *d* shown). These holes are small as we go down the impactor.

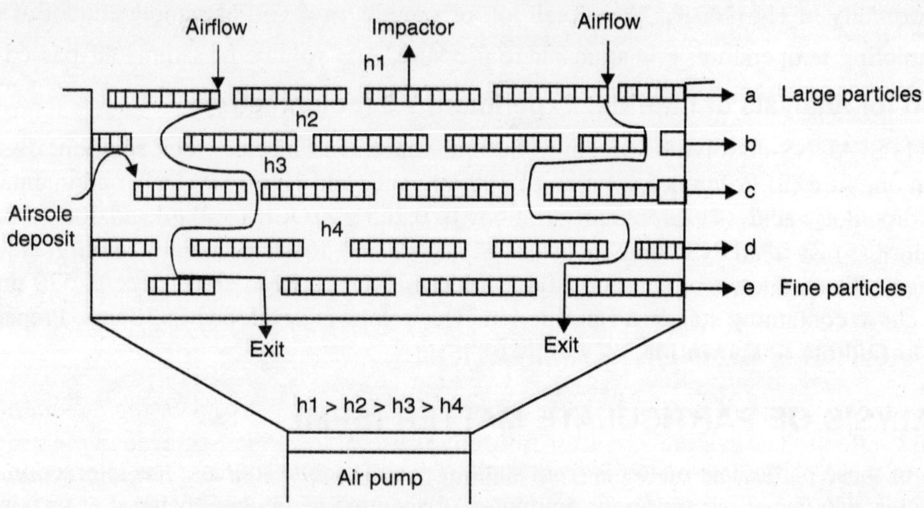

Fig. 7.2 Cascade impactor

The rate of drawing air is 1 cubic ft/mt. Large particles are deposited in stage (*a*) while finest particles are deposited in stage (*e*) in Fig. 7.2. Velocity in each stage rises due to aerodynamic effect. This is because aerodynamic impaction property depends upon density, shape, hollowness of species. Too long sampling gives erroneous results. The model with four stages is called volume fractionating impactor used with high volume sampler.

(3) *Light scattering method:* This method involves the use of nephelometer which is more useful for analysing particulate of very small size. It permits automatic monitoring in network. One generally

measures the extent of scattering of the radiation. The clean air and half scale signal provides countercheck for calibration. Another device used is aerosol particle counter. The third device is the directional sampler. This performs the task of controlling sampling by wind direction.

7.10 SAMPLING OF DUST FALL

The deposit gauges, monitor the deposition of particulate matter. For dust angle of fall is of vital importance e.g., with average density of 2, the fall at angle of 75° lead to 61% degree. We have variety of gauges to measure dust fall. They include British Standard, French Standard, Norwegian standards with minor variation from one another. These gauges are used in recognising as trouble shooters source of nuisance of dust. The measurements are made for long period to get conclusive data e.g. sampling for one month duration, short survey are possible to know level of dust fall. If wind speed exceeds certain limit sampling must be stopped (see Fig. 7.3). Disc sampling can be done at spot which location is at least 1.515 metre above ground level. Further there should not be building or vertical structure in the radius of 3 metre of dust collector. Exposer is made for 8 hours. The change in weight is measured in evening.

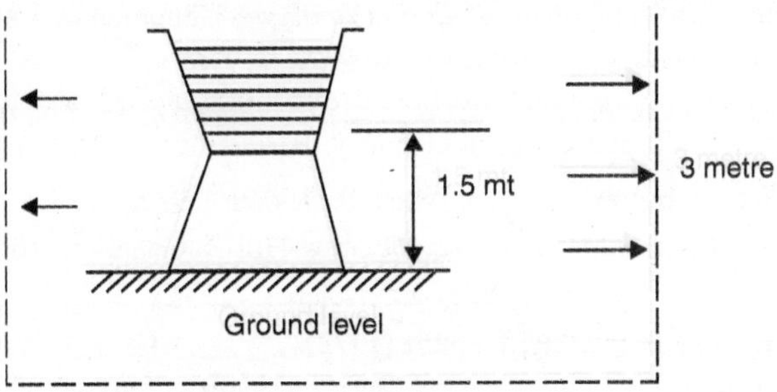

Fig. 7.3 Short survey of dust fall location

In various ways we can classify these particulates but expected support for such classification should be justified. One can design density gradient separation and density gradient liquids for such purpose of classification.

7.11 PARTICLE SIZE DETERMINATION

The assessment of particle size is of great importance. Here one has to measure the density and size to relate toxic effects. The cascade impactor fruitfully is useful for initial segregation of particulate matter. The coarse particles can be segregated by ranging from mesh size of 40–1000μm. Microscopic investigations can be carried out for small sized particulates. Martin diameter is used to specify the size and shape of particulates. Micrometer can be also used to gauge the size. Granules are used to assess size.

So far we have considered the theoretical as well as the practical aspects of air pollution. In subsequent chapters we propose to discuss air pollution monitoring and in multifaced characteristics. It will be noted that many laboratory methods used for the analysis are somewhat closer to monitoring technique.

We would try to summarise various methods used for the analysis of air pollutants. There are three methods for the analysis of SO_2 but PRA is most preferred method of analysis. For analysis of mercaptans, H_2S have limited methods. For analysis of oxides of nitrogen specially NO and NO_2 have three methods including Saltzman method of acid dissociatives. There are limited methods for the analysis of ammonia or nitrate. For analysis of CO_2 and CO we have as many as four methods while for CO nondispersive infrared spectrometry is the ideal choice.

We had also considered number of methods for the analysis of ozone, fluoride and HCl vapour from air. For ozone spectrophotometric method is more reliable. The bleaching of colour of the complex with SPADNS or Alizarnic is best method for the characterisation fluoride. OTA continues to be best method for the analysis of halides or halide vapour in the atmosphere. The fluorocarbons are seriously objectionable pollutants which are analysed by spectrophotometry. Finally for the analysis of the suspended particulate matter, mostly reliable method is gravimetry provided if the particles are large or we resort to nephelometry if particles are finally divided. Cascade impactor facilitates the collection of suspended matter before analysis.

In the next chapter we would consider monitoring methods. They involve methods for continuous measurement of the pollutants. It will be noted that several of the laboratory methods are also used for monitoring of the pollutants.

ADDITIONAL READINGS

7.1 R. Perry and R.J. Young, *Handbook of Air Pollution Analysis*, Chapman & Hall, New York (1977).

7.2 P.O. Warner, *Analysis of Air Pollutants*, John Wiley & Co. (1976).

7.3 A.C. Stern, *Air Pollution—Measuring Monitoring and Surveillance of Air Pollution*, Academic Press (Indian Ed.) Vol. 3 (2001).

7.4 J.O. Ledbetter, *Air Pollution Analysis*, Part A, Marcell Dekker (1972).

7.5 C.E. Hall, *Introduction to Electron Microscopy*, McGraw-Hill & Co., New York (1953).

7.6 R.D. Cadle, *Particle Size Determination*, Wiley-Interscience (1955).

7.7 P.W. West, *Determination of Air Quality*, Plenum (1972).

7.8 P. Drinker and T. Hatch, *Industrial Dust*, 2nd Ed., McGraw-Hill (1974).

7.9 S.S. Butcher and R.J. Charison, *An Introduction to Air Chemistry*, Academic Press (1972).

7.10 R. Belcher, T.S. West and C. Wilson, *New Methods of Analytical Chemistry*, 2nd Ed., Chapman & Hall (1964).

7.11 S.M. Khopkar, *Environmental Pollution, Monitoring and Control*, New Age International Publishers (P) Ltd., New Delhi (2007).

7.12 S.M. Khopkar, *Basic Concepts of Analytical Chemistry*, 3rd Ed., New Age International Publishers (P) Ltd. New Delhi (2008).

CHAPTER 8

Air Pollution Monitoring

8.1 INTRODUCTION

In earlier chapters we had considered very reliable and accurate methods for the analysis of air pollutants. However what is significant of such methods is the reproducibility. However, such methods can be carried out only with accuracy in laboratory but may totally fail if such pollutants are measured in outdoor environment. It may also be necessary to carry out analysis on continuous basis to get periodical results where conventional laboratory methods fail in spite of that they are very good and selective but unfortunately cannot be performed at frequent intervals. This problem necessitated to discover such methods which can only be carried out in outdoor surroundings but also one can measure the level on time to time basis. Such analysis which involved continuous analysis was termed as 'monitoring'.

The special instruments were developed to measure some vital physical properties of the pollutant like electrical conductivity, optical absorbance, or redox potential to carry out measurements and note their periodic variation. Hence many sophisticated analytical instruments like spectrophotometer, conductometer, fluorescence analyser, gas chromatograph, polarograph were freely used provided if these devices are available as portable models. For instance, the phenomena of chemiluminescence was largely used in SO_2, NO-NO_x and H_2S analyser while difference in conductance was readily measured in SO_2 monitor, the effect of light scattering was used in nephelometer for assaying quantum of dust and suspended particulate matter in atmosphere. As already discussed various sampling devices like cyclone, impinger, electrostatic precipitator were fabricated to achieve reproducible results. Several efforts led to design and development of pollution monitoring instruments. One such effort to design H_2S and SO_2 analyser are discussed in greater details. These prototype instruments were designed and developed by us in our laboratory.

This chapter therefore embarks on the consideration of various air pollution monitoring instruments, their principles of working limits of analysis. The growing efforts are made to fabricate these instruments in our country which are ordinarily imported from abroad. With the advances in field of analytical instruments, there lies greater challenge to our scientists to design and develop simple monitoring devices which can stand test of time, accuracy and speed of analysis.

In the earlier chapters we have considered laboratory methods of air pollution analysis. When the pollutant is emitted ceaselessly and especially when it is necessary to measure it at site such as in a field, or in any other outdoor location, several of the laboratory methods fail to furnish reliable information. Many a times it is difficult to collect the sample and bring it to the laboratory for further analysis. In addition there is a good possibility that we have to take several measurements on regular basis at constant and regular intervals. In such a case we need rapid methods by which we can analyse air pollutants on a continuous basis. Fortunately, few of the laboratory methods can be used on the outdoor locations, sometimes it is always possible to use basic principles underlying laboratory methods for testing of pollutants to devise a novel method for continuous measurement. Such continuous measurement is termed as 'monitoring'. Therefore, we would consider in this chapter air pollution monitoring methods.

8.2 PRINCIPLES OF MONITORING METHODS

Since it is necessary to measure air pollution on a regular basis, we need to identify some physical property of the pollutants for such measurements. The component of instrument which measures such physical parameters is called a 'Transducer'. A specific property, characteristic of a particular pollutant such as light transmittance or absorbance, electrical conductivity; change in potential, generation of material by passage of electricity, nucleation or fog formation or radio-activity is usually measured. Such properties are summarised in Table 8.1. The examples of gaseous pollutants which can be measured are cited. The monitoring instrument is composed of three components, viz. air remover, transducer and recorder. The air remover consists of a flow meter and a pressure gauge. They measure rate of flow of air passing and the pressure under which gas pollutants exist. Transducer in fact is the heart of the instrument, which measures a physical property, while the recorder notes change in physical property of gaseous pollutant, e.g., flow colorimeters or spectrophotometers are used to measure the concentration of the pollutant. The instrument should be checked for response time, specificity, sensitivity, noise level, maintenance and downtime and overall accuracy. Very few instruments are available for aerosol analysis in this country at economic rates.

Table 8.1 Physical principles involved in air pollution monitoring methods

S. No.	*Physical principle of measurement*	*Example*
1	Molecular absorbance (UV/visible/IR)	SO_2, NO_2, H_2S, CO
2	Atomic absorption (AAS)	SO_2, Hg
3	Atomic emission (ICP-AES)	SO_2 metal pollutants
4	Molecular luminenscence (Fluorescence)	SO_2, NO_2
5	Light scattering (Nephelometry and Turbidimetry)	Particulates
6	Magnetic resonance/Electron spin resonance (NMR/ESR)	Organics
7	Polarographic, amperimetry	Pesticide
8	Conductance, High frequency titration	SO_2
9	Potential–pH measurement, ionselective electrodes	NO–NO_x
10	Coulometry–anodic stripping voltametry (ASV)	SO_2
11	Radioanalytical (NAA/IDM)	Radioactive pollutants
12	Gas chromatography (GC)	Organics, HCl

8.3 AIR POLLUTION MONITORING INSTRUMENTS

The common air pollution instruments for measurement are summarised in Tables 8.2 and 8.3. The limits of analysis are specified in Table 8.2, while the technique of measurement is given in the last column of Table 8.3. There are a good deal of instruments for measurement of SO_2, CO, CO_2, NO–NO_x and H_2S. They are based upon measurement of various physical properties listed in Table 8.1, e.g., for SO_2 monitor one can use either conductance measurement or coulometric technique. On the contrary, for H_2S we have only one method involving flow colorimetry (Table 8.2). The other uncommon pollutants like oxygen, ozone, smoke and dust, organic pollutants as well as metal pollutants, can also be monitored on a continuing basis. In such a technique, apart from conventional methods one can use measurement of light scattering phenomena (for smoke), fluorescence quenching (for organic pollutant), emission and absorption spectroscopy (for metal pollutants) for quantitative analysis.

Table 8.2 Air pollution monitoring instruments and their limit of analysis

S. No.	*Instrument*	*Pollutant to be measured*	*Limit of analysis ppm/m³*
1.	IR gas analyser	CO, CH_4 and HC	10–1000
2.	NDIR analyser	CO, CO_2	Non-dispersive IR
3.	Car exhaust meter	Auto-exhaust analysis	CO, CO_2 meter
4.	CO monitor	Catalytic oxidation	0–50
5.	SO_2 monitor	Conductivity	0–100
6.	SO_2 monitor	Coulometric titration	0–40
7.	O_3 monitor	Chemiluminescence	0–10
8.	NO-NO_x-monitor	Chemiluminescence	0.05
9.	H_2S analyser	Colorimetry	0.40

Table 8.3 Air monitoring instruments and techniques

S. No.	*Instruments*	*Pollutant measured*	*Technique of analysis*
1.	Conductometric analyser	SO_2	Conductometric measurement
2.	Continuous SO_2 monitor	SO_2	Coulometric measurement
3.	Gas liquid chromatograph	CO, NO hydrocarbons	Gas chromatography with different detectors or hyphenated with spectrometer
4.	Nitric oxide monitor	NO_2	Fluorescence technique
5.	O_2 analyser	O_2 in atmosphere	Measurement of emission spectrograph
6.	Ozone monitor	O_3	Chemiluminiscence technique
7.	Infrared gas analyser	CO, CH_4 and other hydrocarbons	Infrared absorption limit of 10–1000 ppm for CO, NDIR technique
8.	Fluorescence spectroscope	Organic pollutant	Fluorescence, spectroscopy
9.	Smoke meter	Smoke, smog	Light scattering technique, nephelometer
10.	Atomic absorption spectrophotometer	Trace metal analysis	Atomic absorption spectroscopy
11.	Spectrograph	Metal pollutants at tracer levels	UV visible spectroscopy, ICP-AES

8.4 STATUS OF MONITORING INSTRUMENTS IN INDIA

It is worthwhile to review the situation of monitoring instruments available in India. In the last two decades, we have made impressive progress in enactment of pollution control measures. To some extent we have succeeded in clamping reasonable standard for pollutants in the country.

On the front of development of monitoring instruments, we have failed. While preparing standards for pollutants we were aware of our limitations, but while prescribing methods for analysis, we have never thought of whether a particular instrument is easily and cheaply available. Table 8.4 lists status of monitoring instruments in India. It will be interesting to note that out of 21 basic instruments six are not fabricated in India, while 50% of listed instruments need refinement; and of this only 25% instruments are in the process of development. Table 8.4 gives such approximate idea of important instruments. This is based upon reports prepared by the special committee on environmental monitoring instruments of the Government of India. One encouraging factor is, although simple instruments like pH meter, and spectrophotometer were also imported some decades ago, now they are available indigenously. Similarly, since one decade sophisticated instruments like AAS, GLC, C, H, N analysers and all meteorological instruments are also readily available. The day will not be too far when NMR, GC-MS, NDIR ICP-AES, Capillary Electrophoresis, SFC, SFE; ESCA; & ESR shall be also designed and developed within our country. Thus we have infrastructure, facilities and know how available. With little effort and proper utilisation of expertise and manpower, a day is not too far when we will be self-supporting on this count of design and development of these instruments. We should fabricate only such kind of monitoring devices which best suits Indian conditions and environment. Table 8.5 lists monitoring instruments so far developed in India.

Table 8.4 Cost of air monitoring set-up as in 1984

S. No.	*Instrument*	*Parameter to be measured*	*Cost (in lakh) as in 1990*
1.	High volume sampler	Dust load	0.5
2.	Microscope	Particulate size	0.2
3.	SO_2 analyser	Conductance	2.5
4.	NO–NO_x meter	Absorbance	2.5
5.	O_3 analyser	Absorbance	1.5
6.	O_3 analyser	Chemiluminescence	2.5
7.	CO analyser	Infrared	2.5
8.	NDIR analyser	CO, CO_2, HX	2.7
9.	Hydrocarbon	FID-ionisation	1.0

8.5 METEOROLOGICAL INSTRUMENTS

While measuring air pollution, it is absolutely necessary to consider meteorological and topographical effects of air movements in the atmosphere. The important parameters considered for climate, is direction of wind, frequency of certain type of weather (like periodical cyclones in the ocean), lapse rate and humidity in air. The topographic considerations include regional surface, local surface and urban structure. Variation in air temperature decides dispersal of pollutant at ground level, but when temperature of surrounding air decreases with height, the pollutant cannot accumulate. The rising parcel of pollutant

Table 8.5 Monitoring instruments so far developed in India

S. No.	*Type of instrument developed in India*
1.	Smoke meter
2.	Recording stack smoke density meter
3.	Stack monitoring kit
4.	High volume sampler
5.	Size selective gravimetric dust sampler
6.	Portable multigas sampler
7.	CO monitor including NDIR model for both CO, CO_2
8.	SO_2 monitor
9.	H_2S monitor
10.	O_3 monitor
11.	NO–NO_x monitor
12.	NDIR CH_4 monitor
13.	Fluorescent SO_2 monitor
14.	Electrodeposited lead dioxide plate
15.	Wind speed recorder
16.	Wind direction recorder
17.	Temperature sensor
18.	Humidity meter
19.	Anemometer and wind vane meter

Table 8.6 Meteorological instruments

S. No.	*Name of instrument*	*Property to be measured*	*Analytical technique*
1.	Temperature recorder	Temperature	Temperature sensor by diode
2.	Anemometer and wind meter	Wind velocity and wind direction	Velocity threshold
3.	Tide and wave meter	Wave and tide	Hydrostatic pressure
4.	Salinity meter	Salinity of system	Conductivity cell
5.	Telemeter	Measurement of depth	Hydrostatic sensing probe
6.	Temperature logger	Temperature recorder	Probe with thermistors
7.	Water speed measurement	Water flow meter	Signal conditioners and loggers
8.	Water level meter	Water level	Resistance variation
9.	Capacitance bridge	Height and level of water indicator	Capacitance sensors measurement of difference
10.	Telemetry design	Data transmitter	UHF radiolink used
11.	Rain gauge	Rainfalls	Rain gauge measurement
12.	Humidity indicator	Humidity measurement	Measurement with thermistor probes

when air expands it rises above and cools also. When there is inversion of temperature with height at ground level, vertical movement either upward or downward is strongly arrested and consequently pollution remains stagnant near the level at which it had emitted. Since lapse rate is negative, it leads to 'inversion'. When ground is cool, specially at night, low level or noctural inversion occurs. Then mist or fog is formed. The inversion level extends up to 30 m above in early morning. A subsidence inversion involves high atmospheric pressure, when air in high pressure system descend, leading to stability and inversion phenomena to a height of 2–3 km. The elevated inversion consist of inversion layer sandwiched between two neutral layers at cold polar air slides. Sea breeze is an important meteorological factor of consideration, as after sunrise cold sea air forms at ground based layer of air, which is 300–400 mm thick and moves under warmer inland air at a speed of 15 km/hr. The meteorological parameters form an important segment of various monitoring systems (Table 8.6). The important instruments required for such purpose are temperature, humidity, rain, wind direction and wind velocity instruments. Fortunately, this is one of the fields where meteorological department, Indian Space Research Organisation defence labs, have made spectacular development and progress, with the result that all kinds of instruments listed in Table 8.6 are manufactured in India. We have excellent temperature sensors, hydrostatic pressure and sensing probes, thermistors, etc. The evaluation of these parameters would help us to measure gases, pollutants, in atmosphere and furnish reliable and reproducible results on their nature.

8.6 ENVIRONMENTAL MONITORING PROGRAMMES

Monitoring programmes can be classified into various groups. The priority is given in the order as ambient environmental monitoring: source monitoring; inplant monitoring and personal monitoring. The ambient monitoring covers the monitoring of ambient air. The source or stack monitoring is carried out by enforcement agencies to conform to standards led down by controlling authorities. Last two groups, viz. inplant and personal monitoring is useful for ascertaining occupational health. The ambient monitoring is carried out to achieve proper enforcement of control strategies. The fixed station ambient monitoring can be done by mobile monitoring van. The stationary or stack monitoring is carried out in many process industries at the place of origin. At source the concentration of pollutant is obviously high. The auto-exhaust emission falls under the category of mobile source monitoring.

The measurement of pollution in working area of worker is inplant monitoring. Keeping track of exposure of an individual leads to personal monitoring. As a matter of fact it was possible to fix, TLV, *i.e.,* threshold limiting value of toxic pollutant from the knowledge of personal monitoring. Whatever strategies are used for air pollution monitoring can also be extended for water pollution monitoring. Broadly the air pollutants can be classified due to presence of particulate matter, sulphur, hydrocarbons; nitrogen compounds, halogens, carbon compounds, inorganic compounds and organic pollutants in the atmosphere. The monitoring of air quality parameters is done according to importance of pollutants like suspended particulate matter, SO_2, hydrocarbons, CO, NO–NO_x and O_3. Therefore, we would consider air monitoring methods for each important pollutants in the atmosphere in the following sections.

8.7 MONITORING OF SULPHUR DIOXIDE

The laboratory methods were already described earlier. The various methods used for the monitoring of SO_2 are summarised in Table 8.7. It will be seen that important methods are based upon measurement of colour or electrical conductance, or use of electrochemical transducer, by chemiluminescence technique, NDIR measurement or gas chromatography. In simple barium perchlorate

Table 8.7 Monitoring methods for SO_2

S. No.	*Method*	*Remarks*
1.	Ba $(ClO_4)_{2-}$ Titration	Valid till 500 ppm but many metals interfere
2.	Flow colorimetry	Barium chloroanilate used as the chromogenic ligand but NO–NO_x interference
3.	PRA-colorimetrically	Very sensitive
4.	Conductometric method	Commonly used
5.	Coulometric titrations	KBr reduction with the use of three electrodes
6.	Electrochemical transducer	Semipermeable membrane used
7.	Pulses fluorescence spectroscopy	Very sensitive (0.1–100 ppm)
8.	NDIR technique	*In situ* monitoring
9.	GLC-with FID	Specific for SO_2 (but not for H_2S)

Table 8.8 SO_2 monitor – developed in I.I.T. Mumbai

1.	Working principle	Conductance measurement
2.	Methods of analysis	Photometry, coulometry, chemiluminescence
3.	Reacting solution	Acid, water, H_2O_2
4.	Absorbing liquids	Monoethanoal amine, formaldehyde and triethanolamine and hydrogen peroxide
5.	Solid absorber	Glassfibre filters
6.	Interference	NO_2, NH_3, H_2S, CO
7.	Special features	High accuracy, minimum maintenance IC used, reading is automatic
8.	Level of analysis	0.1–80 ppm
9.	Threshold limit (TLV)	5 ppm
10.	Advantages	Low cost (Rs. 15,000 approximately)
11.	Cost of imported model	Rs. 40,000
12.	Weight of model	3 kg. portable
13.	Atmospheric SO_2 level	4–40 ppm present in air
14.	Toxic effects	Asphyxiation, dermatitis, pneumoconisis, renal and hepatic disorder
15.	Source of SO_2 pollution	Foundry, oil refineries, furnace oil, coal gas, heavy chemical units, and fertiliser industries.

titration technique, SO_2 is reacted with H_2O_2 or H_2O to form H_2SO_4, which in turn is reacted with $BaCl_2$ to form $BaSO_4$. This method is valid when high concentrations of SO_2 are encountered, but many metals like Pb show strong interference. Flow colorimetry is the best technique. One involves interaction of Ba-chloroanilate with SO_2 to give coloured compound, while in another p-rosaniline (PRA) reacts with SO_2 to give coloured species, which can be measured spectrophotometrically at regular intervals. In widely used conductometric method, conductance of an absorbing solution is measured at regular intervals. The increase in the conductance is caused by the ions formed, as SO_2 combines with the absorbing solution like H_2O_2. The increase in conductivity is proportional to the concentration of SO_2 in the air. In fact this principle forms the basis of SO_2 monitor developed indigenously by us (Table 8.8).

The liquid absorber monoethanol amine, triethanol amine, H_2O_2 and formaldehyde are used to eliminate interference of other gaseous pollutants like CO, H_2S, NO–NO_x and NH_3. The SO_2 monitor permits analysis in the range of 0.1–80 ppm (Fig. 8.1), when levels of SO_2 in atmosphere is 4–40 ppm. Other working details are also provided in Table 8.8. In coulometric technique, in alkali halide solution like KBr, a constant halogen-to-halide ratio is maintained by electric current. It measures the current required to regenerate the halogen depleted by SO_2, this current being proportional to the quantity of SO_2 which is converted from halogen to halide. The only limitation of this technique is the periodic replacement of the reagent and constant cleaning of electrodes. Thus a cell containing acidic potassium bromide with four electrodes is taken, SO_2 is passed and free Br_2 liberated, causes imbalance. The signal is amplified and current is passed to generate enough free bromide coulometrically to restore balance in the circuit. The generating current which is indirectly proportional to SO_2 present is continuously measured. In electrochemical transducer, a sealed coulometric devise is used. SO_2 selectively diffuses through a semipermeable membrane and as a dissolved gas through thin-film electrode. Current generated by oxidation of SO_2 is a measure of its concentration. Selectivity is determined by the choice of the

Fig. 8.1 SO_2 analyser

membrane electrolyte and electrode materials of construction. The most efficient technique is pulsed fluorescence method. A pulsed ultraviolet source excites the SO_2 molecules, which emit a characteristic fluorescence with intensity proportional to the quantity of SO_2 present. e.g.

$$SO_2 + h\upsilon \rightarrow S^*O_2;\ S^*O_2 \rightarrow SO_2 + \text{fluorescence}$$

The intensity of such fluorescence is detected by a photomultiplier tube and after suitable signal processing given to the meter and recorder. The only limitation of this method is humidity, which can quench fluorescence intensity. In nondispersive IR method, a broad spectral adsorption is measured and sensitised for a particular gas by a detector. Limitations are H_2O, CO, aromatic compounds show strong interferences, in the same region of SO_2 absorption. We can use filters to eliminate such interference. Finally, in gas chromatography, flame ionisation detector technique is useful for specific measurement of SO_2, but not useful for H_2S. There is a dire need to develop these instruments in India.

8.8 MONITORING OF SULPHURETTED HYDROGEN

Most simple and reliable methods for monitoring of H_2S is flow colorimetry. Instead of using any kind of chromogenic ligands, it would be worth to explore simple chemistry of second group of inorganic qualitative analysis. Amongst metals tested, viz. Cu, Cd, As, Sb, Bi, Sn, arsenic gives reproducible results by formation of As_2S_3, which conforms to Beer's law. The details of such H_2S analyser, which was indigenously developed by us in our laboratories is furnished in Table 8.9 and Fig. 8.2. The tape

Fig. 8.2 H_2S analyser

coated with arsenic chloride is slowly moved, K_2CO_3 is impregnated on glassfibre filters to remove SO_2 contamination; a sponge containing dilute HCl wets the tape before its interaction with H_2S, to undergo reaction as

$$2AsCl_3 + HCl + 3H_2S \rightarrow As_2S_3 + 8H^+ + 8Cl^- \text{ (yellow coloration).}$$

Table 8.9 H_2S analyser – details

1.	Working principle	Flow colorimetry
2.	Laboratory methods	Acid base and redox titration, lead candle method
3.	Monitoring methods	Molecular luminescence, colorimetry, coulometry
4.	Transducer	Phototube sensor, transmittance measurement
5.	Threshold limiting value	5 ppm (TLV)
6.	Limit of analysis	1–800 ppm
7.	Power supply	230 V AC or 6V DC
8.	Recorder	10 mV recorder possible
9.	Digital display	Can be adopted with 3 1/2 digit meter

The lowest detection limit is 5 ppm, with working range of monitoring of 1–800 ppm. The instrument is provided with multiple switch for measurement even in high concentration. This model works both on AC (230 V) and DC (6 V). The instrument can be attached to 10 mV recorder (x-y type). It is worthwhile to consider cost of components as shown in Table 8.10. Except pump working on DC (often used in musical horn), all components are indigenously available. The cost indicated was as it existed in the market in 1980. An imported model cost more than Rs. 14,000 and our model would cost Rs. 4,000. Even if we assume there is substantial price rise since 1980 till date, the instrument is no doubt inexpensive and is best replacement for imported model.

Table 8.10 Cost of components of H_2S analyser (1980)

		Rs.
1.	Pump working on DC	400.00
2.	Phototube	250.00
3.	Analogue meter	200.00
4.	IC—chips	75.00
5.	Gears	75.00
6.	Tape recorder motor	80.00
7.	Ball bearing and resistances	120.00
8.	Dry cells (9 volt)	30.00
9.	Spools, knob, cover, etc.	100.00
10.	Miscellaneous components	300.00
	Total	2,000.00

8.9 MONITORING OF NO–NO_x

Various methods are recommended for monitoring of oxides of nitrogen, which are listed in Table 8.11. Of these the best method is colorimetry involving Saltzman reaction. An air to be sampled is bubbled through a mixture containing NEDA, *i.e.*, N-(1-napthyl) ethylene diamine and acetic acid. NO_2 present in the sample reacts with the solution to form a dye. The absorbance of this dye is a measure of NO_2 concentration. Only limitation is that colour development is not instantaneous. The second best method is chemiluminescence, which is highly specific, e.g.,

$$NO + O_3 \rightarrow NO_2 + O_2;\ NO_2 + h\upsilon = N^*O_2;\ N^*O_2 \rightarrow NO_2 + \text{chemiluminescence}$$

Table 8.11 Monitoring methods of NO–NO_x

S. No.	*Method*	*Remarks*
1.	Colorimetric technique	Saltzman method, needs somewhat longer time for colour development
2.	Chemiluminescence technique	O_3 used to convert NO_2 to NO
3.	Ion selective electrodes	Nitrate in aqueous phase is measured
4.	Electrochemical transducer	Slow response time
5.	Non-dispersive infrared/ UV method	Interferences from SO_2, H_2S resonance fluorescence show good specificity

Anion reacting with O_3 gives positive interference, provided such ion should overlap the spectral region of (NO + O_3) emission; fortunately no such molecule exists in the atmosphere. Ions showing quenching of chemiluminescence show negative error. O_2 and N_2 can show quenching, but their concentration in the atmosphere is fixed and is constant. A thermal conversion of $NO_2 \rightarrow NO$ is commonly utilised. Gold, stainless steel, carbon impregnated metals are used for such conversion. In ion selective electrode method nitrite formed is oxidised to nitrate and is measured with ion selective electrode. Electrochemical transducers are slow to work, as response time is very poor. In NDIR method, SO_2 and H_2S seriously interfere, one has to use resonance fluorescence method in such a situation. For *in situ* monitoring, gas cell correlation technique is favourable. In an interferometer one can use Raman scattering for detection of NO in *in situ* analysis. Finally in Jacob's Hocheiser method, NO–NO_x are collected over NaOH as $NaNO_2$ measurements followed by colorimetric analysis. No measurement can show large variation if dye is not properly formed. NO, SO_2, interference are present. This was the basis

Table 8.12 NO–NO_x meter – details

1.	Principles	Flow colorimetry
2.	Transducer	Phototube sensor, transmittance
3.	Level of analysis	0.02–600 ppm
4.	Power supply	230 V AC or 6V DC accumulator
5.	Recorder	10 mV recorder possible
6.	Digital display	Can be adopted with 3.5 digit meter
7.	Reading solution	NEDA ((1-napthyl)-ethylenediamine-hydrochloride and anthranilic acid
8.	Interferences	NO_2^- ion interferes
9.	Weight of analyser	8 kg roughly
10.	Recorder print out	10 mV can be coupled
11.	Cost of model	Rs. 9,000 approximately
12.	Special feature	High accuracy, less time for analysis, no maintenance, trouble free working, low cost
13.	Toxic effects of NO–NO_x	Irritation of nose and throat, eyes and skin, dermatitis, lower respiratory track affected, vertigo, vomiting
14.	Source of pollution	Atmosphere burning of nitrogenous matter and fuel, lightening conductor conversion, auto-exhaust emission

of NO–NO_x meter developed in our laboratories. The details are given in Table 8.12. It can detect 0.02–600 ppm of NO_2. The weight of instrument is just 8 kg. The estimated cost of various parts (in 1980) is given in Table 8.13. All items except pump are locally available.

Table 8.13 Cost of components of NO–NO_x meter (1988)

S. No.	*Components*	*Price (Rs.)*
1	Pump working on DC	600
2	Phototube	500
3	Analogue meter	500
4	IC chips	200
5	Gears	100
6	Tape recorder motor	200
7	Dry cell 9 V	50
8	Spools, Knobs, cover	200
9	Miscellaneous component	500
10	Tungsten lamp	500
11	Digital display	1,000
	Total	4,350

8.10 MONITORING OF CO, CO_2

Various methods for monitoring CO are given in Table 8.14. These are all valid for CO. We rarely need methods for monitoring of CO_2, as the conventional classical titration technique is superior to any instrumental method. In NDIR technique, the extent of absorption or transmittance of infrared energy by a column of sample gas in a specified narrow wavelength region (v–1800 cm^{-1}) is determined. IR-

Table 8.14 Monitoring methods for CO, CO_2

S. No.	*Method*	*Remarks*
1	Non-dispersive IR	H_2O vapour interferes, but it is the most reliable method used as reference method
2	NDIR spectrometry	Specific, sensitive, reliable technique
3	IR fluorescence method	Sensitive, no interference by water vapour
4	AAS (Reduction of HgO–Hg)	Need charcoal filter, however H_2O and hydrocarbons interfere
5	Electrocatalytic method	Needs selective scrubbing of SO_2, NO_2 but H_2 interfere
6	Electrochemical transducer	Ethylene and acetylene interfere seriously, battery operated, simple, inexpensive instrument
7	FID (after CH_4 formation)	Very sensitive method as compared to NDIR
8	Hopcalite based CO monitor	Temperature rise for converting to CO_2 is measured

pneumatic detector is used in conjunction with interference filters of about 0.1 mm. Sample needs no preconditioning. However, dust and moisture must be scrupulously absent. Gas flow rate need not be controlled. In NDIR correlation, spectrometer specification and selectivity is satisfactory. The IR fluorescence method is very sensitive, but excess water vapour creates problems. AAS method measures Hg after reduction of HgO to Hg by cold vapour AAS. Hydrocarbons and H_2O vapour however show strong interference. In electrocatalytic method, one has to use scrubber, while in electrochemical transducer ethylene and acetylene show strong interference. The instrument can work on battery, the model is portable and also inexpensive; but shows excessive drift which is a serious problem. GLC with FID gives good results where CO is hydrogenated to CH_4, non-methane hydrocarbon can also be measured by this method.

A novel method for CO monitor is Hopcalite based on CO monitor. The temperature rise is measured. Such rise is due to heat generated in the exothermic conversion of CO to CO_2 in the presence of Hopcalite catalyst. Oxides of Mn or Cu serve the purpose of such a catalyst. Humidity affects measurement. Hence it is necessary to maintain catalyst at a temperature of 100°C.

8.11 MONITORING OF HYDROCARBONS AND OZONE

Different methods for monitoring of hydrocarbons, ozone and oxidants are summarised in Table 8.15. Gaseous hydrocarbon may be reactive, *i.e.,* ethylene and also non-reactive methane. Gas chromatography is a commonly used technique of analysis with FID or TCD analysers. Hydrocarbon level is expressed as methane equivalence hydrocarbon. Preconcentration of hydrocarbon gives reliable results. Silica gel column is useful for lighter organics. The reactive hydrocarbon can be measured with chemiluminescence detector. Less than 0.1 ppm of ethylene can be detected by the chemiluminescent detector at ppb level on silica gel coloumn. The polyaromatic hydrocarbons (PAH) are collected with glassfibre filters. Various instruments including mass spectrometer can be used for such analysis. PAH can be best extracted in benzene and then analysed by GLC. TLC can also detect them, but HPLC offers best results. NDIR method is used for hydrocarbons emitted from heavy duty vehicles. The iodometric method involving wet chemical technique is used for the analysis of oxidant, but SO_2 and NO_2 show interference. See also Experiment 16.1 to 16.7 in section 16.

Table 8.15 Monitoring of hydrocarbons and ozone

S. No.	*Method*	*Remarks*
1.	TCD gas chromatography	Good for mobile source, light organics
2.	FID detectors	Coupled in gas chromatography useful for reactive hydrocarbons
3.	Chemiluminescence for hydrocarbon	$CH_2 = CH_2$, measured, reactive HC
4.	Nondispersive IR spectroscopy	Ideal for heavy duty vehicles
5.	Iodometric method	Wet chemical technique
6.	UV absorption technique	Change in beam on passage through O_3 is measured
7.	Chemiluminescence for CO_3	$CH_2 = CH_2 + O_3$ gives light which is measured
8.	Thermistor catalyst	O_3 passed, heat generated measured by imbalance in bright circuit.

The chemiluminescence method can also be used with O_3 and Rhodamine B probe on silica gel and NO in gas phase. Photochemical reactions can measure level of ozone. Photolysis of NO_2 followed by chemiluminescence detection of NO helps to analyse ozone. Ozone is formed by photochemical decomposition of NO_2 by sunlight in the lower atmosphere. Ozone is monitored with wet chemical methods. In this method, total oxidant is bubbled through neutral buffered KI reagent and liberated iodine is titrated and measured as indicator of ozone concentration. In chemiluminescene technique, ozone is reacted with ethylene and light produced is measured. In UV absorption technique, proportional intensity changes of UV beam as it travels through ozone is measured. Finally ozone is passed on two thermistors, one being coated with a catalyst. The heat of decomposition of ozone is then measured on thermistor bridge circuit (See Experiment 17.1).

8.12 MONITORING OF SUSPENDED PARTICULATE MATTER

Various methods for monitoring of SPM are summarised in Table 8.16. The most reliable method is high volume sampler with gravimetry. SPM is collected on filter paper, millipore glass fibres or nucleopore using air, such as pumps and residue is weighed. Pump can work between 12 and 24 hours. Moisture must be removed. Calibration flow meter measures total volume of air sampled; β-absorption analysis is also a good method. The air is sampled through filter tape, which passes through radio-active source detector like GM counter with PM-147 or C-4 detector. Decrease in count rate is recorded and is related to level of SPM. In piezoelectric balance, two quartz crystals oscillating at same frequency are used. One crystal is encapsulated, while the other crystal is open, coated with selective chemical, which absorbs SPM. Rise in weight is measured. In thimble sampling, change in weight on deposition of SPM on cellulose filter (150°C), boroglass filter (300°C) or ceramics filter (> 300°C) is accurately measured (See Experiment 4.2).

Table 8.16 Monitoring methods for suspended particulate matter (SPM)

S. No.	*Method*	*Remarks*
1.	Gravimetry	High volume sampler (12.24 hours) filter paper weighed
2.	β-gauge analyser	β-rays used to note loss in weight of tape dilution tunnel
3.	Piezoelectric balance	Sensitive method, change in weight noted
4.	Thimble sampling	Change in weight of cellulose, borosilic or glass ceramics filter is measured
5.	Nephelometer	0.1–1 μm diameter can be measured
6.	Lidar method	Measures reflection of light with laser beam
7.	Acoustical particle	*In situ* analysis (5 μm)
8.	Transmissometer	Measures optical density; low cost, cannot detect mass
9.	Electrostatic precipitator	Measures particle flow

In nephelometric technique, beam of light is focussed on spot and air is sampled through. Light is scattered, which is measured on photocell. The scattering of right angles is utilised for measuring particle density. In Lidar method, laser beam is used as light source and it is used for measurements. Other important methods include accoustical particle counter, which involves *in situ* measurement of individual particles above 5 μm. It needs sample conditioning. In transmissometer, optical density of suspended particles is measured, which has a low cost of operation. Only limitation of this technique is, it cannot detect mass of the particles. In electrostatic precipitator, it measures particle flow rate, the system has low potential as mass monitor. To evaluate particle size, electronmicroscopy, light scattering, impaction, minicyclones, laser doppler velocity meter and charged particle drift spectrometry are usually used.

Table 8.17 summarises some more methods for analysis of particulate matter. They are based on the same principle as narrated in the earlier section. Only difference is, some of them have names of persons who have discovered these techniques.

Table 8.17 More monitoring methods for particulate matter

S. No.	*Name*	*Mode of operation*	*Technique*
1.	High volume	Filtration	Gravimetry
2.	Paper taps	Filtration	Transmission densitometer
3.	Durham	Impaction	Counting
	Anderson	Impaction	Counting
	Cascade	Impaction	Microscopic examination
4.	Midget impinger	Impingement	Gravimetry
5.	Cyclone	Centrifugal separation	Gravimetry
6.	Electrostatic and	Precipitation	Non-quantity
	thermal precipitator	Precipitation	Gravimetry

8.13 MONITORING OF TRACE METAL POLLUTANTS FROM AIR

The methods for monitoring metal pollutants or trace elements from the atmosphere involves use of instruments like AAS, ICP-AES, ion selective electrodes, fluorescences, plasma spectroscopy, neutron activation analysis, TGA techniques, etc., e.g. Hg is analysed by flameless cold vapour AAS from air, Hg, Cd, Se, Pb are measured by electric discharge; for XRF the great advantage is the nondestructive method of analysis. Cd, Be and other trace metals can also be measured by ICP-AES. Spark source mass spectrometer is used to measure elemental composition. TGA coupled with mass spectrometry gives analysis of complex molecules. Important elements encountered in air are Hg, Be, Pb, Fe, Cd, Se, As, Ni, V, Cr, Mn, Cu, Sb, Zn, Ba, Sn. The typical methods used for their monitoring are Hg (AAS, UV, XRF), Fe (XRF), Be (AAS and spectrophotometry), Pb (NAA, XRF; AES; spectrophotometry) Cd (AAS, AES, spectrophotometry), Se (Spectrophotometry); Ni (AAS; spectrophotometry); Cu (AAS, AES; Sb (spectrophotometry), Zn (AAS, XRF and AES); Ba (AES) so also Sn (AES). As more research and advancement takes place, newer and better methods would come in the forefront. Table 8.18 to 8.20 indicate experimental results of the analysis of metal pollutants from air, e.g. Table 8.18 indicates how a simple ring-oven technique can be used for analysis of Pb. Se, V and Cd from air with almost cent percent accuracy. While Table 8.19 indicates analysis of trace elements from aerosols. It is necessary to take a number of samples and an average value of the results is considered. Finally Table 8.20 indicates analysis of Pb and V individually. The sampling time was varied from 4 to 18 hours.

Table 8.18 Toxic metal pollutant analysis by ring-oven technique (SCOPE)

Metal ion	*Amount taken µg/litre*	*Amount found µg/litre*	*Amount found by spectrophotometry µg/litre*	*Chromogenic ligand in spectrophotometric analysis* (λ_{max})
Pb (II)	340.0	340.0	360.0	Dithizone (520 nm)
	220.0	220.0	230.0	
V (V)	78.0	78.0	73.0	N-benzoyl-N-phenyl-
	62.0	62.0	68.0	hydroxylamine (525 nm)
Se (VI)	26.0	26.0	27.0	3,3′–diaminobenzidine
	24.0	24.0	28.0	(420 nm)
Cd (II)	13.5	13.5	14.0	2,2′ bipyridyl (526 nm)
	16.0	16.0	19.0	

Table 8.19 Trace elements in the aerosols in typical fertiliser plant
Flow rate = 150 litre/min

Metal ion	*Vol. of sorption liquid taken ml*	*Amount found, µg/10 ml*			*Average µg/10 ml*	*Average µg/litre*
		(I)	*(II)*	*(III)*		
Pb (II)	2.0	0.20	0.25	0.20	0.21	21.0
	5.0	0.45	0.45	0.50	0.475	47.5
V (V)	5.0	0.25	0.25	0.25	0.263	26.3
	10.0	0.50	0.45	0.50	0.483	48.3
Hg (II)	3.0	0.30	0.40	0.25	0.31	31.0
	5.0	0.45	0.55	0.40	0.475	47.5
Se (VI)	2.0	0.10	0.15	0.10	0.113	11.3
	6.0	0.30	0.25	0.30	0.288	28.8
Cd (II)	5.0	0.45	0.55	0.50	0.50	50.0
	10.0	1.10	0.90	0.90	0.967	96.7

Table 8.20 Analysis of particulates, lead and vanadium in the atmosphere of typical refinery

S. No.	*Sampling duration (hr)*	*Wt. of paper sample (g)*	*Wt. of filter paper (g)*	*Wt. of particulate (g)*	*Mean flow rate 1 pm*	*µg of Pb/m³ of air*	*µg of V/m³ of air*
1.	4	0.787	0.049	0.038	142.5	1378.5	193.0
2.	6	0.789	0.750	0.039	135.0	763.96	152.7
3.	16	0.771	0.719	0.052	135.0	343.78	114.59
4.	18	0.803	0.724	0.079	135.0	3667.0	73.34

8.14 FUTURE DEVELOPMENT IN MONITORING INSTRUMENTS

The research and development activities in this area are restricted to established institutions. Attempts were made to indigenously develop some monitoring instruments. Such instruments are inexpensive. However, coordinated approach is lacking in such development. Most of the efforts have been directed towards air quality monitoring instruments. Development should lead to standardisation of procedure in data collection, analysis and calibration. A considerable expertise for development is available within the country and import of know how is not at all necessary. Personal monitoring instruments must also be developed in future. Improved aerosol sampling methods, that are size selective, are essential in future year; fine particles should be measurable with absolute accuracy. Analytical chemistry has moved in the past decade from manual one-by-one sampling procedure to modern automatic techniques; but such methods unfortunately are not extended to routine environmental monitoring. Methods for analysis organics at microgram level will have to be considered in future. Good gas chromatograph, specially portable, battery operated models, will be required for field work. Instruments based on remote sensing will have to be developed within the country. Low cost techniques for detection of gas in indicator tubes will also have to be promoted in years to come. LIDAR technique will have to be improved with tunable dye laser to monitor upper atmosphere.

ADDITIONAL READINGS

8.1 A.C. Stem, H.C. Wohlers, R.W. Boubel and W.P. Lowry, *Fundamentals of Air Pollution*, Academic Press, New York (1973).

8.2 J.H. Seinfeld, *Air Pollution: Physical and Chemical Fundamentals*, McGraw-Hill Co., New York (1975).

8.3 A.C. Stem, *Air Pollution*, Vol. 1 to 5, 3rd Ed., Academic Press, New York (1977).

8.4 J.H. Seinfeld, *Atmospheric Chemistry and Physics of Air Pollution*, Wiley-Interscience, New York (1977).

8.5 R.C. Flagan and J.H. Seinfeld, *Fundamentals of Air Pollution Engineering*, Wiley Inc., New York (1988).

8.6 J.O. Ledbetter, *Air Pollution,* Part A and B, Marcel Dekker Inc., New York (1972).

8.7 R.D. Ross, *Air Pollution and Industry*, Van Nostrand Reinhold Company, New York (1972).

8.8 H.C. Perkins, *Air Pollution*, McGraw-Hill Kogakusha Ltd., Tokyo (1974).

8.9 W.L. Faith and A.A. Atkinson Jr., *Air Pollution*, 2nd Ed., Wiley-Interscience, New York (1972).

8.10 W. Strauss, *Air Pollution Control*, Wiley-Interscience, New York (1971).

8.11 A.K. Ganguly, Report of the panel on *Environmental Monitoring Instruments*, Electronics Commission, Government of India, New Delhi (1976).

8.12 T.E. Harvey, Notes on *Environmental Pollution, Monitoring and Control*, Tottenham College of Technology, London (1978).

8.13 B.L. Deekshatulu, Report on *Environmental Monitoring Instruments*, Department of Environment, New Delhi (1984).

8.14 S.M. Khopkar, *Environmental Pollution, Monitoring and Control*, New Age International Publishers (P) Ltd., New Delhi (2007).

CHAPTER 9

Introduction to Water Pollution

9.1 INTRODUCTION

The environmental science largely covers the study of the air, water, soil, etc. However, out of these areas the one which has been extensively investigated is the area of water pollution. The public health or sanitary engineering is one of the classical discipline which was in existence since ages. Under the programme of water supply, water management or disposal of sewage extensive studies were carried out by civil engineerings. While air pollution was not considered a serious problem earlier but it has reached alarming proportion now-a-days.

However, with the advent of time and specially with the emergence of waste water, it had become imperative to consider all aspects of quality of water. These aspects included physical examination, chemical characterisation and biological investigation of the water. Such examination needed a thorough knowledge of physics, chemistry or biology. Consequently, the entire field spread to the involvement of experts from all interdisciplinary fields.

Therefore, this chapter provides a glimpses of theory of the quality of water. Such studies include colour, conductivity, temperature, odour, turbidity, hardness as well as presence of several interfering metals and anions like fluoride as well as dissolved oxygen, BOD and COD of water. However, the present chapter introduces these important characteristics which decide the quality of water. While detailed experimental procedures listing important ingredients as pH, temperature, need careful consideration while carrying out the experiments. These are listed in the next chapter. Apart from the consideration the method of analysis colour, turbidity, odour, acidity/alkalinity, hardness, conductivity, presence of suspended solids are considered in greater details in next chapter. Finally some important precautions to be taken during collection and interpretations of collected data are given at the end of next chapter. Therefore, first let us consider important physical, chemical and biological properties of the water.

We have considered so far the analysis of toxic pollutants as well as the analysis of air pollutants. We propose to consider water pollution including analysis of quality of drinking water. This will be followed by the analysis of water after its treatment, which is being used for domestic and industrial purposes. The fresh water cannot be used directly for drinking purposes due to pollution. A treatment is essential before it is used for drinking, or for industrial processes. The quality of water depends upon its source or history. The word history of water signifies the terrain through which water is flowing, its origin and most important the extent to which it is contaminated on its way to down. The solubilisation and fragmentation of the terrain through which it flows affects its quality. The extent of dissolved solid materials present in water decides its quality. Such quality is also decided by the solubility of the geological deposits, contact of water with sediments, time of interaction and special factors related to environment. To ascertain suitability of water for human consumption, it is necessary to undertake examination of quality of water. Such quality is ascertained in three ways, viz.

(*i*) Physical examination of water
(*ii*) Chemical characterisation of water
(*iii*) Biological investigation of water

Table 9.1 indicates the characterisation of water from Powai Lake, wells and municipal taps in the city of Mumbai, while Table 9.2 describes important characteristics of well water in typical coalfields in north east India, where most of the water comes from wells. Both the tables show status of the quality of water. For instance, there is too much of chromium, hardness and dissolved solids in Powai Lake water. Also the MPN count is alarmingly high, viz. 430–2400. This shows Powai water is unfit for drinking. The second table (Table 9.2) throws much light on the quality of well water in north east India.

Table 9.1 Characterisation of water from different sources

Source	*pH*	*Hardness* $CaCO_3$ *ppm*	*Dissolved oxygen*	NO_2^-	F^-	*MPN per 100 ml*
Powai lake water	8.1	74.4	7.0	0.6	2.0	2400
	7.9	71.6	7.2	0.8	1.0	430
	8.3	80.0	8.1	0.6	2.0	430
	8.2	74.4	5.6	1.3	1.0	2400
	8.0	74.8	8.0	0.6	2.3	2400
	8.2	78.2	8.2	0.7	1.6	1100
	8.2	76.8	8.4	0.6	2.4	2400
	7.0	12.5	12.5	0.1	0.2	100
Well water	7.2	30.1	1.4	1.2	–	1000
	7.9	95.2	1.4	0.7	–	1000
	7.1	124	1.4	1.1	–	1000
	7.4	251	1.8	0.9	–	1000

9.2 PHYSICAL EXAMINATION OF WATER

Much information on the quality of water was ascertained by physical examination (Table 9.3). We consider characteristics such as colour, conductivity, temperature, odour, turbidity and hardness.

(a) Colour of Water

Colour of the water is due to presence of minute amounts of humus materials from decaying vegetable matter. Industrial waste also contributes to colour of water. We classify colour into two categories, viz. true colour and apparent colour. True colour is the real colour of water seen after filtration. Such colour does not depend upon background. While apparent colour is colour due to reflection of background e.g., sky, blue colour of sea water or reflection of trees imparts green colour to well water. The classical

Table 9.2 Characteristics of well water of coalfield area in north east India

Parameters	*Average values (mg/L)*
pH	7.0 units
Dissolved solids	672.0
Total hardness	617.0 of $CaCO_3$ hardness
Nitrate	78.0
Sulphate	47.0
Chloride	62.0
Oil and Grease	0.038
D.O.	6.0
B.O.D.	0.82
C.O.D.	4.5
Iron	0.10
Manganese	0.003
Fluoride	0.23

Table 9.3 Physical characteristics of water

S. No.	*Physical characteristics*
1	Acidity (no definite data)
2	Alkalinity–10–500 mg/litre as $CaCO_3$ (100) (total)
3	Colour–50 platinate units (10)
4	Hardness–5–1000 mg/litre $CaCO_3$ (150)
5	pH 6.0–9.0 (7.5)
6	Specific conductance 30–1000 microohms (400)
7	Temperature 0–30°C (20°C)
8	Turbidity 0–1000 JU (100 JU)
9	Odour-3 TON (3.0)

method of measurement of colour consists of the comparison of colour with one prepared by dissolving known weight of dipotassium hexachloroplatinate and cobalt chloride. This is called as the platinum-cobalt method of comparison. The unit of colour is one produced by dissolving 1 mg platinum/litre in the form of chloroplatinate ion. Thus 1.25 gm of K_2PtCl_6 and 1 gm of $CoCl_2$, $6H_2O$ are dissolved in 100 ml of concentrated hydrochloric acid and made up to one litre with distilled water. The resulting stock solution has colour of 500 units. This method of analysis involves personal error. However, the spectrophotometric method is the most reliable method, wherein the absorbance of water is measured between 400 and 700 nm. The tristimulus colorimetric method was also used for measurement of colour. The experimental details are provided in next chapter (Expt. 10.1. For details of analysis refer to Expt. No. 10.7).

(b) Conductivity of Water

The equivalent conductance of various ions is listed in Table 9.4. The conductance is the measure of concentration of mineral constituents present in water. One can get an idea about quality of dissolved solids in water. The conductance is reciprocal of resistance, which is measured between two electrodes kept at distance of one cm apart, with area of cross-section as one cm^2. The value depends upon the presences of total concentration of ionised substances. It checks the reliability of mineral analysis or quality of distilled water. The average value should be less than 2 μohms. The self-contained conductivity instruments are available for such measurements. In conductivity cell, first it is necessary to evaluate cell constant. The conductivity is measured at a fixed temperature. Table 9.5 gives conductivity as well as other hydrological parameters of the potable water from various suburbs of Mumbai, as well as that of supply of water from reservoirs like Tansa Vaitarna, Bhatsa, Tulsi, Vihar and Powai (Fig. 9.1).

Table 9.4 Equivalent conductance of various ions

S. No.	*Anion*	*Conductance μmho/mg/litre*	*S. No.*	*Cations*	*Conductance μmho/mg/litre*
1	Cl^-	2.14			
2	SO_4^{2-}	1.54	6	Ca^{2+}	2.60
3	CO_3^{2-}	2.82	7	Mg^{2+}	3.82
4	HCO_3^{2-}	0.71	8	Na^+	2.16
5	NO_3^-	1.15	9	K^+	2.00

Table 9.5 Characterisation of water from various lakes in Mumbai as in 1990

Location		*Place*	*pH*	*[$\bar{F}$] mg/litre*	*[DO] mg/litre*	*Hardness $CaCO_3$ mg/litre*	*Turbidity NTU*	*Conductance μmhos/mg/litre*
Lakes:	1	Tansa	7.52	0.219	5.56	0.50	0.60	131.0
	2	Vaiterna	7.72	0.134	6.66	0.45	0.55	126.9
	3	Bhatshai	7.36	0.177	6.80	0.37	0.40	98.27
	4	Vihar	7.77	0.156	5.60	0.68	3.23	190.3
	5	Powai	7.09	0.112	5.95	0.93	0.50	335.8
Sea Side:	6	Juhu	7.39	0.909	4.0	18.3	2.76	5794.2
	7	Chowpati	7.11	0.376	3.70	6.81	2.36	4955.3
	8	Dadar	7.81	0.109	5.45	0.60	1.20	233.3
Mumbai (Eastern)	9	IIT, Powai	7.24	0.141	5.53	0.93	0.55	163.7
	10	Mulund	7.22	0.139	4.75	0.68	0.60	188.3
	11	Dombivali	7.29	0.136	5.75	0.80	2.50	221.1
Mumbai (Western)	12	Grant Road	6.81	0.121	5.50	0.70	0.76	122.8
	13	Vile Parle	7.29	0.116	5.45	0.55	1.03	114.6
	14	Goregaon	6.83	0.113	5.20	0.56	0.90	137.0

The results show low conductivity for less polluted water is due to absence of ionic impurities, e.g., sea water at Chowpati in Mumbai has the highest conductivity due to presence of large proportion of inorganic salts, while water from Bhatshai lake which is less polluted has the minimum value for conductance as it is located about 80 km away from human habitat.

(c) Temperature of Water

It is measured immediately after collecting the water sample. The depth temperature is required for limnological studies, which may be measured with a reversing thermometer or thermophone or thermistor. The thermistor is the most reliable instrument for measurement of temperature.

(d) Odour of Water

The method for the measurement of odour is latter given (Expt. No. 10.4 in next chapter). The volatile compounds produce odour. The dilution of sample at which odour is detectable is a measure of its

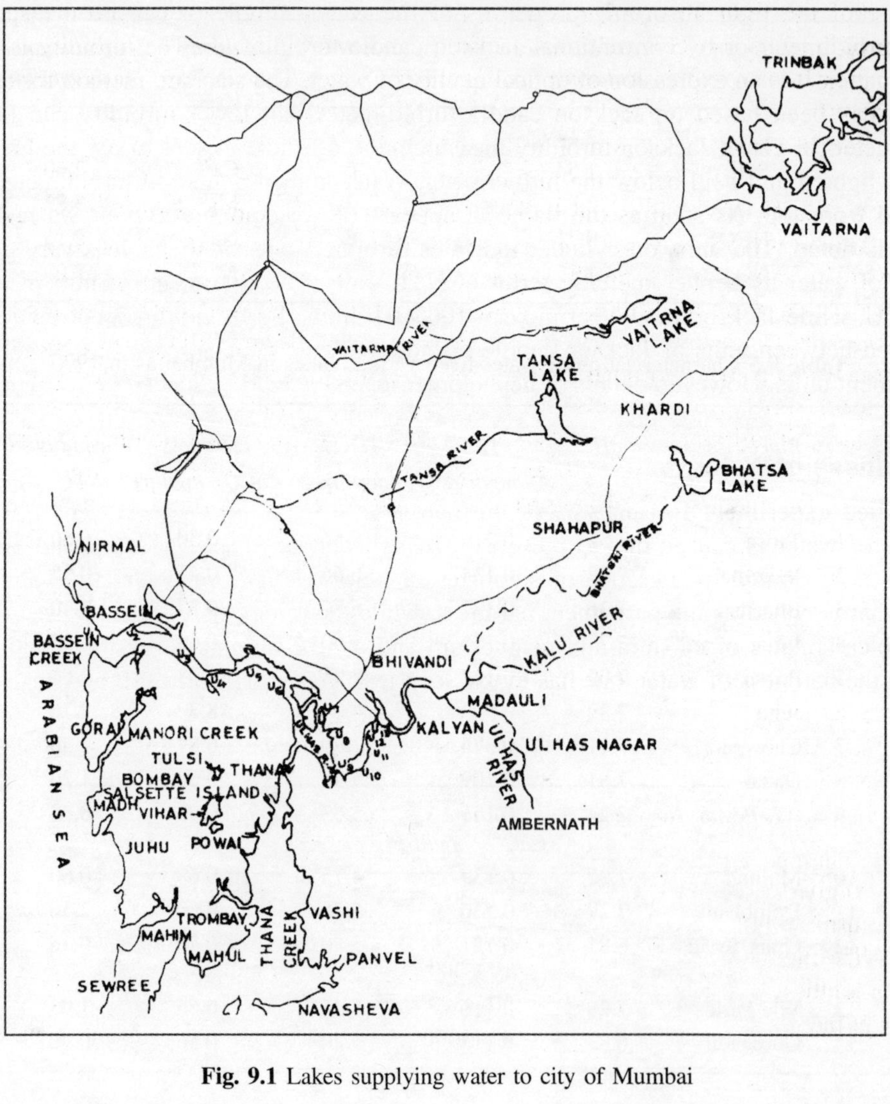

Fig. 9.1 Lakes supplying water to city of Mumbai

intensity and thus odour can be quantitatively measured in the laboratory. The odour is qualitatively described in words as medicinal (phenolic) fishy (due to algae), earthy (decaying matter) or chemical (chlorine).

So far no instrument has been developed for the measurement of odour. Such measurement largely depends upon contact of a stimulating substance with appropriate human receptor cell. Such stimuli are chemical in nature and hence termed as chemical senser. The details of odour, its sampling, measurement and control is very well described in next chapter. It is measured in terms of TON *i.e.,* threshold odour number, indicating the number of times the dilution one should carry out with odour free water, in order to get no perceptible odour, e.g., TON = 4.0 indicates if we dilute it with water in the ratio of 1:4, no odour will persist. Obviously, smaller the value of TON, better is its quality of water. Accepted average value of TON = 3.

(e) Turbidity of Water

The detailed experiment for measurement of the turbidity is given latter (Expt. 10.3, Chapter 10). It is a measure of the light absorbing properties of the water sample. It can be measured either with nepheloturbidimeter or by conventional Jackson candle turbidimeter. The turbidity is caused by suspended matter. It is an expression of optical quality of water. The standard method for determination of turbidity has been based on Jackson candle turbidimeter, but lower turbidity can be measured by nephelometer. In visual Jackson turbidity measurement, a candle is kept below the ground glass. The candle is lighted and held below the turbidimeter. Water is then poured in turbidimeter and the flame is viewed from top. As soon as the flame disappears (if we continue to pour water) reading on the cylinder is noted. The units on cylinder indicates turbidity units of water. It is very easy to measure turbidity of water by nephelometer in terms of NTU units. The instrument permits measurement from 1000 NTU, while Jackson model permits 25–1000 JU limits. So far no direct correlationship has been established between units of Jackson candle turbidimeter and nephelometer, *i.e.,* in terms of turbidity measurement units. However, details of nephelometric analysis are presented in next chapter (Expt. No. 10.3).

(f) Hardness of Water

The detailed experiment for analysis of the hardness is given latter (Expt. No. 10.6, Chapter 10). Hardness of water is caused due to presence of sulphates and chlorides of calcium and magnesium. Hardness is attributed to presence of sulphates or chlorides of iron, manganese and aluminium. It causes what is called as the permanent hardness, while temporary hardness is caused due to the presence of bicarbonates of calcium and magnesium and can be eliminated by mere boiling. In order to ascertain the hardness of water, one has to use soap test. Water will form soft and beautiful lather with

Table 9.6 Classification of hardness

S. No.	*Hardness* $CaCO_3$ *mg/litre*	*Quality*
1	0–40	Soft
2	40–100	Moderately hard
3	100–300	Hard
4	300–500	Very hard
5	500–1000	Extremely hard

soap if water is soft. However, if water is hard then calcium and magnesium will combine with stearates in soaps to form curd or insoluble lather of stearate of alkaline earth. The alkali metal stearates are soluble. The hardness is expressed in terms of $CaCO_3$/litre. Table 9.6 lists typical hardness of water in words for practical purposes.

Thus a water with 25 mg/litre $CaCO_3$ is considered as soft water, while water with 500 mg/litre of $CaCO_3$ is termed as hard water. The method for analysis of hardness is titration method, involving $KMnO_4$ solution, while other method is involving the use of EDTA for the purpose, which is more dependable. The reactions involved in permanganate titrations are as follows:

$$CaCl_2 + H_2C_2O_4 \rightarrow CaC_2O_4 + 2HCl \text{ (precipitate filtered)}$$

$$CaC_2O_4 + H_2SO_4 \rightarrow CaSO_4 + H_2C_2O_4 \text{ (in solution)}$$

$$2MnO_4^- + 5H_2C_2O_4 + 6H^+ \rightarrow 2Mn^{2+} + 10CO_2\uparrow + 8H_2O.$$

The most commonly utilised method is EDTA titration (See details in Expt. No. 10.6, Chapter 10).

$$CaCl_2 + 2EDTA \rightarrow Ca(EDTA)_2 + 2NaCl$$

Thus we use mureoxide as the metallochromic indicator. During titration the pH is maintained up to 4.5. If magnesium is present, then the reaction is as follows:

$$MgSO_4 + 2EDTA \rightarrow Mg(EDTA)_2 + Na_2SO_4$$

The indicator used is Eriochrome black T, where pH is maintained at 10.0. The EDTA combines first with calcium from water; when magnesium can be first precipitated as $Mg(OH)_2$. So first titration volume gives the amount of calcium only. Solochrome dark-blue can also be used instead of mureoxide indicator, with equal efficiency. When EBT is used, one gets reading for both calcium and magnesium. EBT is grounded with solid NaCl in the ratio of 1:1, and is used during titration. A new indicator called Patton Reeder indicator is best for determination of calcium in hard water. It permits simultaneous analysis of Ca and Mg. The details of calculations are furnished latter (chapter Expt. 10.6).

9.3 CHEMICAL CHARACTERISATION OF WATER

Most important aspect of water quality assessment is the chemical characterisation of water. Table 9.7 lists important chemical constituents of water, which are analysed to assess its quality.

The table also lists important physical properties of water, which we have already discussed in the preceding section. The biological factors which are investigated are also listed in the same table. The various components which are present in major concentrations in potable water are calcium, magnesium, sodium, potassium, chloride, sulphate, carbonates, bicarbonates and dissolved solids. The important minor constituents of potable water are aluminium, barium, boron, fluoride, bromide iodide, nitrate, nitrite, phosphates, polyphosphates, iron, manganese and trace elements. We do consider other important properties such as acidity/alkalinity, dissolved oxygen (DO), biochemical oxygen demand (BOD), chemical oxygen demand (COD), for waste water and finally nitrogen from ammonia, nitrite and nitrates.

Table 9.8 describes the major components of water and their corresponding concentrations. The values in bracket indicate mean level of the component present in sample of water. The metals constitute an important portion of drinking water. Table 9.9 shows, WHO limit of metals, which is in agreement with that prescribed by public health services or pollution boards in India. It gives guidelines for us to control quality of drinking water, by devising proper water treatment process. It is worthwhile at this stage to consider two important aspects of chemical constituents. First is their significance in determining quality of water and second is their analysis in the laboratory, details of such analysis is provided in next chapter (Chapter 12, Expt. 12.1 to 12.13).

Table 9.7 Physical, chemical and biological examination of water

Chemical characteristics	*Physical properties*	*Biological investigations*
Ca^{2+}	Colour	Total coliforma
Mg^{2+}	Conductivity	Faecal coliforma
Na^{+}	Temperature	Faecal streptococci
K^{+}	Odour	Crenothrix
Cl^{-}	Turbidity	Plankton
SO_4^{2-}	Hardness	Algae
CO_3^{2-}		Nitrosomonas
HCO_3^{2-}		
Total dissolved solid (TDS)		Diatomaceous algae
Al^{3+}		
Ba^{2+}		Protozoa
Fe^{3+}		
Mn^{2+}		
B^{3+}		Rutifers
F^{-}		*Escherichia coli*
NO_2^{-}		(*i.e., E. coli*)
NO_3^{-}		(See details in Chapter 17.0)
PO_4^{3-}		
SiO_2^{2-}		

Table 9.8 Major components of water

Component (anions)	*Concentration (mg/L)*	*Component (cations)*	*Concentration (mg/L)*
SiO_2^{-}	0–100 (25)	Mg^{2+}	0–100 (15)
CO_3^{2-}	0–10 (< 5)	Mn^{2+}	0–0.2 (0.25)
HCO_3^{2-}	0–300 (100)	Fe_3^{3+}	0–0.5 (0.25)
Cl^{-}	0–500 (50)	K^{+}	0–25 (15)
Dissolved solids	0.0–750 (300)	Ca^{2+}	0–250 (150)
F^{-}	0.20 (0.5)	Na^{+}	0–100 (25)
NO_3^{-}	0–5 (1)	–	–
SO_4^{2-}	0–500 (100)	--	–

Table 9.9 Major elements present in drinking water (mg/litre)

Element	*PHS–limit*	*WHO–limit*
As	0.05	0.20
Ba	1.0	None
Cd	0.01	Nil
Cr	0.05	0.05
Cu	1.0	1.5
Fe	0.03	1.0
Pb	0.05	0.10
Mn	0.05	0.50
Se	0.05	0.05
Ag	0.05	None
Zn	5.0	15.0

(g) Calcium in Water

This is a major constituent to natural water. It is dissolved from rocks and soils which cause hardness and forms scales which is deposited in conduit tubes and pipes. The concentrations up to 100 mg/litre of calcium is capable of forming scales in pipes and boilers. But it has no adverse physiological manifestation on human system, and in fact it is believed that a person consuming water containing certain amount of calcium is likely to have less chances of cardiac disorder, than a persons drinking soft water without any calcium contents. The method of analysis of calcium is gravimetric estimation, if present in large concentration. It is weighed as CaO (see details in Chapter 12, Expt. No. 12.3).

$$Ca^{2+} + H_2C_2O_4 \rightarrow CaC_2O_4 + 2H^+; \; CaC_2O_4 \rightarrow CaCO_3 \rightarrow CaO + CO_2\uparrow$$

If the level is small, one can use $KMnO_4$ titration by dissolving the oxalate in sulphuric acid and carrying our redox titration. Similarly, the EDTA titration is the another method which is narrated (Expt. 12.3). One can also use indicators like calcein purpurate, etc. Fe, Mn if they present are masked at pH of 10.0 with alkali cyanide.

(h) Magnesium in Water

For details of analysis refer to next chapter (Expt. No. 12.11 in Chapter 12). In comparison to calcium, magnesium is present in a smaller concentration. It is determined by titration with EDTA with EBT (Eriochrome Black T in NaCl) at pH 10.0. If a large amount is present, gravimetrically it is estimated as $Mg_2P_2O_7$. The reaction involved is:

$$MgCl_2 + (NH_4)_2HPO_4 \rightarrow 2MgHPO_4 \rightarrow Mg_2P_2O_7 + H_2O$$

In very few instances, if amount is small, it is determined by spectrophotometer, by making use of the brilliant yellow solution.

(i) Sodium in Water

The next chapter (Expt.12.11 in Chapter 12) describes method for the analysis of alkali metals present in water. It is a major component of potable water. The average level of sodium is > 100 mg/litre. If high amount of sodium is present, it will render boiler operations difficult. No doubt sodium has medicinal effect on humen health. However, high concentration in blood leads to hypertension. The best method for analysis is by flame emission spectroscopy at 590 nm, potassium does not interfere, as it has absorption maxima at 770 nm. One can also use standard addition and internal standard method during flame photometric analysis. One can estimate 5×10^{-3} ppm of sodium or potassium in the sample (For details see Chapter 12).

(j) Potassium in Water

It is not present in high concentration. The ratio of sodium to potassium is 10:1 or 20:1. The method of analysis is by flame photometry at 770 nm. If moderate levels are encountered, one can resort to colorimetry with sodium cobaltnitrate as chromogenic ligand.

(k) Chloride in Water

A method for the analysis of chloride is provided in next chapter (Chapter 14, Expt. No. 14.3). The concentration of chloride varies from sample to sample. It has no adverse effects on health, and it imparts bad taste to drinking water. With this increase in level of chloride in water, its corrosiveness on metals also increases. Two methods for the determination of chloride are Mohr's and Volhard's methods (See details in Chapter 14). In such method important reactions involved are:

$$NaCl + AgNO_3 \rightarrow AgCl + NaNO_3 \text{ (excess of } AgNO_3 \text{ added for back titration)}$$

$$AgNO_3 + NH_4SCN \rightarrow AgSCN + NH_4NO_3 \text{ (Equivalence point)}$$

$$3NH_4SCN + Fe^{3+} \rightarrow Fe(SCN)_3 + 3NH_4^+ \text{ (colour after end point)}$$

We add excess of $AgNO_3$ to water containing chloride, and titrate excess of unreacted $AgNO_3$ with KSCN or NH_4SCN. After end point, blood red colour appears due to excess of iron thiocyanate.

In Mohr's method, we use K_2CrO_4 as an indicator for indirect titration. The reactions are:

$$NaCl + AgNO_3 \rightarrow AgCl + NaNO_3 \text{ (till end point)}$$

$$2AgNO_3 + K_2CrO_4 \rightarrow Ag(CrO_4)_2 + 2KNO_3 \text{ (end point)}$$

Both methods are reliable and excellent for analysis of dissolved chloride.

(l) Sulphate in Water

The detailed method for the analysis is provided in next chapter (Chapter 14, Expt. No. 14.12). This is a major anion present in water. It is formed by biological oxidation of sulphur compounds if present in water. It has no adverse effect upon health. If the amount is large, it can be gravimetrically estimated as $BaSO_4$. If the amount is moderate, it can be determined by the nephelometric method. It is also possible to estimate it colorimetrically using sodium rhodiazote as chromogenic ligand. Gravimetry is useful if $SO_4^{2-} > 10$ ppm. The titration method involves interaction of $BaCl_2$ with thorin as an indicator. The method is best if $SO_4^{2-} \sim 5–150$ mg/litre. Formaldehyde is added if interfering ions are present in sample of water.

(m) Carbonates and Bicarbonates in Water

The next chapter provides details of the analysis of these anions (Expt. 14.13 in Chapter 14). Their presence in water renders it temporarily hard. They are formed due to dissolution of CO_2 in water at lower pH (bicarbonates) or at higher pH (carbonates). Their level is 40–400 mg/litre. They form a scale in pipes and boilers. The temporary hardness can be removed by the process of boiling or addition of lime water to the sample and water is rendered soft as:

$$Ca(HCO_3)_2 \rightarrow CaCO_3 + H_2O + CO_2\uparrow \text{ (by heating)}$$

or

$$Ca(HCO_3)_2 + Ca(OH)_2 \rightarrow 2CaCO_3 + 2H_2O \text{ (by addition of lime)}$$

(n) Dissolved Solids in Water

The next chapter (Expt. 10.8 in Chapter 10) provides details of the analysis of TDS. They are present in the range 10–20 mg/litre. Certain water samples may have as much as 1000 mg/litre. The excess level in drinking water can cause stomach upset, fever and impart salty taste to water. Total solids are determined by evaporating a known volume of water to dryness and then by weighing its percentage in water. (See chapter 10.7, Expt. 10.8.)

9.4 MINOR COMPONENTS OF WATER

We have considered the major components of water. The important minor components are aluminium, barium, boron, bromide and iodide, which are of less importance. Aluminium passes through clays and minerals in water. There level is 1 mg/litre. Fortunately, they have no ill effects on health. The details of experiment for analysis of Aluminium is described elsewhere (Expt. No. 12.1 also for analysis of barium in Expt. No. 12.11).

Table 9.10 Fluoride in drinking water in various states in India

S. No.	*State*	*Name of district*	*Average fluoride level, mg/litre*
1	Tamil Nadu	Chinglepet	4.5
		Salem	4.5
		Madurai	4.5
		Trichinapalli	4.5
		Coimbatore	4.5
2	Karnataka	Belgaum	0.8–1.2
3	Andhra Pradesh	West Godavari	0.8–1.2
		Nalangonda	0.8–1.2
		Vellore	0.8–1.2
		Anantpur	0.8–1.2
4	Gujarat	Mehsana	4.0
		Banaskantha	4.0
		Sabarkantha	4.0
		Panch Mahal	4.0
		Surat	4.0
5	Punjab	Bhatinda	3.0
		Ferozpur	3.0
		Amritsar	3.0
6	Haryana	Gurgaon	3.6
		Hissar	3.6
		Karnal	3.6

Aluminium is determined spectrophotometrically with aluminion as the chromogenic ligand. While barium if present in concentration of 1 mg/litre; has toxic effects upon nerves and heart and blood vessels. It is determined flame photometrically at 554 nm. Boron is another minor element used as micronutrient in the soil. It helps in farming, specially the growth of citrus fruits like oranges and lemons. It has no physiological effects on health. It is analysed spectrophotometrically with caramine red as the chromogenic ligand. The phosphates and polyphosphates are present in the range of 0.01–0.5 mg/litre. The aquatic life, flora, fauna, cannot survive in water if there is no phosphate in water. The method for its analysis is interaction with ammonium molybdate. A canary yellow solution can be reduced with $SnCl_2$ to molybdenum blue and then measured spectrophotometrically. As small as 0.002 μg/ml of PO_4^{3-} can be analysed. In place of $SnCl_2$, one can also use aminonaphtholsulphonic acid as the reducing agent. The anions of importance among minor constituents are: fluoride, nitrite, nitrate and cations as iron, manganese, hence their analysis is considered. [The experimental details are given in Expt. No. 14.4 (Fluoride) Expt. No. 14.8 (nitrate) Expt. No. 14.9 (nitrate) Expt. No. 12.7 (iron) Expt. No. 14.9 (manganese) resp. mostly in Chapter 14.0.]

(a) Fluoride in Water

The analysis of fluoride is described in next chapter (See Expt. No. 14.4). It will be noted from Table 9.10 that out of 28 states of Indian Union, six states have considerable amount of fluoride in potable water posing a problem. Although inadequate concentration it causes dental crevices, acute poisoning of central nervous system; excess of it can be deposited in bones, as it has a great affinity for calcium. Dust containing fluoride if inhaled causes irritation of nasal passage, gastric troubles, loss of appetite, nausea and vomiting. In concentrations greater than 74 mg/litre, it is considered toxic. It shows diffuse osteoporosis of bone with calcification of ligaments and restriction of mobility of columns and thorax as confirmed by X-ray therapy. There is a growing controversy regarding role of fluoride in tooth paste and drinking water. In the USA they allow its use freely, while in UK a controversy exists as to whether one should use fluoridised potable water. In India, in most of cities we do not fluoridise our water supplies; About 0.8–1.2 mg/litre is a safe concentration for potable water, as it is good for growth of teeth of children. The best method for analysis of fluoride is shown in Fig. 9.2 (Experiment No. 14.4 in Chapter 14).

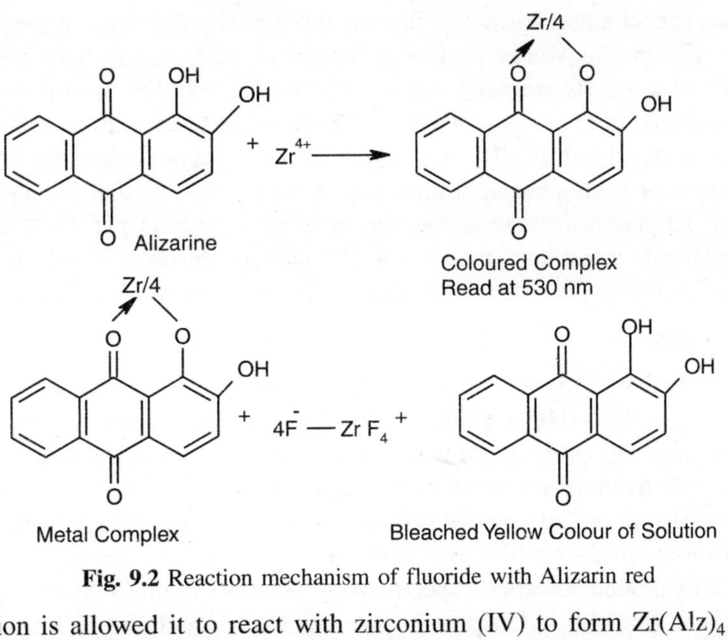

Fig. 9.2 Reaction mechanism of fluoride with Alizarin red

Alizarin solution is allowed it to react with zirconium (IV) to form $Zr(Alz)_4$ complex, which is violet in colour. This complex $Zr(Alz)_4$ subsequently reacts with excess of fluoride and bleaches colour to form ZrF_4 or ZrF_6^-. The decrease in intensity of colour indicates concentration of fluoride. Erichrome Cyanine R or SPADNS (*i.e.*, 2-parasulphophenylazo), 1, 8 dihydroxy 3–6 naphthalene sulphonate are also best reagents, but in such analysis $AlPO_4$ and SO_4^{2-} show strong interferences at 570 nm. However, the best method for analysis is Ion analyser with specific electrode or Ion chromatography (such method is described in Expt. No. 14.1 in Chapter 14).

(b) Nitrogen in Ammonia, Nitrite and Nitrate and Organic Nitrogen in Water

The nitrogen in water undergoes oxidation to form ammonia, nitrite or nitrate. The nitrogen originates from vegetable materials from land runoff, due to sewage pollution. Infants should not drink water containing nitrogen content greater than 45 mg/litre, as it will cause methaemoglobinaemia. The synthesis of protein in body and amino-acids is adversely affected by nitrogen. Nitrate occurs in trace quantities in surface water. A limit is 10 mg nitrate nitrogen in mg/litre of drinking water. Nitrate is actual etiologic agent for methaemoglobinaemia. Of this ammonical nitrogen, NH_3–N is determined colorimetrically by the Nessler's reagent. (The experimental details for the analysis of ammonia are in (Expt. 14.7) nitrite (Expt. 14.8) nitrate (Expt. 14.9) and cyanide (Expt. 14.14)). It will facilitate the actual analysis of these ions. In high concentration, it is titrimetrically analysed. Ammonia selective electrodes are also available. Secondly, nitrate-nitrogen, *i.e.,* NO_3–N is determined spectrophotometrically with brucine and phenoldisulphonic acid. The chromotropic acid *i.e.,* (4, 5–dihydroxyl–2, 7 naphthalene disulphone acid) is also used for spectrophotometry. The best method for nitrite nitrogen (NO_2–N) is Saltzman dizotisation colorimetric method. At pH 2.2–2.5 a red dye is formed with NEDA (*i.e.,* N–(1–naphthyl) ethylene diamine hydrochloride) and sulphanilamide reagent. All absorbances are measured at 543 nm. For determination of organic nitrogen, best method is macro-kjeldahl method of titration. (For details of analysis refer to Chapter no. 14 which provides analysis of anions in water with all details.)

(c) Iron in Water

A large amount of iron is encountered in water flowing through Basaltic rock region, similar to Deccan plateau in India. Few places like Murud in district Raigad of Maharashtra have high level of iron in potable water specially in wells. Its average level is 1 mg/litre *i.e.,* 1 μg/ml. It causes staining of laundry and ceramics ware. The growth of iron specific bacteria raises the level of iron in water. On account of aeration, iron (II) gets oxidised to iron (III). It is determined by spectrophotometry by its complexation with 1–10 phenanthroline or Bathophenanthroline. Bis or tris pyridyl ion can also be used for a similar purpose in place of 1–10 phenanthroline as the chromogenic ligand; at pH 3.2–3.4. The presence of phosphate or polyphosphate interferes in analysis. Bathophenanthroline is selective for iron in the presence of copper as impurity. (The details of analysis is provided in Expt. No. 12.7.)

(d) Manganese in Water

Water samples if collected from Vidarbha region of Maharashtra or in Goa are likely to be contaminated with manganese. Average level is 0.05 mg/litre, *i.e.,* 0.05 μg/ml. Manganese imparts objectionable and tenacious stain to laundary. The presence of manganese in water gives metallic taste to tea, as several complex compounds of manganese are formed with ingredients of tea. The best method for analysis is the persulphate method. In this $K_2S_2O_8$ or KIO_3 is used to oxidise Mn(II) to Mn(VII), which in turn is determined spectrophotometrically at 525 nm. One can also determine manganese spectrometrically with formaldioxime or by atomic absorption spectroscopy. (For more details refer to Expt. 13.3 for AAS analysis. The experiment No. 14.9 in Chapter 14 provides expt. details for its analysis.)

(e) Silica in Water

This is the common impurity in drinking water. It is determined gravimetrically if the amount is greater than 20 mg/litre but a small concentration can be estimated by spectrophotometric method involving reduction with 1-amino 2-naphthol-4-sulphonic acid. However, all operations must be carried out in polythene vessels, as glass contains silica. In molybdosilicate technique, colour formed with $(NH_4)_6 Mo_7O_{24} \cdot 4H_2O$ is measured spectrophotometrically at 410 nm. The heteropoly blue method is very sensitive.

(f) Alkalinity/Acidity or pH in Water

This is a measure of the capacity of water to neutralise H^+ and is expressed in terms of equivalent amounts of $CaCO_3$. It inhibits corrosion. This is determined by titrating 50 ml of water with 0.02 N NaOH to pH 8.3 or 50 ml of H_2O with 0.02 N H_2SO_4 till a pH value of 4.5. Alkalinity is expressed as phenolphthalein and total alkalinity. During titration, CO_3^{2-} is converted to HCO_3^{2-} at pH 8.3. The end point of pH 8.3 is called phenolphthalein alkalinity. However, if same titration is continued further, HCO_3^{2-} reacts with acid to form free H_2CO_3. The reaction is complete at pH 4.5. The amount of acid required to react is termed as total alkalinity. The alkalinity is expressed in terms of $CaCO_3$. The acidity originates due to dissolution of CO_2 from atmosphere, CO_2 is produced in water also due to biological oxidation of organic matter. Mineral acidity is present in many industrial wastes. The salts of trivalent metals (e.g., Fe, Al) hydrolise to release mineral acidity. The mineral acidity is measured by titration to pH 4.5. When methyl orange is used as an indicator, it is called methyl orange acidity. The titration of sample to phenolphthalein end point of pH 8.3 is called total acidity or phenolphthalein acidity. This is also expressed in terms of $CaCO_3$. As $CaCO_3$ has equivalent weight of 50, N/50 NaOH when used will give 1 ml = 1 mg of acidity in $CaCO_3$. (The expt. 10.5 spells out all details for the analysis of acidity.)

(g) Other Parameters

In chemical characterisation of water, certain parameters like dissolved oxygen, BOD, residual chlorine, pH, odour and bacteria are determined immediately, while some aspects like hardness, Na^+, K^+, Ca^{2+}, Mg^{2+}, F^-, Cl^-, SO_4^{2-}, colour, turbidity, specific conductance, alkalinity, dissolved solids can be

Table 9.11 Analysis of water sample from Jorhat in North East India

Parameter	*Sample A (Surface)*	*Sample B (bottom of well)*	*Standard WHO*	*Remarks*
Colour	70 Hazen	70 Hazen	5 Hazen	Very high
Turbidity	75 JU	85 JU	5 JU	Very high
pH	6.5	7.0	7.8–8.5	Slightly acidic
Hardness	105	110	100	Slightly high
Aluminium*	0.3	0.2	1	Tolerable
Chromium*	0	0.1	0.050	High
Iron*	2.5	2.0	0.1	Very high
Manganese*	0.5	0	0.05	Very high
Chloride*	80	85	200	Tolerable
Fluoride*	1.5	2.0	0.70	High
Nitrate*	25	20	22.6	Slightly high
Oxygen* dissolved (DO)	0	0	1.5	Too low

* Values of constituents are in mg/litre

determined at leisure. Finally, a few components like nitrogen ammonia, nitrate-nitrogen, suspended solids or heavy metals need addition of some preservative, before commencing analysis in laboratory. How we use the knowledge of chemical characterisation is demonstrated in Table 9.11, where a typical analysis of well water collected from city of Jorhat North East India adjacent to Civil Hospital is given. This speaks about the quality of health of people in the region. Accepted standard for potable water indicating maximum, minimum and average limit of different constituents for various countries is presented in Table 9.12. Such characterisation is of extreme importance in deciding water treatment process in next few chapters.

Table 9.12 Comparison of standards for drinking water by different agencies (ppm)

S. No.	*Parameter*	*WHO*	*EEC*	*USA*	*Canada*	*France*	*Acceptable value*
1	Dissolved solid	–	–	–	–	–	500
2	Colour	–	5	3–15	15	5	5
3	Turbidity	–	5	1–5	5	5	5
4	pH	–	6.5–8.5	–	6.5–8.3	7–8.5	7.0–8.5
5	Conductivity	–	400	–	–	2000	–
6	Hardness	–	35	–	18	–	100
7	Dissolved oxygen	–	5	–	–	1–1.5	–
8	Sulphate	200–400	5	250	500	250	200
9	Chloride	–	5	250	250	200–600	200
10	Free chlorine	–	–	0.2–0.3	–	–	–
11	Nitrate	45	–	45	10	50–100	–
12	Cyanide	0.05	0.05	0.01–0.2	0.01–0.2	0.05	0.05
13	Fluoride	1.5	0.7–1.5	1.4–2.4	–	1.5	0.70–1.0
14	Ammonia	–	0.05	–	0.5	0.05	–
15	Boron	–	–	–	5	–	–
16	Phosphorus	–	0.3	–	0.002	–	–
17	Calcium	–	100	80–100	200	–	5
18	Magnesium	–	30	80–100	150	–	30
19	Silver	–	0.01	0.05	0.05	–	–
20	Arsenic	0.05	0.05	0.05	0.01–0.05	0.5	0.05
21	Barium	–	0.1	1.0	1.0	–	–
22	Cadmium	0.01	0.005	0.01	0.01	0.005	0.01
23	Chromium	0.05	0.05	–	–	–	–
24	Copper	–	0.05	0.2–1.0	1.0	0.05	0.05
25	Iron	–	0.3	0.05–0.3	0.3	0.1	0.1
26	Mercury	0.001	0.001	0.002	0.05	0.01	0.001
27	Manganese	–	0.05	0.01–0.05	–	0.05	0.05
28	Nickel	–	0.05	–	–	–	–
29	Lead	0.1	0.05	0.05	0.05	0.05	0.1
30	Antimony	–	0.01	–	–	–	–
31	Selenium	0.01	0.01	0.001	0.01	0.01	0.01
32	Zinc	1.5–5.0	0.1–0.2	5	2	–	5

9.5 BIOLOGICAL INVESTIGATION OF WATER

Apart from direct bacteriological examination, indirect proof for the presence of micro-organism and biodegradable organic matter can be found out from the knowledge of dissolved oxygen (DO) and biochemical oxygen demand (BOD). (For extensive discussion please refer to Chapter 17.)

(a) Dissolved Oxygen of Water

All living organisms are dependent upon one another, to maintain the metabolic process that produce energy for growth and reproduction. N_2 and O_2 dissolve poorly in water, but their solubility increases with temperature as per Henry's law. At saturation, dissolved gases in water has 38 per cent oxygen. The solubility is 14.6 mg/litre at 0°C and it rises to 7 mg/litre at 35°C at one atmosphere pressure. Further, the rate of dissolution of O_2 in polluted water is much less than in pure water. Thus, DO determines whether biological changes are brought out by aerobic or anaerobic organisms. Former use free oxygen for oxidation to produce innocuous end products. DO is one single test to indicate how aquatic life is supported; as all aerobic process depend upon DO. Samples of water for DO analysis is collected in fields. (The detailed method for the analysis of DO is given in Expt. No. 15.2 in Chapter 15). The dissolved oxygen from water is determined by Winkler's method. The important reactions are:

$$MnSO_4 + 2KOH \rightarrow Mn(OH)_2 + K_2SO_4$$

$$2Mn(OH)_2 + O_2 \rightarrow 2MnO(OH)_2$$

$$MnO(OH)_2 + 2H_2SO_4 \rightarrow Mn(SO_4)_2 + 3H_2O$$

$$Mn(SO_4)_2 + 2KI \rightarrow MnSO_4 + K_2SO_4 + I_2$$

$$2Na_2S_2O_3 + I_2 \rightarrow 2NaI + Na_2S_4O_6$$

Thus, we use manganese hydroxide, which is oxidised to manganic hydroxide by dissolved oxygen. This manganic hydroxide is in turn converted to manganic sulphate which is reacted with KI to liberate free iodine, which in turn titrated against standard solution of thiosulphate. The amount of iodine consumed represents the dissolved oxygen in water used in oxidation. The oxidising agents in water will produce positive error, while the reducing agents produce negative error. Thus NO_2^-, Fe^{3+} can oxidise I^- to I_2° and give high results; also reducing agents like Fe^{2+}, SO_3^{2-} or S^{2-} also give negative results, as they reduce I_2° to I^-. The Winklers reaction can be also given in ionic form as:

$$Mn^{2+} + 2OH^- \rightarrow Mn(OH)_2 \text{ white ppt}$$

$$Mn^{2+} + 2OH^- + 1/2O_2 \rightarrow MnO_2 + H_2O \text{ (fixation of oxygen)}$$

$$Mn(OH)_2 + 1/2O_2 \rightarrow MnO_2 + H_2O$$

$$MnO_2 + 2I^- + 4H^+ \rightarrow Mn^{2+} + I_2 + 2H_2O$$

$$2Na_2S_2O_3 + I_2 \rightarrow 2NaI + Na_2S_4O_6$$

In presence of NO_2^- (Azide method) if excess of nitrite is present in water, we have to make modifications to Winkler's method. It is called Azide modification. This involves addition of azide to destroy free nitrite as:

$$2NO_2^- + 2I^- + 4H^+ \rightarrow I_2 + N_2O_2 + 2H_2O \text{ (Nitrite presence gives wrong results)}$$

$$N_2O_2 + 1/2\ O_2 + H_2O \rightarrow 2NO_2^- + 2H^+ \text{ (Excess iodine is consumed)}$$

Such interference is eliminated by addition of NaN_3 as:

$$NaN_3 + H^+ \rightarrow HN_3 + Na$$
$$HN_3 + NO_2^- + H^+ \rightarrow N_2(g) + N_2O(g) + H_2O$$

Thus, nitrite interference is eliminated. The Rideal-Stewart modification eliminates many interferences caused by reducing agents. $KMnO_4$ is used for pretreatment. Excess $KMnO_4$ is removed with $K_2C_2O_4$ e.g.,

$$5NO_2^- + 2MnO_4^- + 6H^+ \rightarrow 5NO_3^- + 2Mn^{2+} + 3H_2O$$
$$5Fe^{2+} + MnO_4^- + 8H^+ \rightarrow 5Fe^{3+} + Mn^{2+} + 4H_2O$$

then $$5(COO^-)_2 + 2MnO_4^- + 16H^+ \rightarrow 10CO_2\uparrow + 2Mn^{2+} + 8H_2O$$

After this treatment, regular Winkler's reaction can be carried out. Apart from Winkler's method, modern trend is to use DO metre or ion analyser with ion specific electrode for measuring dissolved oxygen.

(b) Biochemical Oxygen Demand

BOD is defined as the amount of oxygen required by bacteria while stabilising decomposable organic matter under aerobic conditions. In other words, BOD represents quantity of dissolved oxygen in mg/litre required during oxidation of decomposable or biodegradable organic matter by aerobic biochemical action. The decomposable organic matter at a time serves as the food for bacteria and energy is obtained due to such oxidation. Thus, BOD gives an idea about the extent of pollution. It serves as a test to evaluate the purifying capacity of receiving bodies of water. BOD test is a bioassay test devised to measure oxygen consumed by living organisms, while using organic matter present as waste in water. While evaluating this, samples are protected from sunlight, excessive agitation or shaking, and kept at a fixed temperature in an incubator. This also favours uniform bacterial growth. To complete such growth 20 hours are required. BOD test is a wet oxidation procedure. A 5-day BOD test gives a reliable information on quality of water. (The details for the analysis of BOD (3 days) refer to Expts. No. 15.3 and 15.4 of Chapter 15.) Two 300 ml bottles are filled with water and dissolved oxygen of first bottle is determined immediately by Winkler's method. The second bottle is incubated under 20°C in the dark for 5 days (sometimes 3 days; we need to keep it for 20 days in order to get a complete picture) and dissolved oxygen is once again measured. The reduction in oxygen content is due to demand exerted by the microbiological population and is a measure of oxidisable organic matter in sample value of 5-day BOD. The DO varies from 9.2 to 1.0 mg/litre showing 8.2 mg/litre as BOD value. Typical BOD values of different samples of water is listed in Table 9.13.

Table 9.13 BOD of various samples of water

Type of sample	*BOD_3 mg/litre*	*Remarks*
Potable H_2O	1.0–3.0	Reasonable
River	5.0–20.0	Tolerable
Sewage	50.0–100	Very bad
Industrial waste	100–10,000	Extremely poor

The consolidated study of various hydrological parameters of Ulhas river in Kalyan near Mumbai are given in Table 9.14.

Table 9.14 Water quality parameters of Ulhas river at different sampling stations in Mumbai

Station number (Sewri to Thane)	*pH*	*DO mg/litre*	NO_2^- *mg/litre*	PO_4^{3-} *mg/litre*	SO_4^{2-} *mg/litre*
1	7.8	4.8	0.9	0.09	2.8
2	7.8	4.1	0.4	0.09	2.9
3	8.0	4.3	0.7	0.08	2.7
4	7.4	3.7	2.0	0.05	2.1
5	7.5	4.2	2.1	0.08	2.3
6	7.6	3.1	1.8	1.10	2.9
7	7.6	3.6	1.1	0.05	2.5
8	7.4	3.6	0.8	–	2.7
9	7.3	3.2	4.8	0.06	2.1
10	7.3	2.8	2.7	0.06	2.2
11	7.3	1.9	3.7	–	2.8
12	7.4	2.2	3.2	0.04	2.1
13	7.2	2.8	3.1	0.05	2.2

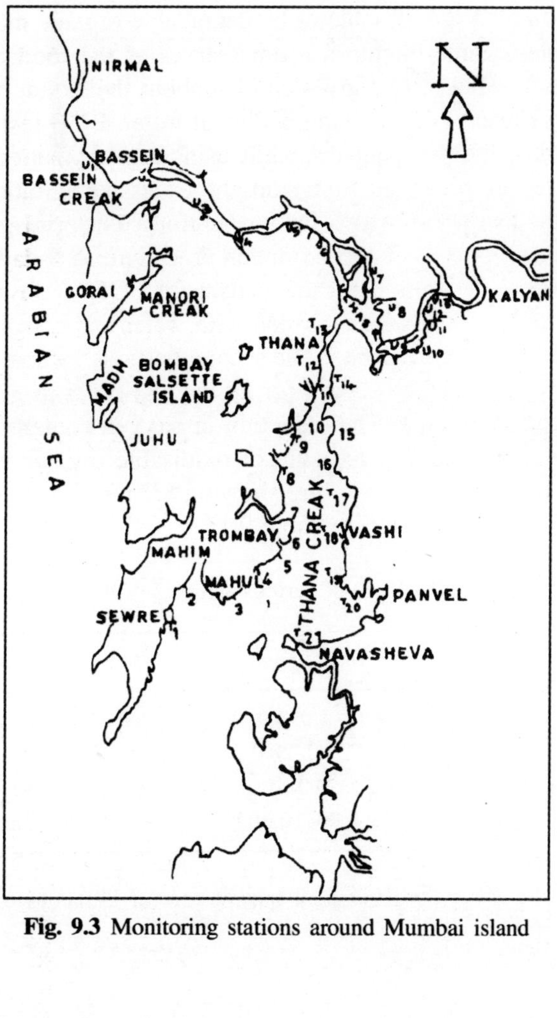

Fig. 9.3 Monitoring stations around Mumbai island

ADDITIONAL READINGS

9.1 D.C. Adriano, *Trace Elements in Terrestrial Environment*, Springer-Verlag, New York (1986).

9.2 D. Purves, *Trace Element Contamination of the Environment*, Elsevier Scientific Publishing Company, Amsterdam (1977).

9.3 D.C. Burrell, *Atomic Spectrometric Analysis of Heavy Metal Pollutants in Water*, Ann Arbor Science Publishers, Ann Arbor (1974).

9.4 M.K. Atkins and J.F. Lowe, *Pollution Control Cost in Industry: An Economic Study*, Pergamon Press, Oxford (1977).

9.5 APHA, *Standard Methods for the Examination of Water and Waste Water*, 20th Ed. (2008).

9.6 C.N. Sawyer and P.L. McCarty, *Chemistry for the Environmental Engineering*, 3rd Ed., McGraw-Hill and Co. (1978).

9.7 H.S. Samant, *Methods for Chemical Analysis of Water and Waste Water*, Surveillance Report EPSS-AR-73-12, Canada (1973).

9.8 I.L. Marr and M.S. Cresser, *Environmental Chemical Analysis*, International Textbook Co., New York, Blacki & Sons Ltd. (1983).

9.9 N.F. Voznaya, *Chemistry of Water and Microbiology*, Mir Publishers, Moscow (1981).

9.10 Mayuri Gandhi and S.M. Khopkar, *Indian Journal of Environmental Protection*, 8, 12 (1988).

CHAPTER 10

Water Pollution Analysis

10.1 INTRODUCTION

We have seen in the earlier chapters the basic principles of physical or chemical examination affecting the quality of the water. Such description did not include details of experimental setup which should be considered during experimentation. Therefore this chapter provides details of experimental conditions for carrying out experiments in the laboratory. This present chapter outlines the experimental details of the analysis of various hydrological parameters influencing the quality of water. While subsequent five chapters encompass the analysis of metals, non-metals, organics, specific organic impurities like pesticides and finally a chapter on the microbiological examination of water. A due care has been taken in all these chapters to provide all minute details of the steps to be looked into after during the experimentation. Even for the preparation of the standard solutions or even buffer solution exact amount of chemicals to be dissolved in water is also specified. Further every section composes of the details of the calculations to be carried out at end of experiment. Also all useful conversion factors are readily provided for rapid analysis and drawing definite conclusion on the data collected. For a reader needing further details of experimentation, he is referred to the additional references provided under each chapter.

When we consider hydrological properties it includes physical, chemical and biological properties of the water. However, there is no hard or fast rule to categorise particular property in specific category. For instance properties like acidity/alkalinity or hardness should be logically under chemical properties of water but they are all considered under physical properties. Similarly dissolved oxygen, biological oxygen demand and chemical oxygen demand have been considered under biological properties although in reality they involve all experimentation from chemistry. Without bothering too much about a class, let us consider experimental details for analysis of various hydrological parameters of water. We would consider such parameters in subsequent sections.

Let us consider each property in details.

10.2 COLOUR OF WATER

Vegetation, iron and manganese compounds impart characteristic colour to the water. The colour may be either true or an apparent colour. True colour is original colour of water while apparent colour is colour due to the background present in the solution. The pretreatment to remove turbidity before colour measurement is most essential. Such treatment includes filtration, centrifugation. The main methods of colour estimation are (a) Visual comparison (b) Absorption spectrophotometry (c) Tristimulus filter colorimetry (Also refer to Section 10.1(a)).

(a) Visual Comparison of Colours: Experiment 10.1: Analysis of Colour of the Water

A comparison is made with known concentration of coloured solution. A platinum-cobalt method is most common as 1mg platinum/L imparts good colour, provided if turbidity is absent. The apparatus and reagents used are as follows: 50 ml Nessler tube and pH meter. Dissolve 1.25 gm of K_2PtCl_6 and 1 gm of $CoCl_2.6H_2O$ in mixture of 100 ml. water and 100 ml. con. hydrochloric acid. Then, make it to one litre. This stock standard solution has a colour of 500 units. By appropriate dilution one can make the solution with 5, 10, 20, 40, 60 colour units from the stock solution which is of 500 units. The water with unknown colour units is taken in test tube and it is then compared visually with colour of varying standards as 5, 10, 20, 40, 80, etc. Look vertically downward through these tubes against background of white surface and measure pH of solution.

Calculation: Colour unit = $\dfrac{A \times 50}{B}$, if A = estimated colour, B = ml of sample diluted. Repeat the experiment several times till satisfactory matching occurs.

(b) Spectrophotometric Measurement of Colour: Experiment 10.2: Spectrophotometry

The colour is expressed in terms of the wavelength and absorbance of the solution. We need spectrophotometer with 10 mm quartz or glass cuvettes. Take 50 ml of sample solution, add sulphuric acid and sodium hydroxide to get pH = 7.6.

The cuvette is filled with coloured solution and the absorbance or transmittance of each sample is measured at 400–700 nm at regular interval of 10 nm against a water blank as the reference solution. The results are shown in Table 10.1.

Table 10.1 Absorbance of sample of water at various wavelengths

Ordinate No.	*Wavelength (x) nm*	*Wavelength (y) nm*	*Wavelength (z)*
1.	435	489	422
2.	461	515	432
3.	544	529	438
4.	564	541	444
5.	577	551	450
6.	588	561	455
7.	599	572	462
8.	610	584	468
9.	624	600	477
10.	645	627	495

In order to get tristimulus values x, y, z add transmittance at particular wavelength and divide by 10 as ordinate. Y is per cent luminescence. The trichromatic coefficient is $x = \left[\dfrac{x}{x+y+z}\right]$; $y = \left[\dfrac{y}{x+y+z}\right]$.

One uses chromatic diagram. State dominant wavelength hue luminescence and purity. One can identify colour hue for dominant wavelength. Visible region (see Table 10.2) is as follows:

Table 10.2 Colour vs range of wavelength

S. No.	*Wavelength (nm)*	*Colour*	*S. No.*	*Wavelength (nm)*	*Colour*
1.	400–465	Violet	8.	587–598	Orange
2.	465–482	Blue	9.	598–620	Orange-red
3.	482–497	Blue green	10.	620–700	Red
4.	497–530	Green		**Also the broad hue range is**	
5.	530–575	Green yellow	11.	400–530	Blue purple
6.	575–580	Yellow	12.	530–700	Red purple
7.	580–587	Yellow-orange			

(c) Tristimus Filter Photometer

The idea is same as was used in earlier experiment in spectrophotometry. The only difference is we measure absorbance with three filters no. 1, 2, 3 as obtained from corning. Filter are CS 3-107 (No. 1), CS 4-98 (No. 2), CS 5-70 (No. 3) instead of varying wavelength.

We obtain transmittance values on three filters and calculate tristimulus values x, y, z from % transmittance for filter 1 (T_1), 2 (T_2) and 3 (T_3). For distilled water with filters T = transmittance

$$\begin{aligned} x &= (T_3 \times 0.06) + (T_1 \times 0.25) \\ y &= (T_2 \times 0.316) \\ z &= (T_3 \times 0.374) \end{aligned} \quad \left| \begin{aligned} x &= 98.09 \\ y &= 100.0 \\ z &= 118.35 \end{aligned} \right\} \text{Tristimulus values of water}$$

Trichromatic coefficients x and y is obtained by following formula.

$$x = \left[\frac{x}{x+y+z}\right] \text{ and } y = \left[\frac{y}{x+y+z}\right]$$

These values are given by the American Dye Manufacture Industries. (ADMI)

10.3 TURBIDITY:

Experiment 10.3: Analysis of Turbidity of Water

Jackson-Candle turbidimeter can be used to estimate the turbidity of water but one can not go below 25 JTU (turbidity units). Below this value we must use nephelometer. Formazin polymer is used as the primary standard or reference suspension.

Apparatus and Reagents: A good nephelometer (90° path of incident light) with 10 mm glass or quartz cuvettes. The reagents to be used during experiment are as follows:

Solution 1: Dissolve 1.0 gm of hydrazine sulphate in 100 ml of distilled water.

Solution 2: Dissolve 10 gm hexamethylenetetramine in 100 ml of water. Then mix solution 1 (5 ml) and solution 2 (5 ml). Allow it to stand for few minutes. It gives solution of 4000 NTU.

Procedure: A solution containing turbidity is subjected to measurement at fixed temperature. Plot a standard graph of turbidence against concentration in the measurement by using standard turbid solutions.

Precaution: Particle flocculation and sedimentation process, must be avoided. The solutions with known turbidity is made up by dilution from the standard solution of 4000 NTU solution for plotting a standard curve.

The results are shown in Table 10.3.

Table 10.3 Turbidity measurements in NTU units

S. No.	*Turbidity (range) (con)*	*Reported NTU*
1.	0–1.0	0.05
2.	1–10	0.1
3.	10–40	1.0
4.	40–100	5.0
5.	100–400	10.0
6.	400–1000	50.0
7.	> 1000	100

To get reproducible results in experiment, repeat the experiment.
(Also refer to section 9.1(c) for principles employed in determination).

10.4 ODOUR OF WATER:

Experiment 10.4: Measurement of Odour of Water

Odour is sensory response experienced by a person by receptor cell like nose. It is quality factor which determines the nature of potable water. One carries out experiment to measure threshold odour number (TON) test. No plastic container should be used as it imparts bad odour. TON varies with temperature at 60°C is standard temperature for analysis at hot threshold.

Apparatus and Reagents: Use odour free glass ware as rubber cork imparts smell to water. A constant temperature bath must be used. The use of odour free water for standardisation from distilled water is required. Activated coal absorbs odour rapidly. A person not contributing his own odour through cosmetics be asked to perform the test at point much less than fatigue limit. The odour reference standard is used to qualify odour (see Table 10.4).

TON is greatest dilution of sample with odour free water. Therefore make total volume of sample and odour free water to volume of 200 ml.

$TON = \frac{A+B}{A}$ if A = ml. sample and B = ml of odour free water added and then the observations are recorded by showing whether odour is noted in each test (see Table 10.5).

Table 10.4 Threshold odour number at various dilution

S. No.	Sample dil. to 200 ml	TON	S. No.	Sample dil. to 200 ml	TON
1.	200	1	9.	12.0	17
2.	140	1.4	10.	8.3	24
3.	100	2	11.	8.7	35
4.	70	3	12.	4.0	50
5.	50	4	13.	2.8	70
6.	35	6	14.	2.0	100
7.	25	8	15.	1.4	140
8.	17	12	16.	1.0	200

Table 10.5 Odour testing response

ml dil. to 200 ml	12	0	17	25	0	35
Response	–	–	–	+	–	+

Calculations: TON is dilution ratio at which odour is just detectable. The smaller value of TON = 1, indicates practically no odour or insignificant odour for the sample. (For basic principles involved in experiment refer to section 9.1(d) Chapter 9.

10.5 ACIDITY OF WATER:
Experiment 10.5: To Analyse Acidity of Water

It is the measure of aggregate property of water. It promotes corrosion, affects chemical speciation and biological process. The volumetric method is followed to ascertain the acidity. Dissolved carbon dioxide in water affects the acidity of water. Hydrogen sulphide and ammonia also affects quality of water and impart offensive odour to water.

Apparatus and Reagents: Titrator, stirrer pipette burette, polythene flask. Electrometric titrator (in reality, it is pH meter). At 25°C titration is carried out in carbon dioxide free water. A buffer of pH 6.0 is used. 0.05 N potassium hydrogenphthalate (10 gm $KHC_8H_4O_4$) is dissolved in one litre of water, 0.5 N NaOH is standardised by titrating against 40 ml phthalate solution. Continue titration below pH 8.7.

Calculations: Normality $= \dfrac{A \times B}{204.2 \times C}$ where A = gms of phthalate solution, B = ml of pthalate solution, C = ml of NaOH. Other reagents used are 30% H_2O_2, bromophenolblue indicator (pH ~ 3.7). Metacresol purple and phenolphthalein as the indicators and $Na_2S_2O_3$ (0.1M).

Procedure: Take known volume of water, measure its pH by adding 0.02 NH_2SO_4. Add few drops of 30% H_2O_2 and boil it and cool. Titrate it with standard alkali to pH 8.3 (Add $Na_2S_2O_3$ if free residual chlorine is present). We can also use potentiometric titration where one must keep electrodes under water. Measure the pH. Add NaOH so that pH change is by 0.2 in units of pH. Use magnetic stirrer for uniform mixing. Note pH at constant reading until pH 9.0. Plot titration curve by noting ml of alkali added against pH attained. We calculate acidity as follows:

$$\text{mg CaCO}_3/\text{L} = \frac{(A\times B)-(C\times D)\times 50{,}000}{\text{ml of sample}}$$

In this equation the symbols A = ml of NaOH used, B = normality of NaOH, C = ml of H_2SO_4 used, D = normality of H_2SO_4.

10.6 ALKALINITY OF WATER:
Experiment 10.5: Analysis of Alkalinity of Water

This add neutralising capacity of water. The measure of the aggregate quality of water CO_3^-, HCO_3^-, OH^- contribute towards development of alkalinity.

Apparatus and Reagents: It is same as described in earlier section, Na_2CO_3 (0.05 N), 2.5 in litre gives this strength standardise against 40 ml of Na_2CO_3 with 60 ml of water till pH = 5.0. Normality $= \frac{A\times B}{53\times C}$ if A = gm of Na_2CO_3 in litre, B = ml of Na_2CO_3, C = ml of acid used of 0.05N H_2SO_4 or HCl with Bromocresol green indicator, Methyl red + Bromocresol mixture is used. To 100 mg bromocresol + 20 mg methyl red in 100 ml. 95% alcohol, or isopropylalcohol. 100 mg bromocresol is dissolved in 100 ml water. Use phenolphthalein and $Na_2S_2O_3$ (0.1 N). There are three ways of titration to be carried out [For detailed discussion of principle employed refer to section 9.2 (f) in Chapter 9].

The potentiometric titration curve can be also obtained as described earlier (At pH 4.5 in standard NaOH). The titration can be carried out at the fixed pH. The end point is obtained by direct addition of acid with small increments. At low alkalinity 0.2 N acid is used at pH 4.3–4.7.

Calculations: In potentiometric titration, we calculate alkalinity as

$$\text{Alkalinity mg CaCO}_3/\text{L} = \frac{A\times N\times 50{,}000}{\text{ml of sample}} \quad \text{if A = ml of acid, N = Normality of acid.}$$

$$\text{Total alkalinity mg CaCO}_3/\text{L} = \frac{(2B-C)\times N\times 50{,}000}{\text{ml of sample}}$$

if B = ml of titrate at first pH, C = Total ml titration, N = Normality of acid.

The carbonate alkalinity is present if phenolphthalein alkalinity is less than total alkalinity while hydroxide alkalinity is present if phenolphthalein alkalinity is more than half of total alkali. Finally bicarbonate alkalinity is present when phenolphthalein alkalinity is less than half of total alkali.

10.7 HARDNESS OF WATER:
Experiment 10.6: Analysis of Hardness of Water

It is of two types viz. temporary and permanent hardness. The former is due to the presence of bicarbonates of calcium and magnesium while latter is due to presence of sulphates or chlorides of calcium and magnesium. The total hardness is sum of calcium and magnesium concentrations expressed as the mg$CaCO_3$/Litre. If hardness is greater than sum of carbonate and bicarbonate alkalinity it is called as "carbonate hardness". The amount of hardness in excess of this is called 'noncarbonate hardness'. While total hardness is defined as sum of calcium and magnesium concentrations expressed as calcium carbonate in milligrams per litre *i.e.*, $CaCO_3$ mg/L. The separate analysis of calcium and magnesium is done to ascertain by EDTA titration. It is best method of analysis.

Apparatus and Reagents: Buffer solution. Dissolve 16.9 gm NH_4Cl in 143 ml ammonical hydroxide. Add 1.25 gm of magnesium salt of EDTA and make it up to 250 ml. In absence of $Mg(EDTA)_2$ salt dissolve 1.179 gm of EDTA and 780 mg $MgSO_4$ or 644 mg $MgCl_2.6H_2O$ in 50 ml of water. The life of such solution is only for one month.

Procedure: An aliquot of water sample is taken and made up to 50 ml in porcelain disc. Add 1–2 ml buffer of pH 10.0. Add 1–2 drops indicator (Eriochrome black T, 0.5 g in 100 ml triethanolamine or 0.10 gm calgamite in 100 ml H_2O) in the form of a solution or powder and titrate with standard EDTA solution. We calculate it as:

Calculations: Hardness (EDTA) as mg of $CaCO_3$/L $= \dfrac{A \times B \times 1000}{\text{ml of sample}}$

if A = ml of titration for sample; B = mg of $CaCO_3$ equivalent to 1 ml of EDTA titrant. (For theoretical discussion on hardness refer to section 9.1(f) of Chapter 9).

10.8 CONDUCTIVITY OF WATER:

Experiment 10.7: Measurement of Conductivity of Water

It is the measure of the ability of solution to carry an electric current. Such ability depends upon total concentration, specific mobility and valence and temperature of sample. Therefore conductance is defined as the function = 1/R if R = resistance in ohms, while C_r is in ohm^{-1} is conductance. AC is used to avoid the polarisation at electrode surface. C_r is directly related to area and distance between two electrodes as by equation (if A = area of electrode surface, L = distance between two electrodes). C = 1/R = K (A/L). It is referred to as specific conductance or conductivity of water. Conductivity is also measured in millisiemans (s) (1 ms/m = 10 μ mhos/cm and 1 μs/cm = 1 μ mho/cm). The equivalent conductivity (λ) is conductivity per unit concentration, (λ = 0.001k/concentration). Table 10.6 gives equivalent conductivity of KCl at different concentration.

Table 10.6 Equivalent conductivity (λ) and conductance (K)

KCl con (M)	*Eq. Conductivity (mho-cm²/eq) (λ)*	*Specific conductance K (μmho/cm)*
0	149.9	–
0.01	141.2	1.412
0.1	128.9	12.890
1.0	111.9	111.900

In laboratory the conductance of standard KCl is measured and then value of cell constant is evaluated by following relation $C = K_s/G_s$. The value for unknown solution is $K_y = CG_Y$. The conductivity of the distilled water 0.5–3 μmhos/cm.

Calculations: Naturally occurring water contains Ca^{2+}, Mg^{2+}, Na^+, K^+, HCO_3^{2-}, SO_4^{2-} or Cl^- and total dissolved solids (TDS) less than 2500 mg/L. Please refer to Table 10.7 for the value of equivalent conductance for ions in water.

Table 10.7 Equivalent conductance λ° for various ions at 25°C

Cation	*λ°*	*Anion*	*λ°*
H^+	350	OH^-	198.6
1/2 Ca^{2+}	59.5	1/2 SO_4^{2-}	80.0
1/2 Mg^{2+}	53.1	Cl^-	76.4
Na^+	50.1	I^-	54.4
K^+	73.5	NO_3^-	71.4
NH_4^+	73.5	HCO_3^-	44.5

For theoretical discussion on conductivity refer to section 9.1(b) of Chapter 9.

10.9 TASTE OF WATER

It refers to gustatory sensation like sweet, sour, bitter, salty. While flavour refers to complex of gustatory olfactory and trigeminal sensation originating from chemical stimulation of the nerve sensory endings. Taste refers to method of sensory analysis. The flavour rating assessment (FRA), flavour profile analysis (FPA) and flavour threshold testing (FTT) are three test conducted by mouth. Water must be clean free from bacteria, viruses and hazardous materials. These tests are identical in procedure as were used to measure odour intensity. Flavour threshold number (FTN) is calculated in following manner.

$$FTN = \frac{A+B}{A}$$ if A = sample volume, ml, B = reference to diluent of water volume.

Table 10.8 gives flavour threshold number of water at various dilution.

Table 10.8 Flavour threshold number of water at different dilutions

S. No.	*Sample volume ml (FTN)*	*Diluent volume (ml)*	*Flavour threshold number (FTN)*
1.	200	0	1
2.	100	100	2
3.	50	150	4
4.	25	175	8
5.	12	188	17
6.	3	197	67
7.	1	199	200

The flavour threshold number is the dilution ratio at which flavour is just acceptable. FTN = 1.0 can be detected also but this is not precise value. In fact flavour and odour are synonymous words.

10.10 SOLIDS IN WATER:

Experiment 10.8: Analysis of Total Solids in Water (TDS)

This represents dissolved or suspended solids in water. With high dissolved solids in water is unfit for drinking (Limit ~ 500 mg/L permitted). The total solids is the material residue left after evaporation of sample of water. When it is dried at fixed temperature, it includes suspended solids also. Dissolved solid pass through the filter of 2 μm pore size. Fixed solids are residual solids. The weight loss on ignition is called "volatile solids". The settleable solid refers to material settling out of suspension. The residues are dried at 105°C to 180°C. Porcelain dish is used for the purpose of evaporation. A temperature of 500°C on muffle furnace is needed. One hour of heating is needed at 103°C. The calculation is as follows:

Calculation: $$\text{mg total solids/L} = \frac{(A-B)\times 1000}{\text{Sample volume}}$$

if A = wt. of dry residue + dish in mg

B = wt. of dish in mg.

By following same procedure except evaporation at 180°C the total dissolved solids can be ascertained. The total suspended solids dried at 180°C can also be measured after simple filtration of sample of water through glass fibre or the membrane filter. Fixed and volatile solids can be ignited at 500°C. The calculations are

$$\text{mg volatile solids} = \frac{(A-B)\times 1000}{\text{Sample volume}}$$

$$\text{mg fixed solids} = \frac{(B-C)\times 1000}{\text{Sample volume}}$$

if A = weight of residue + dish before heating, B = weight of residue + dish after ignition

C = weight of dish or filter (mg).

10.11 TEMPERATURE OF WATER:

Experiment 10.9: To Measure Temperature of the Water

It is useful to calculate alkalinity and stability of $CaCO_3$ in salinity. In Limnological studies, water temperature as function of depth are required. Good mercury filled thermometer is used in terms of 0.1°C marking. For the purpose of depth profile, reversing thermometer is used. Calculations are as follows:

$$\Delta T = \left[\frac{(T'-t)(T'+t)}{K}\right] \times \left[1 + \frac{(T'-t)(T'+V_0)}{K} + L\right]$$

In above expression ΔT = corrected temperature to be added to uncorrected temperature

T′ = uncorrected reading at reversal, t = temperature at which thermometer is read.

V_0 = volume of small bulb end capillary, K = constant (K ~ 6100); L = calibration correction.

To understand significance of temperature measurement refer to section 9.1(c) of Chapter 9.

So far we have considered details of the experimental setup for the determination of various hydrological parameters of water. Let us now consider treatment. What are different methods of the treatment of water to render it fit for drinking or for industrial purposes?

ADDITIONAL READINGS

10.1 American Public Health Association et al., *Standard Methods of the Examination of Water and Waste Water,* 20th Ed., Ed. L.S. Clescerl. A.E. Greenborg and A.D. Eaten (1998).

10.2 S.M. Khopkar, *Basic Concepts of Analytical Chemistry,* 3rd Ed., New Age International Publishers (P) Ltd., New Delhi (2008).

10.3 S.M. Khopkar, *Analytical Chemistry—Problems & Solutions*, New Age International Publishers (P) Ltd., New Delhi (2003).

10.4 W.I. Youden and E.H. Steiner, *Standard Manual of Association Official Analytical Chemistry*, (*AOAC*) Washington DC (1975).

10.5 L.H. Keith, *Principles of Environmental Sampling*, 2nd Ed. ACS reference book, American Chemical Soc., Washington, USA (1996).

10.6 J.P. Dun and R.F. Statzer, *Managing Safety in the Chemical Laboratory*, Van Nostrand Reinhold & Co., New York (1988).

10.7 A.K. Furr, ed. *Handbook of Laboratory Safety*, 3rd Ed., CRC Press, Boca Raton (1990).

10.8 R.W. Manerieff, *The Chemical Sensors*, John Wiley (1946).

10.9 V.I. Sndyink and D. Jenkins, *Water Chemistry*, John Wiley (1980).

10.10 H.V. Svedrop, M.W. Johnson and R.H. Flemming, *The Oceans*, Prentice Hall (1942).

CHAPTER 11

Water Treatment

11.1 INTRODUCTION

In early days, water originating from well, river or brooks was directly used for the purpose of drinking. However with the growth of population as well as various industries, such water hardly remains pure due to contamination and therefore needs preliminary treatment. Even water which is used for the industrial processing also needs proper treatment as impure water may spoil the quality of production and give rise to newer problems. In early days, the potable water was passed through earthen pots containing sand or gravel and charcoal powder before it was utilised for drinking. Such treatment however, lacked the process of disinfection to arrest growth of microorganisms.

The same principle is followed in large scale for water treatment plants. Since water flows from terrain which may be contaminated, it needs sedimentation and coagulation before the process of filtration. If water is to be used for industry it needs softening while if it is to be used for drinking, it must be disinfected by chlorine, ozone or ultraviolet radiation. The fluoridation is considered by some as luxury but it is most indispensable from the point of view of growing infant population in order to protect their teeth. Finally some water may need specialised treatment.

The overall supply of water in world is slowly dwindling due to lack of rains or deforestation of land. The saline water is useless from the point of drinking or industrial consumption. However, some efforts are being made for desalination of sea water by deionisation or treatment involving osmosis. However such water is rarely used for human consumption with exception of few countries like UAR. Unfortunately, the large scale treatment of water for drinking purposes is far from satisfactory in many towns. Most of the metropolitan cities like Mumbai, Kolkata and Chennai do not have all water supplies which are totally treated. This is because of the growth of population and large scale consumption by community. It has become possible either to have partial treatment of raw water or rapid large scale operations involving bear minimum treatment procedure. For disinfection chlorine is good but when operations are to be carried out on large scale, bleaching powder can be used. The treated water has to undergo same steps of analysis as is potable water. One has to guard them from excess of presence of chlorine, PO_4^{3-}, CO_2 and organic pollutants. A large scale testing needs continuous analysis with monitoring instruments. Some efforts are made within our country to design and develop monitoring devices. Such of the devices are discussed at length at the end of this chapter.

Fresh water is rarely used directly for drinking purposes, as the water gets contaminated with impurities during its transportation to the city reservoir. The water which is to be used for industrial process always needs a pretreatment. Water for textile industry should not contain too much of iron, as it causes staining of fabric. Water for food industry should not have manganese, as it causes staining of containers and adversely affects the taste and quality of tea prepared in that water. The boiler feed water should not contain excess of salts of calcium and magnesium, as it will consume more heat due to insulation of containers and pipes by deposited scales. The methods of analysis of treated water are more or less similar to analysis of potable water. However, overall monitoring or analysis of water on regular basis presents a formidable task. In this chapter various treatment processes for water are discussed, followed by an overall review of monitoring instrument used for water quality management.

11.2 IMPORTANT STEPS IN WATER TREATMENT PLANTS

Depending upon the end use of water, the treatment procedure also vary. The overall picture of water treatment, irrespective of end use shall be considered. An endeavour is made in this chapter to understand the chemistry involved in water treatment process. The important stages involved in treatment are: coagulation or flocculation; sedimentation; filtration—slow sand, rapid sand filtration; disinfection—including chlorination and ozonolysis; removal of iron and manganese; softening by lime-soda ash process or deionisation method; control scale and corrosion, taste and odour removal; prophylaxiation treatment, *i.e.,* fluoridisation; and specialised treatment for a specific purposes.

Figure 11.1 indicates steps for water resource management. The main source of water is clouds, rains, melting of ice caps. It gets collected in water-shed areas and goes to reservoir, from where it can be directly let off for distribution, depending upon its quality and end use. For use involve preliminary treatment and if necessary an advanced treatment. Only after this treatment, it is taken to storage tank for regular distribution to city and industries.

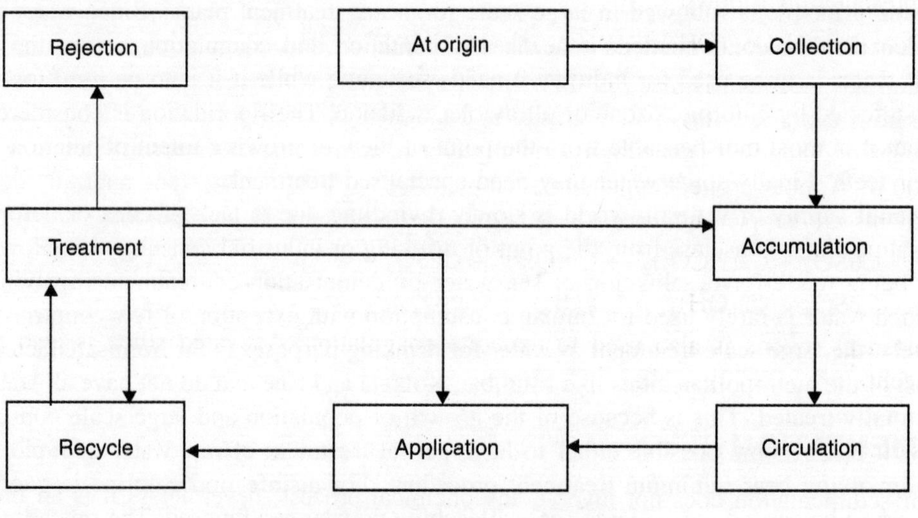

Fig. 11.1 Water supply-management

11.3 COAGULATION

The principal coagulants used in water treatment plants consists of aluminium sulphate or ferric chloride or ferrous sulphate. The important reactions involved in floc formation with alum in lime soda ash treatment involving use of lime water ($Ca(OH)_2$) and soda (Na_2CO_3) is as follows:

$$Al_2(SO_4)_3 + 3Ca(HCO_3)_2 \rightarrow 3\ CaSO_4 + 2Al(OH)_3 + 6CO_2$$

$$Al_2(SO_4)_3 + 3Ca(OH)_2 \rightarrow 3CaSO_4 + 2Al(OH)_3$$

$$Al_2(SO_4)_3 + 3Na_2CO_3 + 3H_2O \rightarrow 2Al(OH)_3 + 3Na_2SO_4 + 3CO_2$$

$$Al_2(SO_4)_3 + 6NaOH \rightarrow \underset{\text{(Floc)}}{2Al(OH)_3} + 3Na_2SO_4$$

This flow of $Al(OH)_3$ absorbs dirt and colour (Floc) during sedimentation. The reaction with ferric chloride coagulation process is as follows:

$$2FeCl_3 + 3Ca(HCO_3)_2 \rightarrow 3CaCl_2 + 2Fe(OH)_3 + 6CO_2$$

$$2FeCl_3 + 3Ca(OH)_2 \rightarrow 3CaCl_2 + \underset{\text{(Floc)}}{2Fe(OH)_3}$$

While with $FeSO_4$

$$FeSO_4 + Ca(HCO_3)_2 \rightarrow Fe(OH)_2 + CaSO_4 + 2CO_2$$

$$FeSO_4 + Ca(OH)_2 \rightarrow Fe(OH)_2 + CaSO_4$$

$$2Fe(OH)_2 + \underset{\text{(Dissolved oxygen)}}{1/2\ O_2 + H_2O} \rightarrow \underset{\text{(Floc)}}{2Fe(OH)_3}$$

The last reaction in the presence of dissolved oxygen is most important. The effectiveness of coagulant is determined by the magnitude of zeta potential. The *zeta potential* is defined as the magnitude of electric charge on ions surrounding suspended particulate matter. The charges are negative and are expressed in terms of millivolts. The agglomeration of floc particles as a result of polymerisation can be represented as follows:

$$2[Fe(H_2O)_5\ (OH)]^{2+} \rightarrow [Fe_2\ (H_2O)_8(OH)_2]^{4+} + 2H_2O$$

A substantial advancement is made in the field of synthetic coagulants or polyelectrolytes. These coagulants are both cationic with positive charge and anionic with negative charge. Those with both positive and negative charges are called polyampholyte or nonionic polyelectrolytes. Their size vary in the range 1–100 μ. Without proper coagulation, sedimentation will be useless. The naturally occurring efficient coagulants are copper sulphate, which favour floc formation as per reaction:

$$CuSO_4 + Ca(HCO_3)_2 \rightarrow Cu(OH)_2 + CaSO_4 + 2CO_2$$

$$CuSO_4 + Ca(OH)_2 \rightarrow Cu(OH)_2 + CaSO_4$$

In few instances, ozone is also used to expedite coagulation. Activated silica is also used as flocculating agent.

11.4 SEDIMENTATION

The process of sedimentation does not involve any chemical process, nevertheless it is an important step of water treatment. This is a physical process of settling of floc. Sedimentation may occur within 15–60 minutes duration after vigorous stirring of water with baffles, which in turn expedites the agglomeration process. In this process the colour of water is also removed, as coloured molecules form basic insoluble salts. The settling tanks are of various sizes and shapes. They have horizontal flow or circular flow or static settling tanks. Sludge clarifier ensures proper circulation with no turbulence, or excess of stirring during process of settling.

11.5 FILTRATION

The main function of filtration is to retain the particles contained in a liquid on the surface of filter media or in-depth. Surface filtration requires supporting media, while in-depth filtration is performed through a filter bed. The dissolved solids and presence of chloride, sulphates of magnesium and sodium increases bed penetration by floc. The real filtration takes place in the first 3–4 cm in total height of 80 cm filter bed. The slow sand filters involve filtration of water without previous treatment, while rapid sand filters are used for water which has been previously treated by coagulation or sedimentation. Anthracite coal is used as filtering media along with sand and gravel. The filter beds are cleaned by back washing. In rapid filtration, 4–50 m/u of water passes through bed. Washing of filter bed leads to expansion of the filter bed. However, simultaneous air and water washing proceeds without expansion of filter bed.

11.6 DISINFECTION

The elimination of taste and smell, the destruction of pathogenic microorganism can be done by chlorination, ozonolysis or ultraviolet light treatment. However, most favoured method is chlorination. This is because chlorine is a cheap material. The treatment with iodine is only used for special swimming pools. The disinfecting gas kills virus bacteria and germs from the water. The source of such disinfectant is chlorine gas or hypochlorates of calcium and sodium, e.g., CaCl(OCl) bleaching powder. The reaction of chlorine with water is as follows:

$$H_2O + Cl_2 \rightleftharpoons HCl + HOCl$$

At low pH, hypochlorous acid predominates, while at higher pH greater than 7.5, hypochlorate ion prevails. HOCl penetrates through organism cell walls and oxidises matter within the cell, but relatively hypochloride is not as effective. Fe^{2+}, Mn^{2+}, Ca^{2+}, Mg^{2+}, NO_2^-, CO_3^{2-}, HCO_3^- and NH_3^- interfere in such a reaction. The most troublesome interference is due to ammonia. It reacts with chlorine to form mono-, di-or trichloroamines at different pH values as:

$$NH_3 + HOCl \rightarrow NH_2Cl + H_2O \text{ (pH > 8.0)}$$

$$NH_3 + 2HOCl \rightarrow NHCl_2 + 2H_2O \text{ (pH = 4.3 – 5.0)}$$

$$NH_3 + 3HOCl \rightarrow NCl_3 + 3H_2O \text{ (pH < 4.3)}$$

Though these chloroamines have germicidal action, they act very slowly. Combined residual chlorination involves chlorination with residual chlorine left behind in water after treatment, since free residual chlorination gives free available chlorine through destruction of ammonia or organic matter. The total chlorine residue remaining after specified contact period is called combined available residual chlorine. Residual chlorine acting as HOCl or OCl^- ion is called free available chlorine. The chlorine demand is defined as the difference between the amount of chlorine added to water and amount of residual chlorine left behind at the end of specified contact period. The demand varies with time of contact and temperature and pH. The germicidal action of iodine is more effective over broader pH range, as elemental iodine and water are active germicidal agents, as chloroiodides like chloroamines are not formed easily. Ozone is also a good germicidal agent. Ultraviolet radiation at 253.7 nm has good germicidal effect on water and is economic.

Break Point Chlorination

Chlorine is mainly added as the disinfecting agent; an excess dose is added to kill microorganisms. Chlorine also removes foul smell. The demand of chlorine is noted by ascertaining the difference between amount of chlorine applied and amount of chlorine left which remains as residue after some contact period. Such demand can be tested in sunlight. The demand depends upon various factors such as time of contact, pH, water temperature and quantum of dose used. The demand suddenly rises if compounds like sulphur mustards, hydrolysis products are present in water. The best method to ascertain concentration of free available chlorine required for killing of microorganism is by plotting graph as given in Fig. 11.2. If a plot of pH versus total residual chlorine gives a peak which indicates the dose of chlorine demand for complete killing of microorganism and pathogenic bacteria.

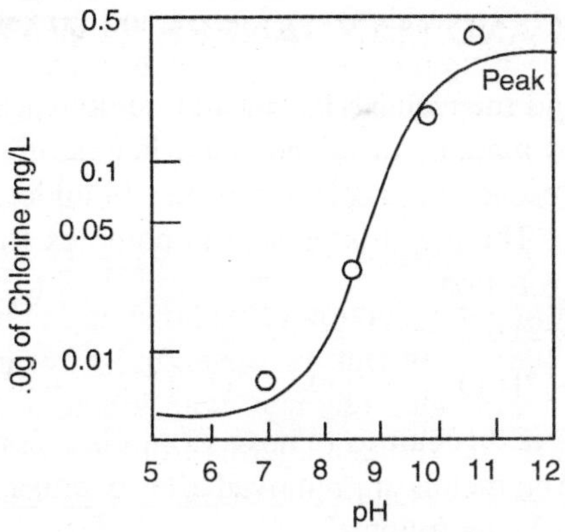

Fig. 11.2 Variation in chlorine with pH

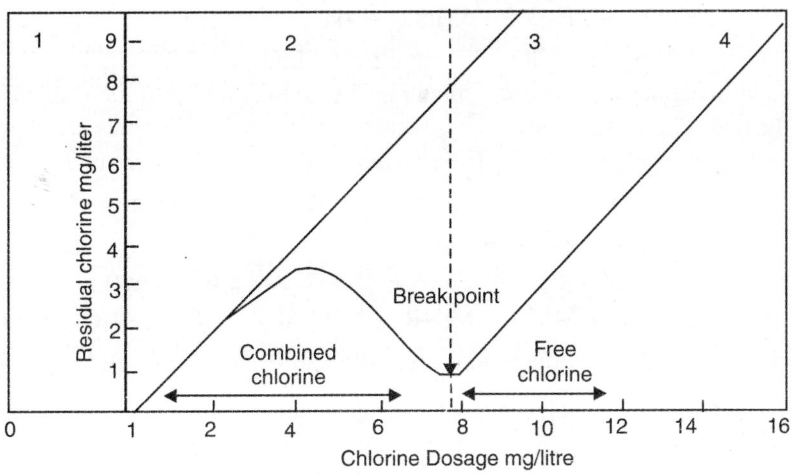

Fig. 11.3 Break point chlorination

Figure 11.3 indicates residual chlorine curve showing a typical break point. A close look at this figure will reveal, in zone (1) all chlorine is consumed for destruction or interaction of reducing compounds. While zone (2) represents formation of chloroorganics and chloroamines, when quantum of chlorine dose is rising. Consequently, in zone (3) the level of residual chlorine tapers off, as all chlorine has been consumed with no residual chlorine. On further injection of chlorine in water, zone (4) shows formation of free chlorine with or without combined chlorine. The inter junction of zones (2) and (3) indicates chlorine demand and it is termed as the break point chlorination. This is the optimum dose of chlorine, which is a must if the disinfecting action has to be effective. Thus, in zone (1) initially no residual chlorine is noted. The break point chlorination is effective for bactericidal action, removal of taste and odour. Ammonia is deliberately added to produce combined chlorine residue. Chloroamines last longer and have bactericidal effect also. It is worthwhile to consider toxicity of chlorine according to dosage, e.g., 1 ppm is the permissible limit, 3.5 ppm gives odour, 15 ppm gives throat irritation, 30 ppm induce cough, 40 ppm is maximum tolerance; 40–60 ppm is injurious to health and 1000 ppm is fatal for a person. Apart from ozone, other disinfecting agents which can be used with advantage are $KMnO_4$, bromine water, chlorine-bromine mixture, ultraviolet radiation at 200–300 nm the ionising radiation of 25 Ci/g and action of silver in ionic form.

11.7 REMOVAL OF IRON AND MANGANESE IN WATER

From drinking water, Iron (III) and manganese(II) must be removed, as they cause corrosion of pipes, affect appearance of water, impart metallic taste and cause laundering problems. The paper and textile mills also cannot tolerate their presence in processing water. In underground water, they are present in reduced form for want of oxygen. The metal is present in complex form. They are converted to stable oxidation state by the following reaction:

$$Fe^{2+} + O_2 + 8OH^- + 2H_2O \rightarrow 4Fe(OH)_3$$

$$Mn^{2+} + 2ClO_2 + 2H_2O \rightarrow MnO_2 + 2O_2 + 2Cl^- + 4H^+$$

The black film of manganese is difficult to remove from glass containers. The prescribed limit for manganese is 0.3–0.05 mg/litre. The metals are removed after pepitisation or oxidation, which proceeds by aeration with Cl_2, ClO_2, $KMnO_4$, as follows:

$$Mn^{2+} + O_3 + H_2O \rightarrow MnO_2 + O_2 + 2H^+$$

$$3Mn^{2+} + 2MnO_4^- + 2H_2O \rightarrow 5MnO_2 + 4H^+$$

A optimum dose is 2.5 mg/litre ClO_2; 1.9 mg/litre $KMnO_4$ and 0.87 mg/litre O_3 is best. If small quantities of iron and manganese are present, then one has to take resort to ion exchange method, at the cost of clogging the bed. The process however is somewhat expensive. Zeolites are cheaper. Zeolites are complex alumino silicates of alkali metals. The reaction with ion exchangers or zeolites are as follows:

$$H_2R + Fe^{3+} \rightarrow Fe\text{–}R + 2H^+ \text{ (cation exchange method)}$$

$$H_2R + Mn^{3+} \rightarrow Mn\text{–}R + 2H^+$$

$$Na_2Z + Fe^{3+} \rightarrow Fe\text{–}Z + 2Na^+ \text{ (zeolite treatment)}$$

$$Na_2Z + Mn^{3+} \rightarrow Mn\text{–}R + 2Na^+ \quad \text{if } Z = Na_2O, Al_2O_3\ xH_2O$$

Activated carbon or alkaline products like dolomite can be used for similar purposes.

11.8 SOFTENING OF WATER

The softening of water is necessary, as water may be hard. Thus $Ca(HCO_3)_2$ and $Mg(HCO_3)_2$ renders it temporarily hard, while $CaSO_4$, $MgSO_4$, $CaCl_2$ and $MgCl_2$ renders it permanently hard. The reliable test to ascertain hardness is soap test; if no lather is formed with soap water, it is considered to be hard. The best process for softening of water are three, viz. (*a*) lime-soda-ash process, (*b*) deionisation or demineralisation and (*c*) calgon process.

(a) Lime-Soda-Ash Process

The important reactions after addition of $Ca(OH)_2$ is as follows:

$$Ca(HCO_3)_2 + Ca(OH)_2 \rightarrow 2CaCO_3 + H_2O + CO_2\uparrow$$

$$Mg(HCO_3)_2 + Ca(OH)_2 \rightarrow MgCO_3 + CaCO_3 + H_2O + CO_2\uparrow$$

while after addition of soda it is

$$CaCl_2 + Na_2CO_3 \rightarrow CaCO_3 + 2NaCl$$
$$MgCl_2 + Na_2CO_3 \rightarrow MgCO_3 + 2NaCl$$
$$CaSO_4 + Na_2CO_3 \rightarrow CaCO_3 + Na_2SO_4$$
$$MgSO_4 + Na_2CO_3 \rightarrow MgCO_3 + Na_2SO_4$$

(b) Deionisation Process

One uses cation exchange resin in H^+ form, while anion exchange resin is used in OH^- hydroxyl form. A strongly acidic resin (like vinyl styrene copolymer with sulphonation) is used as cation exchanger, while basic anion exchanger (also vinyl styrene base with labile OH^- group) is used for removal of anions. The important reactions are:

$$H_2R + CaCl_2 \rightarrow CaR + 2HCl$$
$$H_2R + MgCl_2 \rightarrow Mg\text{–}R + 2HCl$$
$$H_2R + CaSO_4 \rightarrow CaR + H_2SO_4$$
$$H_2R + MgSO_4 \rightarrow MgR + H_2SO_4$$

Then water is passed through anion exchanger RNH_3OH

$$RNH_3 - OH + HCl \rightarrow RNH_3Cl + H_2O$$
$$RNH_3 - OH + H_2SO_4 \rightarrow RNH_3SO_4 + H_2O$$
$$RNH_3 - OH + H_2O \rightarrow RNH_3OH + H_2O \text{ (no change)}$$

The greatest advantage of process is the facility of reusing the resin again and again after regeneration. The regeneration of the resin is carried out as follows:

$$CaR + 2HCl \rightarrow H_2R + CaCl_2 \text{ (cation exchange, } H^+ \text{ form)}$$
$$Mg\text{—}R + 2HCl \rightarrow H_2R + MgCl_2$$
$$RNH_3Cl + NH_4OH \rightarrow NH_4Cl + RNH_3 - OH \text{ (anion exchange } OH^- \text{ form)}$$
$$RNH_3SO_4 + 2NH_4OH \rightarrow (NH_4)_2SO_4 + RNH_3\text{–}OH \text{ (regeneration)}$$

In place of synthetic ion exchanger one can also use naturally occurring sodium zeolite (Na_2O Al_2O_3 SiO_2) *i.e.* Na_2Z. It can also be regenerated by NaCl. The principal reaction involving exchange are as follows:

$$Ca(HCO_3)_2 + Na_2Z \rightarrow CaZ + 2NaHCO_3 \text{ (soluble)}$$
$$CaSO_4 + Na_2Z \rightarrow CaZ + Na_2SO_4 \text{ (soluble)}$$
$$Mg(NO_3)_2 + Na_2Z \rightarrow MgZ + NaNO_3 \text{ (soluble)}$$
$$CaZ + 2NaCl \rightarrow Na_2Z + CaCl_2 \text{ (regeneration of zeolite)}$$
$$MgZ + 2NaCl \rightarrow Na_2Z + MgCl_2 \text{ (regeneration)}$$

During regeneration, chambers containing cation exchange and anion exchange resins must be disconnected to facilitate the reconversion of resins to H^+ and OH^- forms respectively with HCl and NH_4OH as regenerate, respectively. Nowadays there is a growing trend to use mixed bed resins, however only limitation is during regeneration the anion and cation exchangers will have to be separated before physically carrying out regeneration.

(c) Calgon Process

This is largely useful for outdoor treatment for small and limited supply of water. In this process the salt sodium hexametaphosphate $(Na_3PO_4)_2$ is added to water for precipitation of insoluble phosphates of calcium and magnesium, as

$$MgSO_4 + 2NaH_2PO_4 \rightarrow Mg\ (H_2PO_4)_2 + Na_2SO_4$$

$$3CaSO_4 + 3Na_3PO_4 \rightarrow Ca_3(PO_4)_2 + 3Na_2SO_4$$

$$3Ca(HCO_3)_2 + 2Na_3PO_4 \rightarrow Ca_3(PO_4)_2 + 6NaHCO_3$$

However, the large-scale treatment and for softening of water this process is not economical.

11.9 CORROSION AND SCALE PREVENTION

The chemical correction and conditioning of water is carried out in several ways. This also prevents organic growth and fouling of water. The first step is to control pH. The alkaline solutions are not conducive for corrosion of pipes or seams. The other alternative is to inhibit precipitation, e.g., polyphosphate $[M_4P_2O_7n\ (MPO_3)]$, prevents scale formation e.g., if M = Na, we have $Na_4P_2O_7$ $NaPO_3$ or $Na_5P_3O_{10}$ (polymetaphosphate). It is possible to prevent corrosion by reduction of oxygen content in water. One can add corrosion inhibitors like Na_2SiO_3, polyphosphates, nitrites and chromates. Organozinc compounds are also beneficial. The layer of $CaCO_3$ inside a pipeline prevents further corrosion. Soft water promotes corrosion, by the process of recarbonation as:

$$Ca^2 + 2OH^- + CO_2 \rightarrow CaCO_3 + H_2O$$

$$Ca^{2+} + CO_3^{2-} + CO_2 + H_2O \rightarrow Ca(HCO_3)_2$$

$$2Na^+ + 2OH^- + CO_2 \rightarrow Na_2CO_3 + H_2O$$

Hence sodium hexametaphosphate $(Na_3PO_4)_2$ is added to prevent deposition of calcium. The concentration range 0.5–1 ppm of ferric phosphate is ideal, as 5 ppm of $CaCO_3$ forms protective coating of $(Na_3PO_4)_2$ on pipe interiors.

11.10 TASTE AND ODOUR REMOVAL FROM WATER

The dissolved organic substances such as phenols, H_2S, chloroamines and chloroorganics cause odour. Algal decay also produces foul smell. Specially phenols with chlorine form obnoxious products. Odor can be prevented by a number of ways like aeration, coagulation, break-point chlorination and application of certain chemicals like activated carbon, ClO_2, O_3, $KMnO_4$, $CuSO_4$ and others. The activated carbon is best, as it not only removes odour but also removes colour of water; but it cannot be regenerated. Granular carbon is not effective. By the process of adsorption impurities are removed, but dyes, detergents inhibit this reaction.

11.11 FLUORIDATION OF WATER

The term fluoridation is preferable to fluorination, as the element is used in the form of fluoride. (Figure 11.4 indicates relationship between dental caries and fluoride level in drinking water.) The removal of fluoride is called defluoridation. A small quantity of fluorine (0.4–1 mg/litre) in water promotes formation of dental enamel and protects teeth against caries. Too much fluoride leads to destruction of enamel and causes fluorosis–leading to dental disorder, decalcification, mineralisation of tendons, digestive and nervous disorders. The excess of fluorides (1.5 mg/litre) is dangerous to health. Only in the America water is fluoridised. In few places in Australia, South Africa, South America and Europe they treat water with fluorine. Na_2SiF_6, H_2SiF_6, NaF are used as the common fluoridation agents. They impart 0.4–1 mg/litre of fluoride in water. The excess of fluoride can be removed by treating water with $Ca_3(PO_4)_2$, $Al_2O_3SiO_3$, or $Ca(OH)_2$, or filtration through activated carbon.

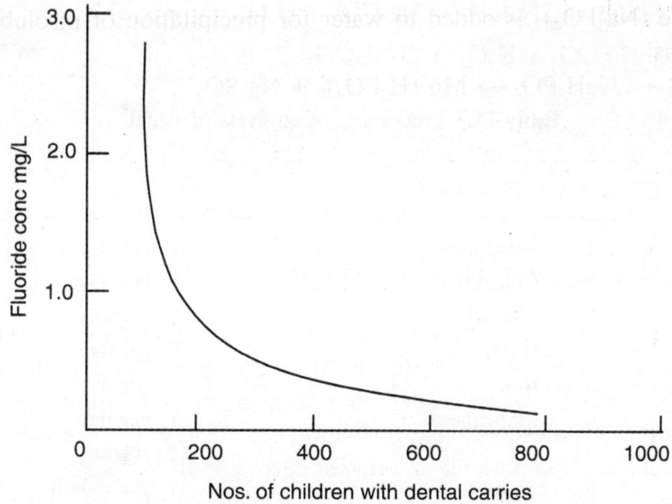

Fig. 11.4 Variation in dental problems and fluoride level

11.12 SPECIALISED TREATMENT OF WATER FOR INDUSTRY

Water which is required for a specific purpose needs a special treatment; e.g. swimming pool, boiler feed water, industrial process, hydrometallurgy. Such treatment depends upon quality of raw water, e.g., for well water no chemical treatment is required; while surface water needs chemical treatment, as described earlier in this chapter. The desalting plants use flash distillation of sea water, or reverse osmosis, wherein the solution passes from less saturated to more saturated solution through membranes. Electrodialysis is equally useful. Swimming pool water needs recirculation, prefiltration, disinfection with bromine or ozone for killing of algae, while boiler feed water needs purification with cooling circuits and antifouling agents, or prevention of scale and corrosion. Industrial process water needs treatment according to end use.

11.13 CHEMICAL ANALYSIS OF TREATED WATER

In order to ascertain quality of water, it is analysed for a fixed number of parameters. The purpose of such analysis is two-folds; it not only gives an idea about quality of water, but also ascertains the success or failure of various water treatment processes so far employed. It is not possible to analyse all the parameters in one single instance. On an average some parameters like alkalinity or acidity or pH needs frequent measurement (Table 11.1). While all trace metals are analysed once in a quarter year, chloride, hardness fluoride, manganese, taste and odour are analysed once a day. Sulphate, iron total solids can be analysed once a week or more. This schedule is valid under normal circumstances; however if there is some emergency due to heavy contamination or pollution of treated water, it will be necessary to undertake more frequent analysis. Thus, common or exotic analysis may be performed during emergency. The details of analysis is already outlined in the earlier chapters.

(a) Carbon Dioxide Analysis of Water

The total concentrations of CO_2 changes periodically. CO_2 should be determined on the spot near the source, because during transportation of water it absorbs more of CO_2 from the atmosphere. The titration of water sample with 0.045 NaOH or 0.023 N Na_2CO_3 is carried out. The reactions encountered are:

$$CO_2 + 2NaOH \rightarrow Na_2CO_3 + H_2O$$
$$CO_2 + Na_2CO_3 + H_2O \rightarrow 2NaHCO_3$$

Table 11.1 Frequency of analysis of water

S. No.	*Parameters*	*Frequency*
1.	Alkalinity	12 per day
2.	Chloride	2 per day
3.	Fluoride	1 per day
4.	Hardness	2 per day
5.	Iron	1 per day
6.	Manganese	1 per day
7.	pH	12 per day
8.	Sodium	1 per month
9.	Sulphate	2 per week
10.	Taste and odour	1 per day
11.	Total solids	2 per week
12.	Trace elements	As 1 per quarter *i.e.* 3 months Cu Pb Se Zn
13.	Turbidity	12 per day

If organic impurities are present, they produce CO_2 by aerobic and anaerobic reactions. Further CO_2 imparts corrosive properties, e.g., concentration of CO_2 > 20 mg/litre is bad. Fortunately, CO_2 has no adverse effect upon health.

(b) Chlorine in Water

All the methods are based on the oxidising power of chlorine gas. Manganese (II) and nitrites have adverse effect on such oxidising property. Simple qualitative tests consist of starch iodide test with reactions as:

$$Cl_2O + 2I_2^- \rightarrow I_2^{\circ} + 2Cl;\ I_2 + \text{starch} \rightarrow \text{Blue colour (qualitative)}.$$

o-toludine method (OTA) is the best method for analysis as shown by the reaction (Fig. 11.5):

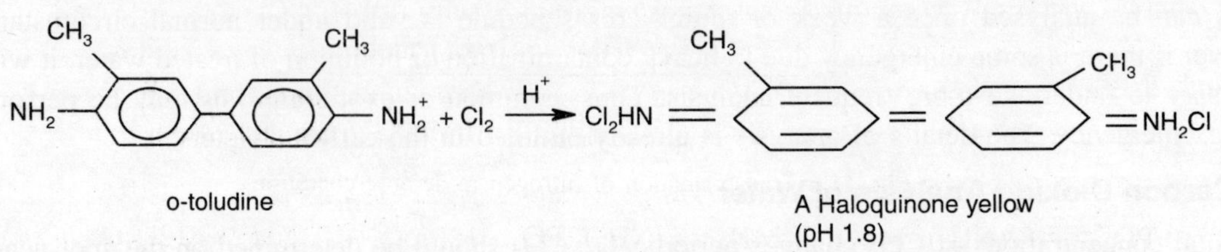

Fig. 11.5 Reaction mechanism of chlorine with o-toludine

Since the nitrites and Mn(II) both oxidise o-toludine to produce haloquinones, they must be removed from water by treatment with sodium arsenite. OTA may give wrong results for free and combined chlorine residual. For such purposes use of amperometric titration, stabilised neutral orthotoludine method, or NN′ diethyl p-phenylenediamine (DPD) reactions; or NN′ dimethyl-aniline techniques (Leuco crystal) phenyl arsenic oxide (C_6H_5AsO) is used as the titratant; at pH 6.5–7.5 in amperometric titration. In second method, slow reaction of OTA at neutral pH is used, buffered at pH 6.5–7.5. A blue colour so obtained is spectrophotometrically measured. In first method, red colour formed is measured when pH for reaction is 6.2–6.5. In leuco crystal, violet method, NN′ dimethyl aniline produces violet colour with chlorine, which is measured spectrophotometrically. The details are spelled out in next chapter.

(c) Coagulation Test of Water

This is also termed as jar test. The idea is to simulate optimum conditions on the laboratory scale and test how much of coagulant is needed for completion of the reaction. This decides its dose at fixed temperature and pH. The flash mixing is done when best appearing floc is noted. It decides relative merit of coagulant or coagulant acid. Zeta potential of floc and colloidal material is determined by electrophoresis cell. The difference should be such that more colloidal material is attached towards floc. Addition of polyelectrolytes should be such that alkalinily also does not increase too much.

(d) Phosphate and Polyphosphates

Some compounds are used for corrosion control. Nitrogen and phosphorus are conducive for growth of algae (Fig. 11.6). The critical levels is 0.005 μg/litre or 5 μg/ml. Sewage contains abundant phosphates. Detergents contain polyphosphates (12–13% P). The method for analysis of phosphate and polyphosphate are relatively simple, e.g., using gravimetry by the formation of pyrophosphate; using volumetry (< 50 mg/litre) by precipitation reaction and colorimetry. The overall reaction in colorimetric method are as follows:

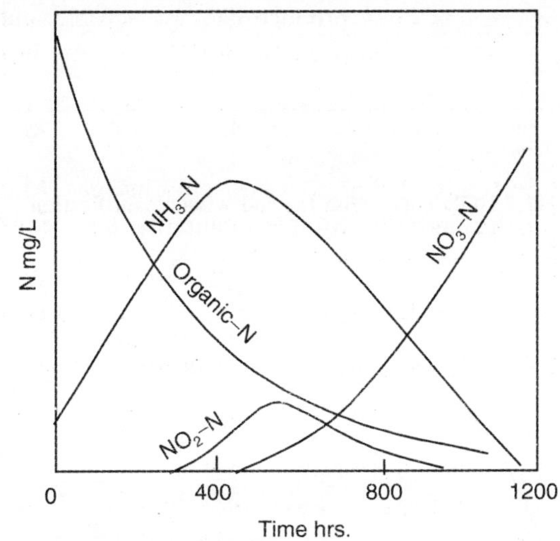

Fig. 11.6 Transformation of nitrogen in aerobic conditions

$$PO_4^{3-} + 12(NH_4)_2 MoO_4 + 24H^+ \rightarrow (NH_4)_3 PO_4 \cdot 12MoO_3 + 21NH_4^+ + 12H_2O$$

$$(NH_4)_3 PO_4 + 12MoO_3 + Sn^{2+} \rightarrow \text{(molybdenum blue)} + Sn^{4+}$$

This colour of molybdenum blue is spectrophotometrically measured. Such a complex can be extracted in benzene or isobutanol.

(e) Analysis for Organic Pollutant

For details of analysis of organic pollutants refer to Expts. 15.1 to 15.11 and 16.1 to 16.7 in Chapters 15 & 16. They are either composed of phenols and derivatives; hydrocarbons; detergents, pesticides and plant health products or micropollutants. Level of phenol is 0.1–0.01 μg/litre. Biodegradability of hydrocarbons is slow. It prevents reoxygenation of water and gives a peculiar smell to it. Surfactants are bad for health. A 4–aminoantipyrene reacts with phenol at pH 10 in the presence of $K_3Fe(CN)_6$ of form coloured antipyrene dye at pH 10.0. This is extracted in chloroform and measured at 460 nm. Pentachlorophenol and 2, 3, 4, 6 tetrachlorophenols are extracted in benzene and analysed by gas chromatography. If HCHO, *i.e.,* formaldehyde is present in water, it can be reacted with acetylacetone in the presence of ammonium salt to form the yellow compound diacetyldehydrotoludine, which is measured colorimetrically at 420 nm. 1–8 ppm of HCHO can be easily analysed. Humic acid if present in water can be directly analysed spectrophotometrically at 250 nm. Amount of 0.1–20 mg/litre can be easily analysed. Finally, lignin if present in water is used to reduce tungstophosphoric acid to produce a blue colour, which can be measured at 624 nm by spectrophotometer. Alkylbenzene sulphonate can also be measured after extraction in carbon chloroform extracts (CCE) photometrically.

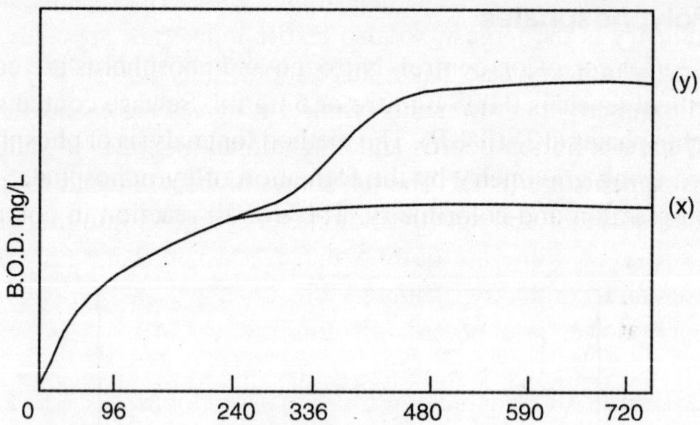

Fig. 11.7 BOD curve with (x) and without nitrification (y)

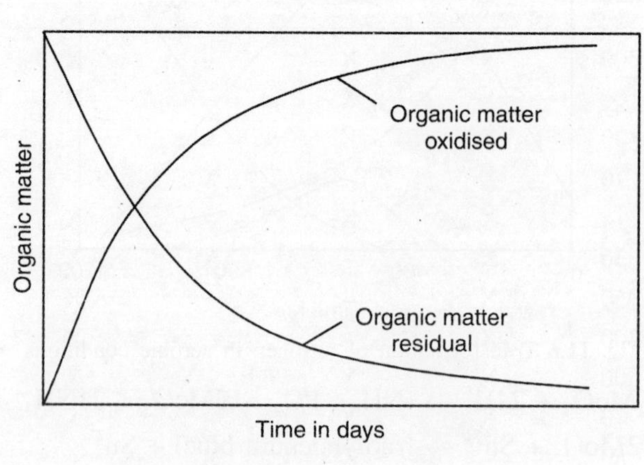

Fig. 11.8 Degradation of organic matter

Figure 11.8 represents BOD curve or oxidation of organic matter and first curve shows influence of nitrification. These are plotted at 20°C for a period of one month. As time passes BOD demand rises. Figure 11.8 indicates changes in organic matter during biological oxidation of polluted waters under aerobic conditions. The good amount of oxygen is used for biochemical oxidation. Hence, we get a parabolic curve (Fig. 11.8). The curve in Fig. 11.7 has characteristics similar to curve in Fig. 11.8 during initial 8–10 days. Then BOD curves decrease subsequently till 30 days.

11.14 WATER QUALITY MONITORING INSTRUMENTS

We have considered various parameters, which are needed to be measured for potable or drinking water or treated water for industrial purposes. In order to keep pace with development and constraint of time, it may be imperative to use rapid methods of analysis on a regular basis, to ascertain the quality of water. Such rapid measurement is possible only by the use of water quality monitoring instruments. Before adjudging such instruments, it is worthwhile to examine requirements of such instrument by various industries. Table 11.2 gives demand of such monitoring devices by industrial enterprises. Eleven important industries are listed, and seven parameters are considered for such a measurement. The average demand for a year is roughly calculated. All industries need to measure suspended solids; BOD and COD while trace metal or oil and grease is determined for thermal plants and refinery. The annual requirement for sugar industry is the largest, so also textile industries, specially in cities like Mumbai, Delhi and Kanpur. The drug industry also has a large input. Food industries, thermal plant, synthetic fibre factory do not need too many measurements. Table 11.3 gives the need for monitoring instruments by research laboratories, state pollution boards and educational institutes in India. In all twenty important water monitoring instruments are listed. The list shows highest demand for pH meter, turbidimeter, and ion selective electrode analyser and spectrophotometer. There is minimum demand for AAS, GC, HPLC and spectrophotometer, as some are quite expensive. Since flame photometer, BOD incubator, BOD meter, DO analyser and conductivity meters are cheap, their demand in boards, educational institutes and research laboratories is expected in future.

Table 11.2 Demand of monitoring instruments by various industries (X = need)

S. No.	*Industry*	*No. of unit*	*Turbidity*	*Suspended solid*	*Temp*	*BOD*	*COD*	*Oil & grease*
1.	Sugar	300	X	X	–	X	–	–
2.	Breweries	100	X	X	–	X	X	–
3.	Soda ash	30	X	X	–	–	–	–
4.	Fertilisers	70	X	–	X	–	–	–
5.	Oil refineries	15	X	X	–	X	X	X
6.	Synthetic fibre	30	X	–	X	X	X	–
7.	Textile mills	950	X	–	–	X	X	–
8.	Paper industry	200	X	–	–	X	–	–
9.	Drugs	300	X	X	–	X	X	–
10.	Pesticides	70	X	X	–	X	X	–
11.	Thermal plants	50	–	X	X	–	–	X

Table 11.3 Demand for water pollution monitoring instruments per year

S. No.	*Instrument*	*R and D labs*	*Pollution boards*	*Educational institutes*	*Total number*
1.	pH meter	4500	100	500	5100
2.	Conductivity meter	–	100	500	652
3.	Dissolved oxygen meter	–	100	500	652
4.	Turbidity meter	3000	100	500	3600
5.	BOD incubator	–	100	100	200
6.	COD meter	–	100	100	200
7.	Mercury analyser	30	40	100	170
8.	Atomic absorption spectrometer	–	20	100	120
9.	Gas chromatograph	–	20	100	120
10.	Spectrophotometer	–	20	100	120
11.	Ion selective electrode	1000	20	100	1172
12.	TOC analyser	20	20	100	140
13.	Gas chromatograph–mass spectrometer	–	5	100	105
14.	High pressure liquid chromatograph	20	5	100	125
15.	Polarograph	–	20	100	120
16.	Ion chromatograph	–	5	100	105
17.	Infrared spectrometer	–	5	100	105
18.	X-ray fluorescence	–	5	100	105
19.	Flame photometer	200	100	100	400
20.	Thin layer chromatograph	–	40	100	140

11.15 TYPE OF WATER QUALITY INSTRUMENTS

Tables 11.4, 11.5 and 11.6 outline the need for water pollution monitoring instruments in small, medium and the advanced research laboratories resp. Table 11.4 lists instruments for the measurement of

Table 11.4 Water quality monitoring instruments for mini laboratory

Quality for measurement	*Instrument*	*Limit of analysis*
Temperature	Thermistor	0–50°C
pH	pH meter	0–14 pH
Conductivity	Conductivity bridge	1 mho to 1 micromho
Dissolved oxygen	DO meter	0–20 ppm
Turbidity	Comparison tubes	1–100 JU
Humidity	Hydrometer	0–100% humidity
Microscopic examination	Microscope	1 × 100 magnification
SPM—measurement	Dust collectors	Large size collection

physical parameters such as temperature, pH, conductivity, turbidity, humidity, suspended particulate matter and sunshine recorder. This set is ideal for a very small laboratory. Table 11.5 gives a set-up for a reasonably big laboratory. It can measure 10–12 parameters. The important chemical parameters are dissolved oxygen, alkali metals, BOD; metal pollutants and plate count (MPN). It gives information on

Table 11.5 Water quality monitoring instrument for medium size laboratory

Instrument	*Quality for measurement*	*Limit of analysis*
Thermistor thermometer	Temperature	0–50°C
pH meter	pH	0–14 pH
Conductivity meter	Conductivity	1 micromho to 1 M mho
Dissolved oxygen meter	Dissolved oxygen	0–20 ppm D.O.
Turbidity meter	Turbidity	0–10,000 JU
Ion analyser	Ionic concentration	10^{-1} mg-μg level
BOD analyser	BOD for 3 days	0.1 mg/litre
Flame photometer	s-block metals	Na, K, Ca filters
Spectrophotometer	Transition metals	325–825 nm
Microscope	Biological investigation	× 500 magnification

Table 11.6 Water quality instrument for advanced analytical laboratory

Instrument	*Quality for measurement*	*Limit*
pH meter	pH	0–14 pH
Conductivity meter	Conductivity	1 mho to 1 micromho
Dissolved oxygen meter	DO	0–20 ppm
Redox monitor	Redox potential	±2000 mV
Turbidity meter	Turbidity	0–10,000 JU
Ion selective electrodes	Ionic concentration	mg–μg
BOD and COD monitor	BOD/COD	0.1–5 μg/litre
Nephelometer	Turbidance	400–700 nm
Spectrophotometer	Molecular absorption	200 to 2000 nm
Atomic absorption spectrophotometer	Metal pollutants at nanogram level	190 to 800 nm
Gas chromatographs	Volatile organic compounds	With FID, ECD, TCD detector
IR spectrophotometer	Organic compounds	With KBr pellet or Nujol technique

the most vital parameters such as conductivity; DO; turbidity metal pollutants, and biological plate count. Table 11.6 contains a comprehensive list of water quality instruments for an advanced analytical laboratory. A dozen instruments should be able to furnish valuable information on water quality. One can get a reliable information not only on physical, but also chemical-inorganic as well as organic pollutants. Table 11.7 lists recommended water quality instruments. Contrary to usual practice, the last column lists principles upon which these methods are based. This set-up is ideal for normal conditions. Figure 11.9 shows water pollution monitoring kit, which was developed by us in our laboratory

Table 11.7 Recommended minimum water quality instruments

Instrument	*Parameter measured*	*Principle*
Water analysis kit	pH, DO, temperature conductivity	—Temperature 0 to 55°C —Conductivity 1.0 mho to 1 mho —Dissolved oxygen 0–20 ppm
Digital pH meter	pH	—Nernst equations 0–14 pH
Chloride meter	Chloride	Coulometry, technique, Faraday law
Chloroscope	Residual chlorine gas	Visual comparison with standard coloured disc (cf OTA method)
Polarograph	Metals and pesticides	Ilkovic equation half-wave potential
Spectrophotometer	Organic and metal pollutants	Beer's Lamberts law 350–700 nm
Turbidity or nephelometer	Scattering or absorbance of suspended particles	Scattering of radiation

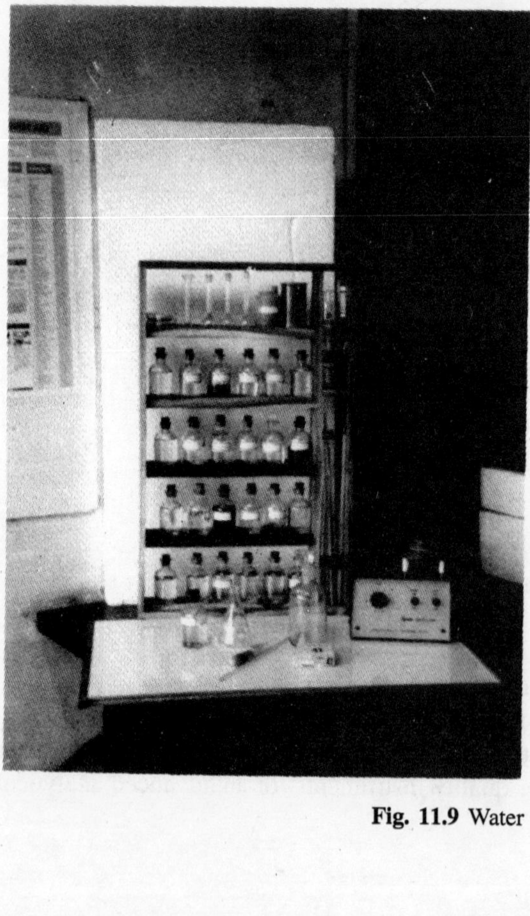

Fig. 11.9 Water pollution monitoring kits

at IIT, Mumbai. This kit can measure as many as eighteen-parameters. They are important for ascertaining the quality of water. This is a portable kit which weights six kg and overall cost of equipment does not exceed Rs. 6000-10,000. The principle of Ring-Oven Technique has largely been used for analysis of metal pollutants. While the method of volumetric analysis is recommended for analysis of acidity, alkalinity, CO_2, NH_3, hardness, one-day rapid test is recommended for plate count and BOD analysis. The extensive field tests must be carried out to assess its suitability for field outdoor work.

11.16 PRESENT STATUS AND PERSPECTIVE PLANNING FOR INSTRUMENTS

Several instruments used for water pollution monitoring are listed in Table 11.8. This table lists fifteen vitally important instruments useful for water quality measurement. Out of these, ten instruments are available in India; while three are in developmental stages. Finally Table 11.9 gives an estimate of the

Table 11.8 Instruments available in India

S. No.	*Monitoring instrument*	*Availability in India*	*Need for modification*	*Progressive development*
1.	Photoelectric photometer	Y	–	Y
2.	Spectrophotometer	Y	Y	Y
3.	Flame emission spectrometer	Y	Y	Y
4.	Atomic absorption spectrometer (AAS)	Y	Y	Y
5.	Turbidimeter	Y	Y	Y
6.	pH meter	Y	–	Y
7.	Ion analyser	N	–	–
8.	Conductivity bridge	Y	–	Y
9.	Gas-liquid chromatograph	Y	Y	Y
10.	Polarograph	Y	–	–
11.	Mass specrometer	N	–	–
12.	Nuclear magnetic resonance (NMR)	N	–	–
13.	X-ray fluorescence (XRF)	N	–	Y
14.	Dissolved oxygen probe (DO)	Y	–	–
15.	Biological oxygen demand monitor (BOD)	Y	–	–

Y stands for yes; N stands for no

approximate cost of instrument, as it existed in 1990. Now it is more 50%. It represents net cost, which is likely to rise in years to come. Out of the 24 instruments considered, about ten will have to be imported while fourteen are locally available in India. Assuming that there is no substantial rise in price, the whole set-up will need approximately Rs. 60 lakhs to have an up to date laboratory. This shows we need perspective planning for the future. We should make vigorous efforts to develop these instruments locally in India. In the initial development stage, manufacturers should not bother too much on cost-benefit analysis, but should give top priority to utility. This will help to give reliable and dependable

Table 11.9 Approximate price of water quality instruments as in 1990

Instruments	*Approx. cost Rs.*
pH meter	3,000 (I)
Conductivity meter	3,000 (I)
Dissolved oxygen meter	2,000 (I)
Turbidity meter	4,000 (I)
Portable water testing kit	4,000 (I)
Biological oxygen demand incubator	3,000 (I)
Chemical oxygen demand meter	6,000 (I)
Mercury analyser	9,000 (I)
Refractometer	4,000 (I)
Atomic absorption spectrometer	3.5 lakh
Gas chromatography with TCD, FID and ECD detectors	3.00 lakh
Spectrophotometer (UV/Vis)	1.5 lakh (I)
Ion selective electrode and specific ion meter	12,000 (F)
Total organic carbon analyser	1.5 lakh (F)
Gas chromatograph-cum-mass spectrometer	18 lakh (F)
High pressure liquid chromatograph	3 lakh (F)
Auto-analyser	1.2 lakh (F)
Polarograph	90,000 (I)
Ion chromatograph	6 lakh (F)
Infrared spectrometer	4 lakh (F)
Plasma scan spectrometer	10 lakh (I)
X-ray fluorescence	6 lakh (F)
Flame photometer	20,000 (I)
Thin layer chromatograph	1,000 (I)

(I - Indian, F - Imported)

data on water quality. This would also help us to adopt exact measures for treatment of water at the source. It will assist us to recover useful materials from waste water. The water quality measurement instruments must be produced within the country. We have expertise, trained manpower and only what we require is a will and enthusiasm to give shape to our plans for a cleaner environment.

ADDITIONAL READINGS

11.1 W.J. Weber, *Physico Chemical Process for Water Quality Control*, Wiley-Interscience (1972).

11.2 *Standard Method for Examination of Water and Waste Water*, 20th Ed., American Public Health Association, New York (2008).

11.3 R.A. Horne, *Chemistry of Environment,* Wiley-Interscience (1978).

11.4 I.M. Kolthoff, P.J. Elving and M.J. Stross, *Treatise in Analytical Chemistry,* Vol. 2, Part III, Wiley-Interscience (1971).

11.5 C.N. Sawayer and P.J. McCarty, *Chemistry for Environmental Engineering,* 6th Ed., McGraw-Hill & Co. (2007).

11.6 L.L. Ciaccio, *Water and Water Pollution Handbook,* Marcell Dekker, N.Y. (1972).

11.7 S.M. Khopkar, *Basic Concepts of Analytical Chemistry*, 3rd Ed., New Age International (P) Ltd., New Delhi (1985).

11.8 *Water Treatment Handbook*, 5th Ed., Degremount Halsted Press of John Wiley & Co. (1979).

11.9 N.F. Voznaya, *Chemistry of Water and Microbiology,* Mir Publishing House (1981).

11.10 F.I. Belan, *Water Treatment,* Mir Publishing House, Moscow (1984).

11.11 I.L. Marr and M.S. Cresser, *Environmental Chemical Analysis,* Chapman and Hall, New York (1983).

11.12 A.K. De, *Environmental Chemistry,* New Age International (P) Ltd., New Delhi, 5th Ed. (2006).

11.13 Mayuri Gandhi and S.M. Khopkar, *Indian J. of Envi. Protection,* **10,** 262 (1990).

11.14 H.S. Samant, *Surveillance Report on Methods for Chemical Analysis of Water and Waste Water,* EPS 5–AR–73–12, Canada (1973).

11.15 A.K. Ganguley, *Report of the Panel on Environmental Monitoring Instrumentation,* Department of Electronics (1976).

11.16 S. Deekshatulu, *Report on Environmental Monitoring Instrumentation,* Department of Environment (1984).

11.17 T. Harvey, *Notes on Environmental Monitoring and Control,* Tottenham College of Technology, London (1978).

11.18 D.B. Shinde, Mayuri Gandhi and S.M. Khopkar—*Ind. J. Env. Prot.* **11**, 801 (1991).

CHAPTER 12

Metal Pollutants Analysis —Specific Methods

12.1 INTRODUCTION

We have so far discussed the analysis of the common hydrological parameters of water. The discussion included the analysis of metals and non-metals which are mainly responsible for influencing the quality of water which is taken up in this chapter. As we have seen in chapter on Toxicology, many of metal pollutants like mercury, lead, chromium are quite harmful to human health if present in sufficiently large concentration. It is, therefore, necessary to analyse them at very low concentrations.

When the problem arises for the analysis of these metal pollutants at microgram concentration, one has only choice to resort to rapid instrumental methods of analysis. Such instrumental methods include atomic absorption spectroscopy, inductively coupled plasma atomic emission spectroscopy or anodic stripping voltammetry. The methods are applicable irrespective of source of metal pollutants. Even those metal pollutants encountered in atmosphere can be analysed by these sophisticated methods.

Apart from these metal pollutants, there are few metals whose presence in the sample of water or air is significant. They need specific methods for the quantitative characterisation. Such metals include aluminium (III), calcium (II), chromium (VI), copper (II), iron (III), lead (II), manganese (II), selenium (IV), vanadium (V) and zinc (II) and s-block metals. Such methods not only include rapid instrumental technique like AAS but also conventional methods like spectrophotometry, ion chromatography or the polarography.

In addition to metal pollutants, organic pollutants pose a serious problem in potable water. Therefore next two chapters are devoted to analysis of organics pollutants by either gas or liquid chromato-graphy as well as by some specialised methods of characterisation. We also have to consider presence of anions which have deleterious effects on the human health. They are characterised by ion selective electrodes or ion chromatography. Such methods are discussed in the separate chapter. Finally microbiological examination of water includes the discussion on the environmental chemistry of water pollution.

Since the last two decades there has been phenomenal growth in process in analytical chemistry for the analysis of metal pollutant. With greater strides in instrumental methods of analysis, newer methods have come to forefront which are rapid, selective and sensitive. Such methods are included AAS, ICP–AES, ICP—MS and ASV. However in spite of the significant strides in instrumentation, some classical methods involving use of organic ligands have remained in use because they are simple, inexpensive and reliable techniques. Many organic ligands were specific for particular metal and could analyse them quantitatively at trace concentration. There were about 12–15 metals which have their own specific techniques for determination. Such metals included common metals encountered in the environment. These methods involved the use of the selective specific sensitive organic reagent for the complexation which was followed by solvent extraction and spectrophotometry. These organic reagents include: Eriochrome cyanine R (Al), Sodium diethyl dithio carbamate (As), sym. diphenyl carbazide (Cr) Neo and Bathocuporine (Cu), 1–10 phenanthroline (Fe), dithizone (Pb), Per sulphate (Mn), diphenyl benzidine (Se), gallic acid (V) and Zincon (Zn). All the remaining metals were determined by either atomic absorption spectroscopy, plasma spectroscopy or by anodic stripping voltammetry. We would consider in this chapter only those of the metal pollutants which were mentioned before and which have very specific chromogenic ligands for complexation.

12.2 ALUMINIUM:
Experiment 12.1: Analysis of Aluminium in Water

The compounds are extensively used in water treatment plants for process of coagulation and floc formation. Al (13) has atomic weight of 26.98. At pH < 4.0, the species Al^{3+} predominates in water. High level of Al is present in patient who is suffering from dermatitis as well as abnormality in brain tissues. Threshold level is 0.05 mg/L as per environment protection agency.

Apparatus reagent: A spectrophotometer with 10 mm matched glass or quartz cuvettes. A stock solution is obtained by dissolving 500 mg aluminium metal in 10 ml hydrochloric acid making it to one litre. The diluted solution is multiple of 5 μgAl is prepared by appropriate dilution. Sulphuric acid (0.002 and 6N), Ascorbic acid (0.1g/100 ml). Buffer solution 136 g CH_3COONa in 40 ml acetic acid and made to one liter. Solochrome cyanine R or Eriochrome cyanine were used as chromogenic reagents—300 mg in 100 ml water, pH adjusted to 9.0 and solution diluted to 100 ml.

The working dye solution is obtained by ten times dilution. Methyl orange and EDTA solution (3.7 g/L) and sodium hydroxide (0.1 and 1N) were used.

Procedure: Prepare standard solutions contains known amount of Al varying in strength from 7 μg–280 μg/L. Add 1 ml 0.002N sulphuric acid, 10 ml of buffer solution (pH ~ 10) and 5 ml of eriochrome cyanine R dye solution, and make it up to 50 ml. Measure the absorbance at 535 nm within 15 mts of complexation. Fluoride if present will interfere in determination. EDTA is used for masking common metals. The phosphate if present must be removed before carrying out photometry. Fluoride if present in sample is measured photometrically with SPADNS a special sensitive reagent.

Calculations: $$\text{mg Al/L} = \frac{\mu\text{g of Al (in 50 ml soln)}}{\text{ml. sample taken}}$$

The method is reproducible free from any interference and very rapid.

12.3 ARSENIC:
Experiment 12.2: Analysis of Arsenic in Water

Silver diethyl dithiocarbamate is used as the chromogenic ligand. At lower concentration, AAS with hydride generator is most popular method. As (33) has atomic weight = 74.92 with As (III) and As (V) in their oxidation states. TLV = 72 μg/L as recommended by EPA and for potable water it is 0.05 mg/L. With silver diethyl dithiocarbamate as chromogenic ligand. AsH_3 is obtained by action of $NaHB_4$ with mineral acid.

Apparatus and reagents: In analysis Cr, Co, Cu, Hg, Mo, Ni, Pt, Ag, Se show interference while generating AsH_3. Pb $(CH_3COO)_2$ is used to complex some of the metals. A device to generate AsH_3 is shown in Fig. 12.1. Rubber or cork should not be used as it reacts with AsH_3. Following reagents are used in analysis viz. Buffer (pH 5.5), 430 ml CH_3COONa and 72 ml 0.2 CH_3COOH, CH_3COONa (0.2 M), 16.4 gm one litre water, CH_3COOH (0.2 M). Dissolve 11.5 ml glacial acetic acid in one litre water. $NaBH_4$ –0.4 gm NaOH + 4 gm $NaBH_4$ + H_2O (400 ml) HCl (2M), $(CH_3COO)_2Pb$ – 10 gm 100 ml; silver diethyldithiocarbamate–1 ml of morphodine 70 ml $CHCl_3$ + 0.3 g Ag-diethyldithiocarbamate – Diluted 100 ml chloroform A 0.1734 gm $NaAsO_2$ in one litre water is used for standard solution (appropriate dilution).

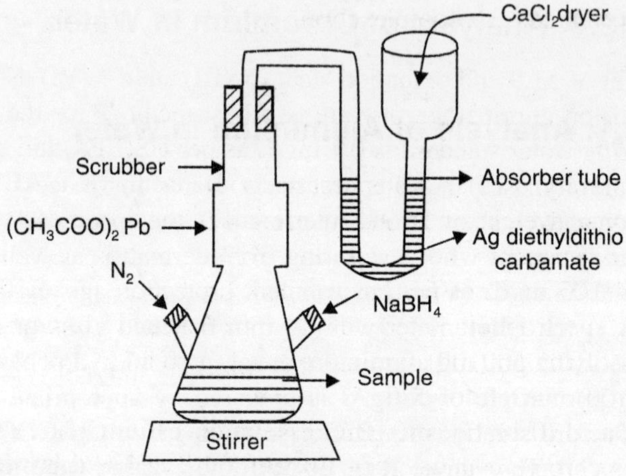

Fig. 12.1 Arsine producer

Procedure: In As generator add 70 ml solution containing 20 μgAs + 10 ml buffer. Then purge nitrogen gas in vessel. Add 15 ml $NaBH_4$. Measure the absorbance of the coloured solution at 520 nm.

Calculations: $\text{mg As/L} = \dfrac{\mu\text{g of As in calibration plot}}{\text{ml. of sample in generator}}$

12.4 CALCIUM:

Experiment 12.3: Analysis of Calcium in Water

In water, we have elements beryllium, bismuth, boron and cadmium, however they are analysed by AAS or ICP-AES technique with greater accuracy. In absence of these metals, Aluminium (Beryllium) and dithizone (cadmium) are used by spectral analysis. Most commonly encountered metals in water are compounds of calcium as carbonates and halides. They render water hard if present as chloride and sulphate. AAS and ICP-AES are no doubt best methods for the analysis of calcium. Otherwise EDTA provides excellent titration method (Ca > 50 mg/L) for routine analysis.

Reagents: NaOH (1N), Mureoxide (150 mg/100 ml glycol). Erichrome Black/Blue 200 mg + 100 mg NaCl–powdered form is used as indicator 0.01M EDTA.

Procedure: An aliquot of solution containing calcium is taken. Add to it NaOH and indicator in 50 ml sample solution and titrate it with standard EDTA (pH < 6.0). Alternatively, take 2 ml NaOH (pH ~ 12–13) 0.1–0.2 g EBT indicator and titrate it against EDTA till end point is reached with colour change from violet to blue.

Calculations: mg Ca/L $= \frac{A \times B \times 400.8}{\text{ml. of sample}}$; calcium hardness as carbonate/L $= \frac{A \times B \times 1000}{\text{ml. of sample}}$

In above expression A = ml. of EDTA added, B = mg $CaCO_3$ reduced to 1 ml EDTA. If mixture (Ca + Mg) is present, EDTA titration end point first reached for calcium as indicated by mureoxide indicator and Mg is precipitated as Mg $(OH)_2$ at pH 10–13. Low concentration of Cu(2), Fe^{2+}(20), Fe^{3+}(20), Mn(10), Zn(3), Pb(3), Al(3), Sn(5) (all mg/Litre) do not interfere during titration. Sr^{2+}, Be^{2+} interfere at high pH in alkaline region.

12.5 CHROMIUM:

Experiment 12.4: Analysis of Chromium in Water

Cr(24) has atomic weight of 51.9 and oxidation state Cr (III) and Cr (VI) as most common. s-Diphenyl carbazide extraction and ion chromatography are two best methods for the quantitative determination of chromium. The drinking water standard is 0.1 mg/L as per US-EPA. Ion chromatography is used for analysis of dissolved chromium (VI). AAS is best method. Plasma (ICP-AES) method is equally good. In colorimetry Hg,Mo, V and Fe show strong interference, however these elements can be masked by use of suitable sequistering agents.

Reagents: Stock solution of Cr (VI). Dissolve 141.4 mg $K_2Cr_2O_7$ to 100 ml, 1 ml contains ~ 500 μg/Cr(VI). HNO_3; H_2SO_4 (6, 18N); H_3PO_4, H_2O_2 (30%) Methyl orange, ammonia $KMnO_4$ (4 gm $KMnO_4$ in 100 ml), NaN_3 (0.5 gm/100) symm-diphenyl carbazide - 250 mg in 50 ml acetone, chloroform cupferron (5 gm/100) NaOH 40 gm/L.

Procedure: From standard solutions (5 μg/Cr) prepare a calibration curve at various dilution of the solution (10–100 μg Cr (VI)]. Then develop the colour of the sample and measure its absorbance at 540 nm with reagent as blank. In case, if sample contains Mo, V, Fe or Cu they are all complexed with cupferron in chloroform. By dry ashing remove droplets of chloroform before estimating Cr (VI) in remaining solution. Cr (III) must be oxidised to Cr (VI) before resorting spectrophotometry. $KMnO_4$ is used for oxidation. Colour can be also developed if 0.25 ml H_3PO_4 + 1 ml 2N H_2SO_4 - at pH 3.0 (1.6–2.2 pH is best) make up to 100 ml, add 2 ml sym. diphenyl carbazide and then within 5 mts. colour is developed which can be measured at 525 nm.

Calculation: mg Cr/L $= \frac{\mu g \text{ Cr (in 102 ml soln)} \times 100}{A \times B}$ if A = ml sample, B = ml digested solution.

However for undigested solution we get amount of answer as per equation.

mg Cr/L $= \frac{\mu g \text{ Cr (in 10 ml final volume)}}{\text{ml of the original sample}}$. One can also use ion chromatographic method.

Experiment 12.5: Analysis of Anions as CrO_4^{2-}, $Cr_2O_7^{2-}$

Ion Chromatography Method: As little as 0.1 μg/L of Cr (VI) can be easily analysed by this method. One can also use ion chromatography. The organic compounds are absorbed in guard before separation. Dissolve 0.1414 gm/500 ml of $K_2Cr_2O_7$ it gives 100 mg (Cr(VI)) per litre. From standard solution prepare series of solutions containing Cr (VI). A 33 gm $(NH_4)_2SO_4$ in 500 ml + 6.5 ml of NH_3 made up to litre is used as the eluent. Dissolve 0.5 gm chromogenic ligand in 100 ml methanol + 28 ml H_2SO_4 made to one litre. The buffer solution 33 gm $(NH_4)_2SO_4$ in 75 ml water + 6.5 ml NH_3 made up to 100 ml.

Procedure: Flow rate be 1.5 ml/ml. A 50 μL loop injection syringe be used. Then construct calibration curve with standard with varying amount of Cr (VI) at pH 9–9.5. Then bring ice-chilled sample in syringe and inject it in chromatographic system.

Calculation: Measure either area under elution curve or measure height of peak in calibration curve. Always use the reference as reagent blank during absorbance measurements at 530 nm. A matrix spike can be also used. For ion chromatography a Dionex model is best and suitable for rapid analysis. The most important component of the instrument is kind of detector used. In general conductivity detector is most commonly used.

12.6 COPPER:

Experiment 12.6: Analysis of Copper from Water

Cu (29) has atomic weight of 63.54 with valency as +1 and +2. As per US-EPA drinking water should not contain more than 1.3 mg/L. Neocuproine is method of choice while for analysis of copper in potable water bathocuporine method is preferred if iron is present.

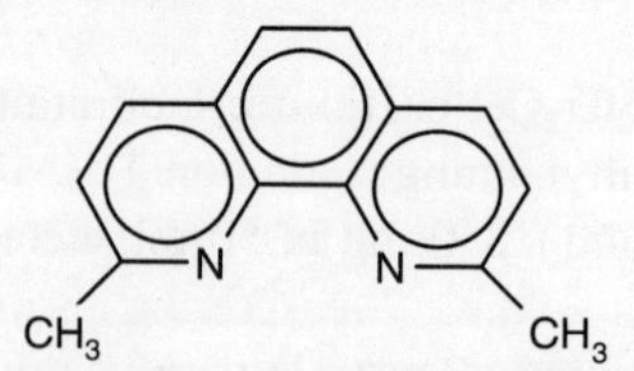

(I)

2–9 dimethy 1–10 phenanthroline

Fig. 12.2 Neocuproine (λ_{max} = 457 nm)

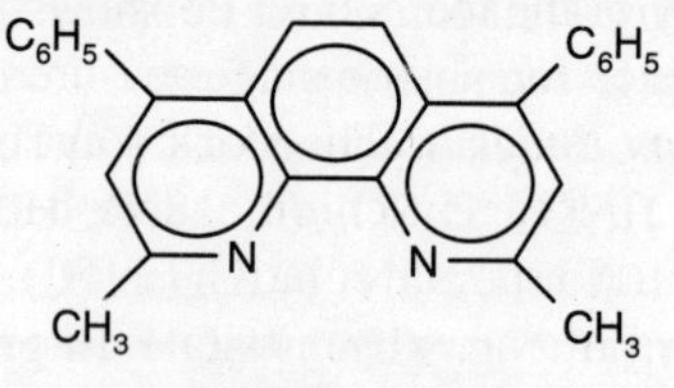

(II)

2-9 dimethyl —
4–7 phenyl 1-10 phenanthroline

Fig. 12.3 Bathocuporine (λ_{max} = 484 nm)

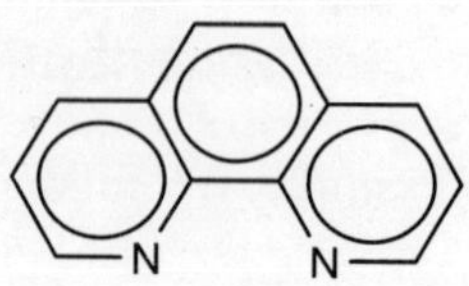

(III)

1–10 phenanthroline (Fe)

Fig. 12.4 1-10 phenanthroline (Fe) (λ_{max} = 510 nm)

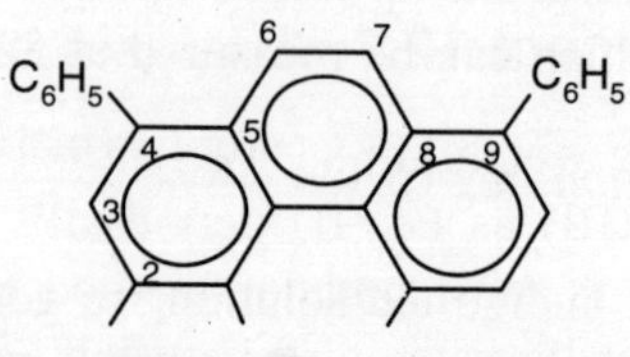

(IV)

4, 7 diphenyl 1-10 phenanthroline

Fig. 12.5 Bathophenanthroline (λ_{max} = 484 nm)

Both reagents provide excellent method for the analysis of copper in water and in waste water. Cr, Sn may show interference. $NH_2OH{\cdot}HCl$ can be used for masking agent.

Reagents: The stock solution of copper is obtained from 200 mg copper wire in 10 ml HNO_3 (con) and made up to 250 ml. About 200 μg Cu/ml is obtained. H_2SO_4 (con), $NH_2OH{\cdot}HCl$ (50 gm/450 ml). Sodium citrate 150 gm in 400 ml + 5 ml $NH_2OH{\cdot}HCl$ + 10 ml neocuproine reagent. 50 ml chloroform extract is used to remove the impurities. NH_4OH (5N), Neocuporine (100 mg in 100 ml CH_3OH), chloroform, HNO_3, methanol and HCl (con).

Procedure: From standard solution prepare different solutions containing varying amounts of copper. Use this to construct calibration curve. Use 1 ml of conc. H_2SO_4 before extraction in chloroform. The extraction is carried out as follows: To an aliquot of solution containing copper add one ml of (con) H_2SO_4 + 5ml (con) HNO_3–boil till solution is colourless. Take a sample solution (~4 –200 μg Cu) add 5 ml $NH_2OH{\cdot}HCl$ + 10 ml sodium citrate adjust the pH to 4.6 with NH_3. Then add 10 ml neocuporine + 10 ml chloroform and extract for two minute, allow it to settle and separate, then measure the absorbance of organic phase at 457 nm. The amount of copper present is calculated as:

$$\text{mg cu/L} = \frac{\mu\text{g cu (in 25 ml soln)}}{\text{ml. portion used for extraction}}$$

One can also use bathocuporine instead of neocuporine to get better results, however many metals like Ca, Na, PO_4^{3-}, ClO_4^-, SO_4^{2-} show strong interferences CN^-, SCN^-, $S_2O_8^{2-}$ and $EDTA^{-4}$ also show interference. We consider this method for the analysis with bathocuporine.

Bathocuporine for Analysis of Copper

The sensitivity of the method for copper is 20 μl/litre. Instead of the spectrophotometer one can also use filter photometer for absorbance measurements.

Experiment Reagent: The stock solution 20 μg Cu/ml, HCl, $NH_2OH{\cdot}HCl$ sodium citrate (300 gm/ Litre of citrate). Reagent chromogenic bathocuporine—1 gm/ L in solution.

Procedure: To sample solution (in 50 ml flask) add 1 ml HCl (1:1), 5 ml $NH_2OH{\cdot}HCl$ + 5 ml citrate solution + 5 ml of chromogenic ligand and make it up to a known volume. Measure the absorbance at 484 nm against a reagent blank. From the calibration curve find out the concentration of copper from the unknown sample.

Calculations: $\text{mg Cu/L} = \dfrac{\mu\text{g Cu (in 66 ml)}}{\text{ml. of sample}}$.

12.7 IRON (III):

Experiment 12.7: Analysis of Iron from Water with Bathophenanthroline

Iron has atomic weight 55.85 with atomic number 26. The colorimetry of iron (II) is reliable but not useful for Fe (III) as Fe (II) gets easily oxidised in air to trivalent state. The method using 1–10 phenanthroline is most popular method for quantitative analysis of iron (III). Bathophenanthroline however can easily distinguish between Fe^{2+} and Fe^{3+} as 1 μg/L can be determined. Little addition of acid prevents deposition of iron (III) on walls of container.

1–10 phenanthroline Method for Analysis

At pH 3.2–3.3 complex is formed. The strong oxidising agents interfere. If solution is initially coloured dry ashing process must be followed. The sensitivity is 10 mg/L. The method is reproducible.

Reagents: HCl (con.), $NH_2OH{\cdot}HCl$ (10 gm in 100 ml H_2O), Buffer solution (pH 3.0). (250 gm CH_3COONH_4 in 150 ml water + 700 ml glacial CH_3COOH), CH_3COONa (200 gm in 800 ml water), 1–10 phenanthroline (100 mg in 100 ml water at 80°C); $KMnO_4$ (0.316 gm in 100 ml H_2O). The stock solution for iron is prepared from $FeSO_4$ $(NH_4)_2SO_4{\cdot}6H_2O$. (Dissolve 1.404 gm salt in 50 ml H_2O + 20 ml H_2SO_4 (con) + 0.1 M $KMnO_4$ - diluted to one litre) 1 ml contains 200 μg/ml of Fe (III). Isopropyl or ether are used as solvents.

Procedure: An aliquot of solution (about 50 ml) is taken, 2 ml (con) HCl is added + 1 ml $NH_2OH{\cdot}HCl$-boil cool and add 10 ml. CH_3COONa buffer + 4ml of ligand. Allow 10 ml for development of colour of complex and then measure its absorbance but ensure complete development of colour before measurement. Fe^{2+} can be also assessed. An aliquot of solution containing Fe (II) is taken, 2ml of (con) HCl is added, then 20 ml of chromogenic ligand is added with buffer solution CH_3COONH_4 (pH ~ 3.0). Make it to 100 ml and measure the absorbance at 510 nm. From total iron found subtract ferrous iron to know amount of Fe (III) ion. The orange yellow colour is measured at 510 nm on spectrophotometer. Fe (II) can be also analysed by extracting coloured complex in propyl ether. You may have to use absorption cell of longer length with optical path of 5 cms, *i.e.,* 50 mm.

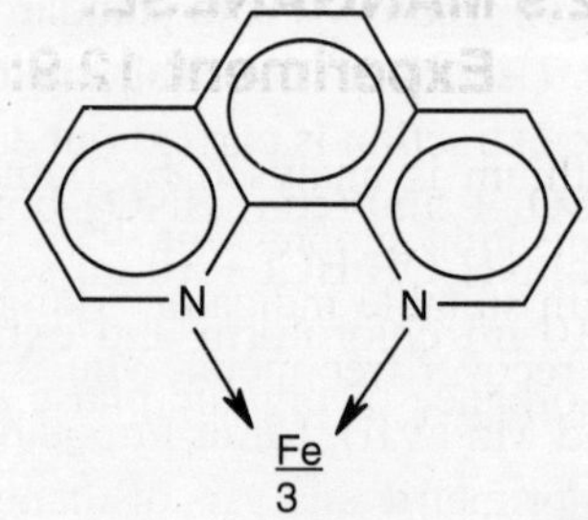

Fig. 12.6 1–10 phenanthroline

Calculations: $\text{mg Fe/L} = \dfrac{\mu\text{g Fe (in 100 ml)}}{\text{ml. of sample}}$ also $\text{mg Fe/L} = \dfrac{\mu\text{g of Fe (in 100 ml)}}{\text{ml. sample}} \times \dfrac{100}{\text{ml. portion}}$

12.8 LEAD:

Experiment 12.8: Analysis of Lead from Water

The best reagent for quantitative analysis of lead is dithizone. Pb (82) has atomic weight of 207.19 with oxidation state of Pb(II) and Pb (IV). The limit for drinking water by US-EPA is 15 μg/L.

Ammonical citrate-cyanide solution containing lead is extractable in dithiozone which is dissolved in chloroform to give cherry red coloured complex. Bi, Sn, Tl show interference. The absorbance maxima is at λ_{max} = 510 nm.

Reagents: The stock solution of lead is prepared by dissolving 0.1599 gm $Pb(NO_3)_2$ is 200 ml water containing 10 ml (con.) HNO_3 and made up to one litre. Working lead solution is obtained from stock solution by suitable dilution. HNO_3 (1:4), NH_4OH (1:9), citrate cyanide reducing solution obtained by dissolving 400 gm ammonium citrate + 20 gm Na_2SO_4 + 10 gm $NH_2OH{\cdot}HCl$ + 40 gm KCN all made to one litre with water. The dithizone solution is also prepared (0.001% in water). The molar absorptivity is 40.6 × 10^3 at 606 nm. Thus 100 mg dithizone in 50 ml chloroform is used. A working solution is obtained from stock solution such that 1 ml contains 40 μg dithizone. Na_2SO_3 (5 gm in 100 ml H_2O). Iodine solution (40 gm KI in 25 ml H_2O diluted to one litre).

Procedure: Add 20 ml (1:4) HNO_3 to 50 ml ammonical citrate-cyanide buffer with 10 ml dithizone-measure the absorbance at 510 nm as blank solution. Then take 100 ml acidified sample (pH ~ 2.0) + 20 ml (1:4) HNO_3 + 50 ml citrate-cyanide buffer + 10 ml dithizone solution and measure the absorbance at 510 nm on photometer. Bi, Sn, Tl interfere however it is eliminated by measuring solution at 465 nm instead 510 nm.

Calculation: $\text{mg Pb/Li} = \dfrac{\mu\text{g Pb (in 10 ml. soln)}}{\text{ml. of sample}}$. (Note absorbance of blank to get real value).

$\dfrac{\text{Corrected absorbance at 510 nm}}{\text{Corrected absorbance at 465 nm}} = 2.08.$ Chloroform gives green coloured solution.

12.9 MANGANESE:

Experiment 12.9: Analysis of Manganese from Water

Lithium is analysed by flame emission spectroscopy at 671 nm with accuracy of 0.1 μg/Litre with ionisation suppressant (95.3 gm KCl in one litre) while magnesium is best analysed by EDTA titration with suitable indicator. Manganese is encountered in areas where haematite or pyrolusite is processed to recover free metal. Mn (25) has atomic weight of 54.93 with common valency of Mn (II), Mn (IV) and Mn (VII). Limit in potable water is 50 μg/L by US-EPA. Persulphate is one of the best method in colorimetric analysis of manganese. $AgNO_3$ is used as catalyst during oxidation of Mn(II) to Mn (VII), but chloride shows strong interference so also Bi^- and I^- interfere. The sensitivity of persulphate method is 42 μg/L or 210 μg/L in 50 and 10 cms cuvettes respectively.

Reagents: 75 gm $HgSO_4$ in 400 ml (con) HNO_3 + 200 ml H_2O + 20 ml of 85% H_3PO_4 and 35 mg $AgNO_3$ all made to 1000 ml as special reagent $(NH_4)_2S_2O_8$. Standard manganese solution from 0.1N $KMnO_4$ obtained by dissolving 3.2 gm $KMnO_4$ in one litre water. The standard manganese solution 1 gm Mn metal in 10 ml HNO_3 – dilute to liter with 1% HCl (1 ml = 1 mg Mn). H_2O_2 (3% HNO_3, H_2SO_4, $NaNO_2$ (5 gm in 95 ml) $Na_2C_2O_4$ and $NaHSO_3$ (10 gm in 100 ml).

Procedure: In case organic matter is present, prior digestion is essential step. To an aliquot of solution 5 ml special reagent is added, with 1 drop H_2O_2 and 1 gm of $(NH_4)_2$ S_2O_8 and boiled. Avoid too much boiling to prevent decomposition of $K_2S_2O_8$ and loss of $KMnO_4$ colour. Dilute to 100 ml and obtain standard from stock solution. Make photometric measurement at appropriate wavelength with water as blank. A correction should be made if solution is turbid preferably by filtration.

$$\text{Normality of } KMnO_4 = \frac{\text{gm } Na_2C_2O_4}{(A - B) \times 0.067}$$

if A = ml of sample

B = ml of titrant

$$\text{mL } KMnO_4 = \frac{4.55}{\text{normality of } KMnO_4}$$

$$\textit{Calculations:}\ \text{mg Mn/L} = \frac{\mu\text{g of Mn/100 ml.}}{\text{ml of sample}} \times \frac{100}{\text{ml portion}}$$

When digested with 100 ml (final) volume then

$$\text{a ml of manganese per litre} = \frac{\mu\text{g of Mn (in 100 ml final) volume}}{\text{ml. sample solution}}$$

The method is selective, sensitive and reproducible as is expected with stable persulphate colour.

12.10 SELENIUM:

Experiment 12.10: Analysis of Selenium from Water

Se(34) has atomic weight 78.96 with valency of Se(II), Se(IV) and Se (VI). The US-EPA drinking water standard for selenium is 50 μg/L. There are many methods for its analysis. AAS with hydride generator is ideal method for rapid analysis. The schemes for speciation are also available. The interference encountered can be eliminated by use of anion exchange chromatography or by use of oxidants in persulphate digestion or by digestion with H_2O_2 or elimination by $KMnO_4$ digestion. The organic selenium is oxidised to hexavalent state. While organic interferences are best eliminated by digestion with the solution of $KMnO_4$. It also oxidises chloride ion to free Cl_2 gas. The process of reduction of Se(VI) to Se (IV) is carried out by HCl digestion. However best method for the quantitative analysis of selenium is colorimetric method utilising 2, 2′ diaminobenzidine or 2-3 diaminonaphthalene as the suitable chromogenic agent for the colour development at pH 1.5–2.0. No inorganic ions interfere. Coloured compound is extracted in cyclohexane. EDTA addition eliminates negative interferences, the sensitivity of proposed method is (10 µg/L of selenium) quite satisfactory.

Reagents: Standard selenium solution. Dissolve 1.9 gm Na_2SeO_3 in 10 ml (con) HCl and dilute to one litre with water (1 ml = 1 mg Se (IV)). Working solutions are obtained from stock solution by appropriate dilution, HCl (0.1 N con), NH_4OH (50%), cyclohexane 2-3 diaminonaphthalene (DAN) solution. Dissolve 0.2 gm in 200 ml of 0.1N HCl, the extraction with cyclohexane (75 ml) and kept it in refrigerator. NH_2OH-HCl, HA-EDTA dissolve 4.5 gm Na_2EDTA in 450 ml water + 12.5 gm of $NH_2OH{\cdot}HCl$ and dilutes to 500 ml.

Procedure: Add 2 ml EDTA solution to 10 ml sample. Adjust pH 1.5 with dil. HCl/NH_4OH. Then add 5 ml DAN and heat it on water bath for 30 mins. till 50°C. cool. Then add 2 ml cyclohexane and shake it for 5 mins. Allow it to settle and separate. Remove red organic phase for absorbance measurement at 480 nm.

Calculations: Prepare as usual calibration curve at 480 nm with H_2O blank. There are sophisticated methods for removing of organic and inorganic volatile selenium. One can use H_2O_2 and NaOH for removing volatile selenium while for non-volatile selenium use methyl sulphonic acid in methanol in the presence of copper sulphate solution. As regards Tellurium $\text{Con } \mu g/L = \frac{\text{Final vol}}{\text{Sample vol}} \times \text{Reading}$ is best analysed by graphite furnace AAS or by plasma emission spectroscopy or by nephelometer as blue sol.

12.11 S-BLOCK ELEMENTS:

Experiment 12.11: Analysis of Alkali-alkaline Earths in Water

The ideal method for the analysis of these elements is flame emission atomic spectroscopy. We would consider here barium, caesium, lithium, potassium, sodium and strontium very briefly as they are rarely encountered in the sample of waste water. Ba (56) atomic weight 137.33 is divalent. Con > 1 mg/L are toxic, is best analysed by AAS or ICP-AES technique. Caesium (55) with atomic weight 132.90 has monovalency. It is analysed by Flame photometry or by Plasma Spectroscopy. Li (3) with atomic weight 6.94 has monovalency. It is best analysed by AAS or flame photometry. ICP-AES-MS can be also used with advantage. The potassium K (19) has atomic weight 39.10 and shows only univalency. Flame photometry, plasma spectroscopy, ion selective electrodes are best methods for analysis. Sodium Na (11) has atomic weight 22.99 and can be analysed with flame emission photometry at 589 nm. This is the best method. Finally Strontium (38) has atomic weight of 87.62 exhibiting divalency. It gets in atmosphere due to the disposal of radioactive and fallout of nuclear waste. Plasma spectroscopy and flame photometry are preferred methods of analysis of these elements.

12.12 VANADIUM:

Experiment 12.12: Analysis of Vanadium from Water

Vanadium (23) has atomic weight of 50.94 with variable oxidation state from V (II), V (III), V (IV) and

V(V) but V^{5+} is most dominant species. Low incidence of heart disease is due to presence of vanadium in water or food (~20–150 μg/L). For drinking water US-EPA limit is 0.1 mg/L. The most preferred method is spectrophotometric analysis with gallic acid. However, apart from gallic acid one can also use ligands like Benzohydroxamic acid (BPHA) or N-phenyl N-Benzoyl hydroxyl amine or N-cinamoyl N-benzoyl hydroxylamine. Gallic acid oxidation is dependent on concentration of vanadium. It is measured at 415 nm spectrophotometrically. Bi^-, I^- interfere but Hg(II) can eliminate their interferences [$Hg(NO_3)_2$] is best for this task. Some ions like Ni^{2+}, Ag^+, U^{6+} can show slight interferences but not Cu^{2+}, Fe^{3+}, Mo^{6+}, Br^- and iodide. The sensitivity of method is 0.025 μg/10 ml or 2 μg/ml of sample solution.

Reagents: In the stock solution dissolve 229.6 mg NH_4VO_3 in 800 ml H_2O with 15 ml HNO_3(1:1) and dilute to one litre (1 ml = 100 μg V), $Hg(NO_3)_2$ solution (dissolve 350 mg in one litre), $(NH_4)_2S_2O_8$ - H_3PO_4 reagent. Dissolve 2.5 gm in 25 ml H_2O add 25 ml H_3PO_4, boil and dilute suitably.

Procedure: From stock solution prepare standard solution by the process of appropriate dilution. In 10 ml sample solution add 1 ml of Hg $(NO_3)_2$ to each and blank solution. Warm up to 25°C on waterbath for 1/2 hr. Add 1 ml ammonium persulphate phosphoric acid mixture and warm and then add 1 ml gallic acid solution, continue the process of addition of gallic acid and the warming after every 30 seconds. After an hour take cold-solution and measure its absorbance at 415 nm on spectrophotometer. Prepare as usual calibration curve and find out amount of vanadium in the unknown sample solution from the calibration curve.

Calculations: $$\text{mg V/Litre} = \frac{\mu\text{g V (in 13 ml of final volume)}}{\text{Original sample volume, ml}}.$$

12.13 ZINC:

Experiment 12.13: Analysis of Zinc from Water

Zinc (30) has atomic weight of 65.38 with divalency. US-EPA standard for Zn in drinking water is 5 mg/L. AAS and ICP-AES are best methods for quantitative analysis. However spectrophotometry utilising zincon is most preferred method of the analysis. Zincon (2 carboxy, 2' hydroxy 5' sulphormazyl) in benzene at pH 9.0 is used for quantitative analysis. Cyanide is added to complex Zn cyclohexane frees Zn from cyanide solution. The interference of Mn is eliminated by addition of sodium ascorbate by process of reduction. Few metals showing interference are Al^{3+}, Mn^{2+}, Fe^{3+}, Cr^{3+} maximum for Ni^{2+}, Cu^{2+}, Co^{2+} and CrO_4^{2-}. The sensitivity of methods is 0.02 mg/L of zinc and method is reproducible.

Reagents: The solution is prepared by dissolve 1000 mg of Zn in 10 ml (1 + 1) HNO_3 (1ml = 1mg Zn). Sodium ascorbate solution, KCN (dissolve 1 gm KCN in 50 ml H_2O) and make it to 100 ml. Buffer (pH ~ 9.6). Dissolve 8.49 NaOH in 500 ml H_2O + 31 gm H_3BO_3 and dilute it to one litre. Zincon (100 mg in 100 ml CH_3OH) is preferably kept overnight before use. HCl (1 N), NaOH (1, 6 N) and cyclohexane (purified).

Procedure: Prepare calibration plot with varying concentration of Zn in solution. To aliquot of solution containing zinc, add 0.5 gm sodium ascorbate, 5 ml buffer, 2 ml of KCN and 3 ml Zincon. Add 1 ml cyclohexane. Measure absorbance of the coloured sample at fixed wavelength. Reagent blank is coloured due to zincon in the cyclohexane. All adjustment if made at pH ~ 7.0 gives reproducible result.

ADDITIONAL READINGS

12.1 APHA, *Standard Methods for the Examination of Water and Waste Water*, 20th Ed., (1998).

12.2 K.J. Irgolic, *Frontiers of Bioinorganic Chemistry,* VCH Publications (1986).

12.3 E.B. Sandell, *Colorimetric Determination of Trace of Metals*, 3rd Ed. Interscience Publishers (1959).

12.4 F. Welcher, *Organic Analytical Reagents*, D. Van Nostrand & Co. Princeton Vol. 3, p. 100 (1947).

12.5 A.M. Ure and R.L. Mitchel, *Flame Emission and Atomic Absorption Spectrometry*, Ed. J.A. Dean, T.C. Rains, Marcel & Dekker (1975).

12.6 K.C. Thomson and R.J. Reynolds, *Atomic Absorption Fluorescence and Flame Spectroscopy*, John Wiley & Sons (1978).

12.7 H.H. Willard, L.L. Merrit, J.A. Dean and F.A. Settle, *Instrumental Methods of Analysis*, 6th Ed., Wordsworth Publishing (1981).

12.8 P.L. Bailey, *Analysis with Ion Selective Electrodes*, Heyden & Sons Ltd. (1980).

12.9 D. Midgley and K. Torrance, *Potentiometric Water Analysis*, John Wiley & Co. (1978).

12.10 J.A. Dean, *Flame Photometry*, McGraw-Hill Publishers (1960).

12.11 S.M. Khopkar, *Basic Concepts of Analytical Chemistry*, 3rd Ed., New Age International Publishers (P) Ltd., New Delhi (2008).

12.12 S.M. Khopkar, *Environmental Pollution, Monitoring and Control*, New Age International Publishers (P) Ltd., New Delhi (2007).

12.13 A.I. Vogel, *Textbook of Quantitative Inorganic Analysis*, Longman & Greens Ltd. (1960).

CHAPTER 13

Metal Pollutants Analysis by Instrumental Methods

13.1 INTRODUCTION

We have considered the analysis of important metal pollutants in water by specific techniques. These techniques largely consisted of absorption spectroscopy. It was possible on account of availability of selective, sensitive specific organic reagents. Most of the metal ions were quantitatively analysed at microgram concentration. Due to access to sequistering agents, many of the interferences by metals were eliminated to the greater extent. Only limitations of these so called specific methods was the time taken for complete analysis. Further at one time it was possible to analyse only one sample at a time. Finally some of the methods had relatively poor sensitivity.

In order to overcome all these limitations, resort was taken to rapid instrumental analysis. With sophisticated instruments consisting of atomic absorption spectroscope with flame or graphite furnace, special modification for the analysis of mercury, selenium etc. was made with hydride generator. Although AAS provided rapid methods for the multi elements analysis (four to six metals at time) their level of analysis was not very satisfactory. Therefore more sophisticated technique of the inductively coupled plasma atomic emission spectroscopy (ICP-AES) was resorted. AAS was absorphometric method while ICP-AES was emission technique with better sensitivity. At later stage it was coupled with Mass Spectrometer to provide a hyphenated method (ICP-AES-MS) for quantitative analysis. Although electroanalytical techniques are very simple, they are very rarely used in metal pollutant analysis. However, anodic stripping voltammetry (ASV) was an exception. It provided rapid and selective method for analysis, DC-polarography, amperiometry, chronopotentiometry, high frequency conductometry are alternate methods available but less frequently used for rapid analysis.

Before analysis it was necessary to pretreat the sample. Such pretreatment included quality control and such preliminary treatment of the sample was followed by various modern methods of analysis. These instrumental methods consisted of AAS, flame AAS, cold vapour atomic absorbance spectroscopy, electrothermal AAS, analysis via hydride generator, plasma emission spectroscopy (*i.e.*, ICP-AES), and finally by anodic stripping voltammetry (ASV).

13.2 SAMPLING OF THE MATERIAL

The presence of metal pollutant in water is objectionable as it affects the qualities of potable as well as industrial water. The following Table 13.1 gives details of the spectral methods for analysis.

Table 13.1 Instrumental methods (largely Spectroscopic) of analysis

Element(s)	*Flame AAS*	*Flame AAS (with extraction)*	*FES*	*Electrothermal AAS with furnace*	*ICP-AES*	*ASV*
Al (III)	✓	✓	×	×	✓	×
Sb (III)	✓	×	×	×	✓	×
Ba (II)	✓	✓	✓	×	✓	×
Be (II)	✓	✓	×	×	✓	×
Bi (III)	✓	×	×	×	✓	×
Cd (II)	✓	✓	×	×	✓	✓
Ca (II)	✓	✓	✓	×	✓	×
Cs (I)	✓	×	×	×	✓	×
Cr (III)	✓	✓	×	×	✓	×
Co (IV)	✓	✓	×	×	✓	×
Cu (II)	✓	✓	✓	×	✓	×
Ga (III)	×	×	×	×	✓	×
Ge (IV)	×	×	×	×	✓	×
Au (III)	✓	×	×	×	✓	×
In (III)	×	×	×	✓	✓	×
Fe (II)	✓	✓	×	✓	✓	×
Pb (II)	✓	✓	×	×	✓	✓
Li (I)	✓	✓	×	×	✓	×
Mg (II)	✓	×	×	×	✓	×
Mn (II)	✓	×	✓	×	✓	×
Hg (II)	✓	×	×	×	×	×
Mo (VI)	✓	×	✓	✓	✓	×
Ni (II)	✓	×	✓	✓	✓	×
Os (III)	✓	×	✓	×	×	×
Pd (II)	✓	×	×	×	×	×
Pt (IV)	✓	×	×	×	✓	×
K (I)	✓	✓	×	×	×	×
Re (VII)	✓	×	✓	×	×	×
Rb (III)	✓	×	×	×	×	×
Se (IV)	✓	×	×	✓	✓	×
Ag (I)	✓	×	✓	✓	✓	×
Na (I)	✓	✓	×	×	✓	×
Te (VI)	✓	×	×	✓	✓	×
Th (IV)	✓	×	✓	✓	×	×
Sn (II)	✓	×	×	×	×	×
Ti (IV)	✓	×	✓	×	×	×
U (VI)	✓	×	×	×	✓	×
V (V)	✓	×	✓	✓	✓	×
Zn (II)	✓	×	✓	✓	✓	✓

(✓) = indicates the metal can be analysed by specified instrumental technique.

(X) = indicates it is not possible to analyse by said instrumental method.

In Table 13.1 metals like Hg, Se, Te need hydride generator with AAS for analysis. While best results are obtained for Mn, Ni, Re, Ag, Th, Ti, V and Zn if extraction precedes AAS analysis. As such ASV is rarely used except in case of Zn, Cd and Pb. Whenever AAS is used it is possible also to use ICP-AES for quantitative analysis.

During quantitative chemical analysis before sampling one must decide as to how much sample is to be actually analysed by proposed method. The sample container should be clean and should not be affected by interfering ions prior to analysis. If necessary one can preserve sample in nitric acid at pH < 2.0 also keep it in refrigerator at 4°C. Thus samples remain safe for 6 months or less. Immediate analysis is required only for trace element analysis. In case of mercury its preservation under 20% $K_2Cr_2O_7$ is essential. To remove organic deposits, use of chromic acid is recommended. Before selecting the method, assess the detection limit (MDL) method as well as maximum range of analysis. Replicate analysis is preferable. To extent possible use reagent blank which has undergone same treatment or as is analytical sample in the absence of metal in question. One can also use spiked samples if necessary. The method of known addition can be also used *i.e.*, standard addition and internal standard method during instrumental analysis. One can also use direct AAS if turbidity is < 1NTU.

Table 13.2 Digestion of metals by acids

Metal	*Acid*	*Whether worthwhile? Yes*	*Certainly not good*	*Volume of acid (mg/L)*
Ag	HCl	Sb Ru Sn	Th Pb	< 0.1 (1000)
Ti	H_2SO_4	–	Ag Pb Ba	0.1-1.0 (100)
–	$HClO_4$	org matter	–	10–100 (10)
–	HF	Silicious matter	–	

The metal after digestion is suitable and ready for AAS or ICP-AES analysis as the organic matter is eliminated earlier. Nitric acid is universally used and best for the digestion of metals. Sometimes addition of acid like $HClO_4$, HCl, HF or H_2SO_4 is required. The process of dry ashing should be avoided to the extent possible to prevent loss of metal by voltisation. If one wishes to analyse by ICP-AES, then microwave digestion is preferable. The ratio of acid volume to metal concentration is given by following equations as:

Calculations:

(*a*) Metal concentration $\text{mg/L} = \dfrac{A \times B}{C}$

(*b*) Metal ion $\text{mg/kg (weight)} = \dfrac{A \times B}{\text{gm sample}}$

(*c*) Metal ion $\text{mg/kg (Dry)} = \dfrac{A \times B \times 100}{\text{gm} \times D}$

In above expression, we have symbol

A = concentration of metal in digested solution mg/L
B = final volume of digestion solution, ml
C = total solids (%) present
D = wt of the sample

The following procedure is used for HNO_3 acid digestion (or in combination with HCl, H_2SO_4, $HClO_4$):

(*a*) A procedure for high level concentration of metal is as follows: A sample + 5 ml conc. HNO_3 is boiled in breaker and evaporated to almost to dryness, till light coloured solution is obtained. Make it to known volume (~100 ml) with addition of water in the procedure (*b*) for digestion trace level metals use $2NHNO_3$ and digest slowly and carefully. The combination of HNO_3-HCl, HNO_3-H_2SO_4 or HNO_3-$HClO_4$ is used depending upon the specific requirements.

13.3 METAL POLLUTANTS ANALYSIS BY ATOMIC ABSORPTION SPECTROSCOPY

Majority of the metal pollutants can be analysed by flame atomic absorption method. The only variation is there may be flamed model or electrothermal or what is termed as graphite furnace atomic absorption spectroscope. Metals as antimony, germanium are analysed by graphite furnace AAS while metals like gold, rhenium, thorium are analysed by inductively coupled plasma atomic emission spectrophy. However, metal as cadmium are also analysed by anodic stripping voltammetry while rare metals like Be, Bi, Cs, Co, In, Ni, Os, Pd, Pt, Rh, Tl, Sn and Ti are analysed directly by flame atomic absorption methods. There is only variation with kind of fuel gas and oxidant used in model. One has the option to use air-acetylene (A-Ac), nitrous oxide-acetylene (N-Ac) as the flame gas. Very small amount can be detected (see Table 13.3).

Table 13.3 Direct flame AAS method for analysis

Metal	*Wavelength (nm)*	*Flame*	*Det. limit (mg/L)*	*Sensitivity (mg/L)*	*Range of analysis (mg/L)*	*Remarks/range*
Be (II)	234.9	NA	0.005	0.03	0.05–2	Good detection
Bi (IV)	223.1	AA	0.06	0.4	0.2–20	Low sensitivity
Co (III)	240.7	AA	0.03	0.2	0.5–10	Good range
Mg (II)	285.7	AA	0.0005	0.007	0.02–2	Good sensitivity
Mo (V) (III)	313.3	NA	0.01	0.5	1–20	Good range of analysis
Ni (II)	232.0	AA	0.02	0.15	0.3–10	Poor sensitivity
Pt metals (IV)	>250	NA	0.1	0.08	~1–20	Average
Sb (V)	217.6	AA	0.07	0.5	1–40	Good range
V (V)	318.4	NA	0.2	1.5	2–100	Good range
Sn (IV)	224.6	AA	0.8	4	10–200	Excellent range
Ti (IV)	213.9	NA	0.3	2.0	5–100	Low sensitivity

NA = Nitrousoxide-acetylene flame

AA = Air-acetylene flame

13.4 FLAME AAS ANALYSIS OF METALS: Experiment 13.1: Direct Flame AAS Analysis with Acetylene Flame

The technique is similar to one with AAS analysis with no special precautions. In such analysis sometimes

-you may encounter chemical interference which in turn can be eliminated by addition of chemical externally e.g., add La to Mg to remove PO_4^{3-} or add Ca in Mn analysis to eliminate interference by silicon. In extractive AAS we have several metals which should be extracted in MIBK with (ammonium pyrroridine dithiocarbamate APDC) followed by AAS analysis used. Such analysis sometimes need background correction. One can use line source with hollow cathode. Zeeman background correction is essential as by magnetic field source is divided into π and σ-lines resulting in absorption in σ-lines, however, sensitivity is thereby reduced. The Smith-Hieftje correction provides more advantages. It eliminates spectral interference. The Table 13.3 represents sensitivity, detection limit, and the range of analysis. (0.0044 absorbance corresponds to 1% absorption to denote sensitivity.) Range represents a region where one can acquire reproducible results. One should dilute solution from stock solution to get calibration curve and hence have dependable results. One always prefers direct flame method with air-acetylene or nitrous oxide-acetylene flame for group of metals analysed by direct AA flame technique in AAS which is rapid and reproducible.

Table 13.4 Analysis of metals by sophisticated instruments

S. No.	*Name of method*	*Kind of metals analysed*	*Section number of this book*
1.	Direct Flame Atomic Absorption Spectroscopy (Direct AAS)	Sb Bi Cd Co Cs Cr Cu Au Zr Fe Pb Li Mg Mn Ni Pt Pd Ru Rh Ag Na Sr Sn Zn Tb Al Ba Be Ni Os Re Sr Ti V	13.3
2.	Extraction followed by Atomic Absorption Spectroscopy (Ext-AAS)	Cd Cr Co Cu Ti Pb Mn Ni Ag Zn (with APDC-MIBK extraction). Al, Be (with 8-Hydroxyquinoline-$CHCl_3$)	13.5 13.7
3.	Hydride-generated Atomic Absorption Spectroscopy (Hydride-AAS)	As (III) Se (IV)	13.10
4.	Cold vapour Atomic Absorption Spectroscopy (Cold vapour-AAS)	Hg (II)	13.8
5.	Electrothermal Atomic Absorption Spectroscopy (Graphite-AAS)	Al Sb Ba Cd Cr Ca Cu Fe Pb Mn Ni Mo Se Ag Sn	13.9
6.	Inductively-coupled Plasma Atomic Emission Spectroscopy (ICP-AES)	Al Sb Be Cd Cr Co Cu Fe Pb Mg Mn Ni Se Ag Ti V Zn (ICP hyphenated with MS)	13.11
7.	Anodic Stripping Voltammetry (ASV)	Pb Zn Cd	13.12

Apparatus and Reagents: A good quality atomic absorption spectrophotometer equipped with air-acetylene (AA) flame. The reagents used are calcium (0.630 gm $CaCO_3$ in 50 ml HCl) cool and dilute it to one litre HCl (1%), lanthanum (58.65 gm of La_2O_3 in 250 ml HCl) and dilute it to one litre H_2O_2 (30%), 2% HNO_3 (aqua ragia (1:3 of HNO_3:HCl) and standard metal solution.

Procedure: The sample containing standard and analyte solution is aspirated in air-acetylene flame and reading for absorbance is noted. A good calibration curve with standard solution is constructed. From this plot concentration of metal in unknown sample is computed as per normal procedure. Use Ca/La solution in case there is chemical interference. Calculate unknown concentration from the calibration curve.

13.5 ANALYSIS OF METALS BY EXTRACTION IN MIBK WITH APDC

This method is useful for analysis of low concentrations of metals. MIBK is used as extractant with 4 gm APDC in 100 ml H_2O, 5% $KMnO_4$, NH_2HCl (10%), Na_2SO_4 are reagents.

Procedure: The different metals extract at different pH shown in brackets e.g. Ag (2-5), Cd(1-6), Co(2-10), Cr(3-9), Cu(0.1-8), Fe(2-5), Mn(2-4), Ni(2-4), Pb(0.1-6), Zn(2-6). Therefore depending upon kind of metal one can adjust pH (or use buffer) and extract metal with APDC in MIBK (40 ml) and then aspirate in air-acetylene flame in AAS and note absorbance. Prepare as usual calibration curve and find out concentration of the unknown sample from its absorbance plotted on standard curve. One can use $KMnO_4$ when chromium is analysed as Cr (III) is oxidised to Cr (VI) with $KMnO_4$ before the analysis. Na_2SO_4 is used to remove moisture from organic MIBK phase.

13.6 ANALYSIS OF METALS WITH AAS IN NITROUS OXIDE FLAME: Experiment 13.2: Analysis of Metals by AAS (Nitrous Oxide Flame)

This group involves several transition metals needing high temperature for atomisation. Therefore these metal pollutants are directly aspirated in nitrous oxide-acetylene flame in AAS determination.

Reagents: HCl (1M), HNO_3 (con), H_2SO_4 (1%), HF (1N), KCl (250 g/L), Al_2 $(NO_3)_3$ 1.39 in 150 ml H_2O add, HNO_3 to make to 200 ml standard metal solutions consisting of Al, Ba, Be, Ca, Os, Re, silica, Th, Ti and V which are dissolved in mineral acids. 2 ml of KCl is added during analysis of Al, Be, Ti, while $Al(NO_3)_3$ is added for analysis of Mo, V, As. Then aspirate standard solution into the nitrous oxide-acetylene flame, note absorbance and prepare calibration curve. Then take unknown sample, aspirate in flame, note absorbance and hence find out unknown concentration from the standard calibration curve.

13.7 ANALYSIS OF METALS AFTER EXTRACTION WITH NITROUS OXIDE FLAME: Experiment 13.3: Analysis of Metals by previous Extraction

Analysis of metals by previous extraction in MIBK and then aspiration in nitrous oxides acetylene flame is common: Low concentration of Be, Al are analysed by this method (~30–1000 μg/L) Fe, Mn, Mg show strong interference.

Reagent: HCl (con), NH_4OH (con), Buffer solution of pH obtained by dissolving 300 gm CH_3COONH_4 in 100 ml ammonia which is made to one litre. 20 gm of oxine in 200 ml water + 60 ml of glacial acetic acid made up to one litre. Prepare standard solution containing Al, Be or any metal to be analysed —1.3 gm $NH_2OH{\cdot}HCl$ and 6.5 gm o-phenanthroline in 500 ml H_2O is used to mask iron after making total volume to one litre. As in previous case a plot standard curve is obtained. Find out concentration of unknown solution from the calibration curve as prepared earlier.

13.8 ANALYSIS OF MERCURY BY AAS: Experiment 13.4: Analysis of Mercury

There is special case for the analysis of mercury by AAS. Since the vapour of mercury are volatile, one has to generate them in vacuum at room temperature before their quantitative analysis. The special kind of attachment consisting of absorption cell, cell support, air pump, flow meter, aeration tube, drying and reaction tube are used (see Fig. 13.1).

Reagents: Standard mercury solution, HNO_3, $KMnO_4$ (50 g/L), $K_2S_2O_8$ (50 g/L), $(NH_2OH)_2$, H_2SO_4 10% in HCl standard tin solution 7 gm/1000 ml.

In Fig. 13.1 is shown cold vapour AAS assembly. Adjust wavelength to 253.7 nm for absorbance measurements. Adjust flow rate of air at 2 L/mt. Prepare standard mercury solution with water blank. Add 5 ml H_2SO_4 and 2.5 ml HNO_3 and 15 ml $KMnO_4$ and allow it to stand for 15 mts. Add 8 ml of $K_2S_2O_8$ solution and heat for 2 hrs. at 95°C cool to room temperature, add NaCl-NH_2OH to reduce $KMnO_4$ add 5 ml $SnCl_2$, one notes that mercury starts volatilising and flows to absorption cell. After reading flush system with water. Prepare calibration curve and from unknown sample ascertain concentration of mercury in unknown sample (Fig. 13.2).

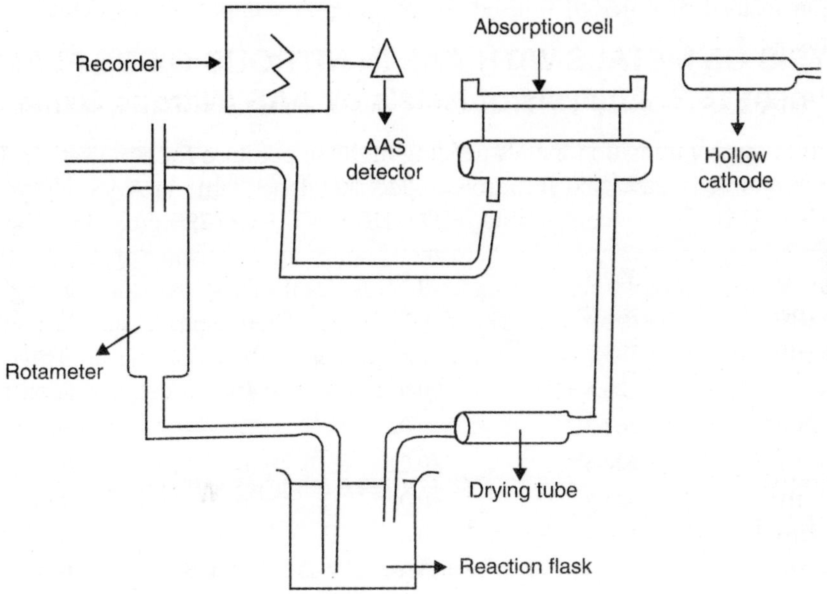

Fig. 13.1 Cold vapour AAS assembly

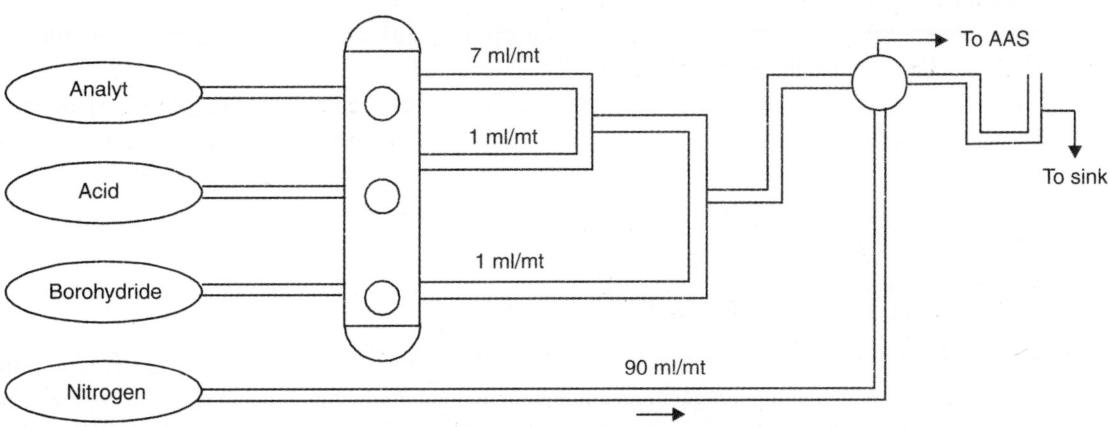

Fig. 13.2 Hydride generator (continuous)

13.9 ANALYSIS OF TRANSITION METALS: Experiment 13.5: Analysis of Group d-block Metals

Graphite Furnace Atomic Absorption Method: Several transition metals having high temperature of atomisation can be analysed faster with electrothermal AAS methods. The sensitivity and detection limit is 1000 times better than the one obtained by Flame AAS. A sample volume is small and as little as 1 μg/L of metal can be analysed. One gets good results with standard addition method. There may be interference by molecular absorption but deuterium lamp helps correction. At high absorbance Zieman or Heistz correction is not valid. Addition of chemical modifier obstructs the correct absorbance measurement. Refractory carbides are formed. Low flow rate or large sample volume improves sensitivity. Argon instead of nitrogen is used for purging. A comparison of detection limit is worth studying in Table 13.5.

Table 13.5 Comparison of Flamed AAS and Graphite furnace AAS, detection

S. No.	*Metal*	*λ (nm)*	*Detection limit (μg/L)*		*Remark*
			F AAS	*E AAS*	
1.	Al (III)	309.3	0.10	3	Poor GF-AAS
2.	Sb (III)	206.8	0.07	3	Poor GF-AAS
3.	As (III)	193.7	50	1	Average
4.	Cd (II)	226.5	0.002	0.1	Sensitive AAS
5.	Cr (VII)	267.7	0.02	2	Poor GF AAS
6.	Co (II)	228.6	0.02	1	Average
7.	Cu (II)	324.7	0.01	1	Best AAS
8.	Fe (III)	259.9	0.02	1	Best AAS
9.	Pb (II)	220.3	0.05	1	Poor GF AAS
10.	Mn (II)	257.6	0.01	0.2	Best AAS
11.	Mo (VI)	202.6	0.5	1	Poor GFAA
12.	Ni (II)	232	0.02	1	Poor GFAA

FAAS = Flamed Atomic Absorption, EAAS = Electrothermal Atomic Absorption

A comparison of values of detection limit (mg/L) (See Table 13.5) indicates that the flame AAS facilitates determination of lower concentration of metal analysis in comparison with graphite furnace AAS method.

Reagents: HCl, HNO_3, Mg $(NO_3)_2$ 10,000 mg/L, $Ni(NO_3)_2$ 10,000 mg/L, H_3PO_4 (10%) $Pd(NO_3)_2$ 4000 mg/L, citric acid 4%, chelating resin in metal free water.

Procedure: Prepare standard solution as per usual procedure. Raise temperature of furnace till atomisation temperature of metal pollutant, and prepare a standard calibration curve. From the knowledge of absorbance of unknown solution its actual concentration can be calculated from the calibration curve.

Calculations: Direct analysis (μg/L) = C × F if C = metal concentration direct reading, F = dilution factor. Method of addition (μg/L = C × F if C = metal ion from plot of addition, F = dilution factor).

13.10 ANALYSIS OF METALS BY USING HYDRIDE GENERATOR: Experiment 13.6: Analysis of Metals by Hydride Generator

The elements like arsenic (III) and selenium (IV) are analysed by AAS with the help of hydride generator. Two methods involve use of hydride generator and is dependable with continuous hydride (Fig. 13.2) generator. Sodium borohydride ($NaBH_4$) is used for preparation of As and Sc hydrides. H_3AsO_3 or H_2SeO_3 are used as the starting materials. With N_2 or Ar, hydrides are purged in instrument in quartz cell which is electrically heated and converted into gas phase atom. The As (III) must be in trivalent stage, while Se (IV) be in quadrivalent state. The hydride generator cell are commercially available. Sulphuric-nitric perchloric acid digestion ensures complete conversion as they destroy organic matter.

Apparatus and reagents: One can use commercial vessel for manual reaction for producing As (III) or Se (IV) hydrides. The reagents used are $NaBH_4$ (8 gm/200 ml, 0.1 N NaOH), NaI (50 gm/500 ml water), KI, H_2SO_4 (18N), H_2SO_4 (2.5 N), $K_2S_2O_8$ 5% (25 gm in 500 ml), HNO_3, $HClO_4$, HCl (con). Prepare standard As (III) solution from As_2O_3 and Se (IV) from Na_2SeO_3 solution.

Procedure: Follow normal procedure for preparation of AsH_3 and H_2Se samples. One can use sodium borohydride also for preparation AsH_3 or SeH_2. After flushing hydrides in quartz cell by purging with argon gas. Note the values of absorbance, then plot standard calibration curve with absorbance versus concentration to ascertain the concentration of As or Se from the unknown sample.

One can also use continuous hydride generation for analysis (Fig. 13.2). It gives reproducible results with low detection limit facilitating high sample volume. Free chloride ion interferes as chloride oxidises hydride. Excess of oxidant can also oxidise the selenium hydride. Nitrite (10/μg/L) is also interfering anion in analysis. Sulphonamide is added to eliminate its interference. One uses HCl (5 + 1), $NaBH_4$, 0.6g + 0.5 g NaOH in 100 ml H_2O, standard solutions of As and Se prepared in 2.5% w/v sulphonamide solution. One has to follow conventional method. Record absorbance values for standard and the unknown solution, plot calibration curve and find out the concentration of unknown sample of selenium (IV) or Arsenic (III) from the standard calibration curve.

13.11 ANALYSIS OF METALS BY PLASMA EMISSION SPECTROSCOPY

The discovery of the ICP-AES is of recent origin. However this technique facilitates multi-element analysis at very low concentration from sample of air as well as water at 6000–8000°C. High temperature completes atomisation. There are spectral and non-spectral interferences like chemical interference. They are eliminated by same methodology as was practised in AAS methods. The change in viscosity or surface tension of the sample solution causes serious interferences. High dissolved solids leads to salt built up at the tip of burner or nebulizer gas orifice. The metals analysed by ICP-AES are as shown in Table 13.6.

Experiment 13.7: The two wavelengths are specified assuming there is spectral interference either due to overlap of wavelength or its proximity of wavelength of other metals. Second wavelength can be preferably chosen for measurement. In spite of detection limits (μg/L) are higher than optimum concentration. The latter should be preferred during routine analysis.

Reagents: Ultrapure reagents are preferred. HCl (1:1), HNO_3 (con), (1:1) also standard solutions of elements to be analysed. The standard calibration plot is made using standards and then absorbance of unknown is noted and concentration in unknown sample is evaluated.

Calculation: There should be correction for the blank by deducting value from experimental figure.

A dilution correction is also essential. It is given as dilution factor = Final wt. or volume/initial wt. or volume. For spectral interference, spectral correction factor (K_d) is evaluated as

$$K_d = \frac{\text{apparent con. of metal}}{\text{actual con. of interfering metal}}$$

Table 13.6 ICP-AES analysis of metals

S. No.	*Metal*	*First wavelength (nm) (λ_1)*	*Second wavelength (nm) (λ_2)*	*Detection limit (μg/L) (D_1)*	*Optimum limit (μg/L) (D_2)*
1.	Al (III)	308	237	40	10
2.	Sb (III)	206	217	30	10
3.	As (III)	193	189	50	10
4.	Be (II)	249	234	0.3	1.0
5.	Cd (II)	276	211	4.0	2.0
6.	Cr (III)	267	206	7.0	5.0
7.	Co (II)	278	230	7.0	10.0
8.	Cu (II)	259	238	6.0	1.0
9.	Fe (III)	220	217	7.0	10
10.	Pb (II)	220	7.0	40	10
11.	Mg (II)	279	279	30	10
12.	Mn (II)	257	294	2.0	2.0
13.	Mo (VI)	202	223	8.0	10.0
14.	Ni (III)	231	221	15	2.4
15.	Se (IV)	196	203	75	5.0
16.	Ag (I)	328	338	7.0	21
17.	Tl (I)	190	371	40	10
18.	V (V)	292	8.0	8.0	1.0
19.	Zn (II)	213	206	2.0	5.0
20.	Si (IV)	212	251	20	21

13.12 ANALYSIS OF METALS BY HYPHENATED ICP-AES-MS TECHNIQUE: Experiment 13.8: The Analysis with ICP-AES-MS

It provides the instrument detection limit from 1-100 μg/L. The ICP-AES-MS is applicable for the analysis of metals like: Al, Sb, As, Be, Cd, Cr, Co, Cu, Pb, Mn, Mo, Ni, Se, Ag, Tl, U, V, Zn. These are very same elements which are analysed by ICP-AES technique. The best results are obtained with the use of internal standard such as Li, Sc, Ge, Tn and In. The instrument detection limit (IDL) varies from 0.003–0.03 μg/L. Method detection limit (MDL) or sensitivity provides the information on the accuracy of the analysis. Linear dynamic range (LDR) is the maximum concentration of analyt above highest point where analyt response is ±10% of the calculated response. Several kinds of interferences are possible. They include polyatomic ion or isotope or physical interferences. Several anions like

OH^-, Br^-, Cl^-, SO_4^{2-}, PO_4^{3-} etc. exhibit positive interferences. The ionisation interferences is also not favourable for the accurate analysis of the metal. Several reagents are used. They are composed of acids (HNO_3) (con) 1:1, 2% and 1% in H_2O. The radio isotopes as ^{6}Li, ^{45}Sc, ^{72}Ge, ^{115}In and ^{232}Th are used as the internal standards. The calibration curve can be obtained with minimum number of standards. It is necessary to use purified argon gas in ICP-torch. It may contain krypton (^{82}Kr) as the impurity. One can assess the method of performance from the values of mean recovery, the standard and relative standard deviation, during continuing calibration and initially verified calibration.

The greatest limitation of ICP-AES method is the necessity of using the inert gas Argon for purging which is much expensive than inert nitrogen gas. The use of Argon gas facilitates its participation in formation of ion with attainment of very high temperature for atomisation of metal ions. The transition metals need very high temperature for complete ionisation followed by the process of the atomisation in the ICP-AES torch flame.

13.13 ANALYSIS OF METALS BY ANODIC STRIPPING VOLTAMMETRY: Experiment 13.9: Analysis of Metals by ASV

This technique is 10–100 times more sensitive than AAS methods, with detection limit in nanograms (10^{-9}) concentration level per litre of sample. One can analyse as many as 4–6 samples simultaneously. However, it is applicable to only those metals which form amalgams with mercury. The method is somewhat slow. One can analyse trace concentration of lead, cadmium and zinc as the metals are electrochemically reduced at anode with negative potential and form an amalgam with pool of mercury. On application of positive potential metal is stripped out. One prefers (HMDE) hanging mercury drop electrode *i.e.*, DME one can also use the thin mercury film electrode (TMCE). Lead, cadmium, zinc readily form amalgam the rate of such formation depends upon pH, temperature, stirring rate and electrode used. Cobalt, nickel show strong interferences by formation of intermetallic compounds. The method is not recommended for analysis of metals at milligram concentration. The organic compounds on account of their adsorption on electrodes also interfere in analysis.

Apparatus: It contains three electrode assembly with a potentiostat for control of potential. One can also use square wave (SW-ASV) and differential pulse (DP-ASV) forms for analysis. The former mode is much faster. SCE is used the reference electrode. Silica or glass cells are used. Good stirrer is useful for homogeneous mixing of solution. Nitrogen gas is purged to remove free oxygen in the cell.

Reagents: Metal free water, HNO_3 (con) and 6N, Ar or N_2 purging gas, standard solutions of metals e.g. Zn , Cd, Pb. Buffer solution of pH ~ 4.5. Dissolve 6.4 gm CH_3COONa in 800 ml of water and then make up to one litre. The citrate buffer (pH ~ 3.0). Dissolve 42.5 gm citric acid in 700 ml water and then dilute to one litre after pH adjustment. Finally phosphate buffer pH ~ 6.8. Dissolve 24 gm NaH_2PO_4 in 500 ml water and make to one litre by addition of NaOH. Amalgamated metal. Dissolve 2 gm of (Zn, Pb, Cd) nitrate in 25 ml. HNO_3, dil to 250 ml. Add Hg $(NO_3)_2$ to solution till amalgam is formed HCl. $(NH_4)VO_3$ soln. (2 gm 250 ml H_2O and HCl), siliconising solution, Alumina, HF, CH_3OH.

Procedure: Prepare as usual standard sample solutions and add 2 ml of (con) HNO_3 to preserve the sample. Prepare the electrodes viz. HDME or TFME as per usual procedure. Let the initial potential be −1.0 V to −1.2 V and have final potential of 0 V, drop size medium, for square wave pulse 25 mV amplitude, 4 mV stop potential, 100 Hz frequency be used. For DP-ASV 25 mV pulse amplitude, 0.5 speed of pulse, 17.50 m sample width with 5 mV/s scanning rate.

Pass current in system as directed. After completion of reaction carry out the anodic stripping of metal by applying positive potential. Plot a graph of applied potential vs. stripping current. Peak current

Table 13.7 ASV of Pb, Zn and Cd (μ/L)

S. No.	Sample	Electrode	Mode	Pb	Zn	Cd	Remarks
1.	Tap water	DME	SW	0.57	–	0.068	Good for Cd
2.	Sea water	TFME	DP	0.008	–	0.012	Good for Pb
3.	Soil water	DME	DP	11.8	–	189	Good for Pb
4.	Pure water	DME	SW	0.79	–	0.13	Good for Cd
5.	Wastewater	DME	DP	74	26	5.2	Good for Cd

DP = Differential pulse ASV, SW-ASV = Square wave ASV, TFMA = Thin mercury film electrode.

indicates quantity of metal present. Follow similar procedure for standard samples. The calculations are as follows:

Calculation: Con. of metal $= C_S = \dfrac{C_1 \times V_s}{V_a} \times \dfrac{i_d}{(i_d - i_a)}$

where C = concentration of metal in sample
C_S = concentration of metal in standard solution
i_a = stripping peak height in original sample, i_d = stripping height in sample.
V_a = volume of sample, V_s = volume of standard solution.

Some interesting results for the analysis of various samples of water are shown in Table 13.7.

ADDITIONAL READINGS

13.1 APHA—*Standard Methods of Examination of Water and Waste Water*, 21st Ed. (2002).

13.2 D.C. Burrell, *Atomic Spectrometric Analysis of the Heavy Metal Pollutants in Water*, Ann Arbor (1975).

13.3 S.L. Sachdev and P.W. West, Env. Sci & Tech, **4**, 749 (1979).

13.4 Z. Grosser, *Technique of GF-AAS*, Perkin Elemer Co. (1985).

13.5 A.G. Jarrel, *High Sensitivity Arsenic Determination by Atomic Absorption Applications*, Lab Bulletin No. 3 (1971).

13.6 K.H. Brainina and F. Nevman, *Electroanalytical Stripping Methods*, John Wiley (1993).

13.7 F.T. Kissinger and W.R. Heinman, *Lab Technique in Electroanalytical Chem.*, 2nd Ed. Marcel Dekker (1996).

13.8 S.M. Khopkar, *Basic Concepts of Analytical Chem.*, 3rd Ed., New Age International Publishers (P) Ltd., New Delhi (2008).

13.9 S.M. Khopkar, *Environmental Pollution, Monitoring and Control*, New Age International Publishers (P) Ltd., New Delhi (2006).

CHAPTER 14

Analysis of Anions in Water

14.1 INTRODUCTION

In the last chapter, we have considered the analysis of metal pollutants by instrumental methods such as Atomic absorption spectroscopy. Inductively coupled plasma-emission spectroscopy and anodic stripping voltammetry. The absorption spectroscopy was largely used for the analysis of metal pollutants utilising specific organic chromogenic ligands for either complexation or complexation accompanied by solvent extraction. All these methods provided reproducible results.

However, analysis of anions was totally neglected field. Very few attempts were made to analyse them by any instrumental methods. In fact use of ion selective electrode or flow injection analysis or the ion chromatography was the right choice for the rapid analysis of these anions. These methods were rapid selective and sensitive. When a chemist was confronted with the problem of ascertaining toxicity of the metal pollutant, he needed an up-to-date knowledge of kind of anion present with the metal cation. The area of environmental speciation is most important from this view point. One can easily cite few examples of pollutants.

Let us consider the illustration of mercury as the pollutant. It has been confirmed that $(CH_3)_2Hg$ or (CH_3CH_2), Hg are toxic but other compounds as Hg_2Cl_2, $Hg(NO_3)$ are not at all toxic. Similarly take a case of chromium (III) which is encountered in effluents from the tanneries. It has been observed that $CrCl_3$ is not toxic while $K_2Cr_2O_7$ is toxic. This shows valency of a metal pollutant plays a significant role in formation of toxic environmental species.

Therefore in this chapter, endeavour is made to characterise anions like nitrate, nitrite by flow injection analysis or analysis of oxyanions like SO_4^{2-}, PO_4^{3-}, NO_3^- by newly invented technique of capillary electrophoresis. The analysis of halides is of vital importance. Although conventional methods like Vohard's or Mohr's method are available they are not well suited for the outdoor analysis. The analysis of sulphur compounds as SO_4^{2-}, SO_3^{2-}, S^{2-} can be also carried out by instrumental methods. While carbonate and bicarbonates, are preferably analysed by acid-base titration technique.

14.2 ION CHROMATOGRAPHY:

Experiment 14.1: Use of Ion Chromatography for Analysis of Anions

It is one of the powerful instrument for the analysis of anions. There are several variation which are possible in this technique viz. one can use chemical suppressor column with conductometric detectors for continuous flow analysis. In fact we have already considered working of the ion chromatograph in earlier chapters. This technique has promise of rapid sequential measuremental method of analysis. It recognises the difference in NO_3^-, NO_2^- or SO_4^{2-}, SO_3^{2-}, or Cl^-, Br^-, I^- in solutions. When weak solutions of carbonate-bicarbonate were used as the eluents the separated anions in the suppressor column were converted into corresponding highly conductive acidic forms which can be measured by conductivity detector, but organic acids interfere in the analysis having low molecular weights. About 1 mg/L of anions can be analysed in 100 μl of the sample. Unfortunately fluoride cannot be analysed by ion chromatography due to water dip. Dianox has produced a versatile model of ion chromatograph with low capacity pellicular anion exchanger resins packed in the column.

Reagents: Deionised water, Eluents: 1.7×10^{-3} M $NaHCO_3$; 1.8×10^{-3}M of Na_2CO_3. Dissolve 0.5712 g of carbonate with 0.7632 g bicarbonate in four litre of water, H_2SO_4 (0.025 N), standard anion solutions (~1000 mg/L concentration). The combined working standard solutions (e.g. 20 ml Cl^- + 80 ml SO_4^{2-}), combined in low range standard solutions.

Procedure: Pass the solution containing mixture of anion in ion chromatograph with flow rate of 2.5–3 ml/mt. The instrument should be calibrated before use and must be compared with standard solution of anions for peaks.

Calculation: C = H × F × D if C = mg of anion/L, H = peak height, F = concentration of stand/height, D = dilution. Table 14.1 summarises in the determination of various anions in potable water. It shows bias for anions like F^-, Cl^-,NO_2^-, NO_3^-, Br^-, PO_4^{3-} and SO_4^{2-} in that order. The bias was highest for NO_3^- (+9.5%) while it was negligible in case of Br^-. It was possible to compare such bias by having samples of water from various laboratories and using them for the interlaboratory comparison.

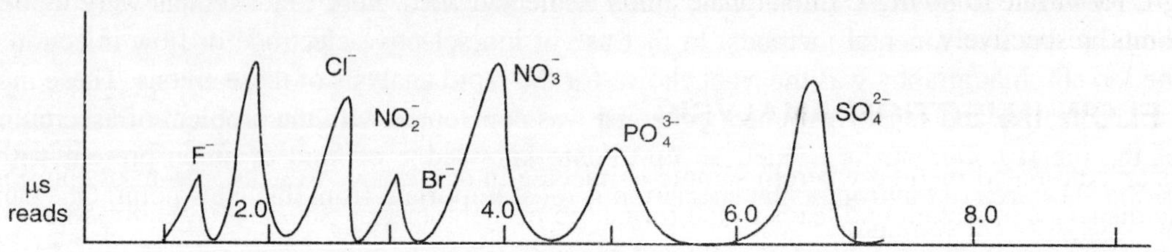

Fig. 14.1 Typical elution curves vs amount

Table 14.1 Determination of anions in potable water

Sample	*Amount added (mg/L)*	*Amount found (mg/L)*	*% Bias*	*Anion detected*
Potable water	0.26	0.24	−7.7	F^-
	0.78	0.54	−30.8	Cl^-
	0.36	0.30	−16.7	NO_2^-
	0.42	0.46	+9.5	NO_3^-
	0.63	0.63	0	Br^-
	0.69	0.70	+1.4	PO_4^{3-}
	2.85	1.12	−60.2	SO_4^{2-}

The anions with same retention time interfere e.g. acetate formate interfere and the technique is unsuitable for analysis of fluoride. A good separation is possible for other anions at pH 5–9. The instrument sensitivity is 0.1 mg/L with 100 μL of sample. A segmented continuous flow analysis method is also used. Table 14.2 illustrates typical separation of anions.

Table 14.2 Separation of anions from sample of water

Anions	*Order elution*	*Elution time (mts)*	*Concentration mg/L*	*Anion*	*Order elution*	*Elution time (mts)*	*Concentration (mg/L)*
Standard peak		–	–	PO_4^{3-}	5	5.7	40
NO_2^-	4	3.1	10	SO_4^{2-}	6	8.3	80
NO_3^-	2	5.3	20	Cl^-	1	2.3	40
F^-	7	0	0	Br^-	3	4.2	40

A flow injection analysis is novel method which facilitates the analysis of anions in several samples. It is wet analysis. Such method is used on industrial scale with continuous analysis of anions. A small volume of sample is needed with no waste of reagents. A typical ion chromatography includes sampling device, liquid transport-peristatic pump, and analytical cartridge where chemical reaction takes place while detector is used to quantify an analyte. In Table 14.2 the order of elution of various anions is as follows:

$$Cl^- > NO_3^- > Br^- > NO_2^- > PO_4^{3-} > SO_4^{2-}$$

The proposed method facilitates the separation of anions at lower concentration viz. from 10 mg/L for nitrate to 80 mg/L for sulphate anion in the solution with the elution time varying from 5.3 to 8.3 mts. respectively.

14.3 FLOW INJECTION ANALYSIS

This is an automated method wherein sample is injected in continuous flowing stream of solution. The sample disperses in stream and leads to the formation of symmetrical Gaussian type elution curve. The concentration gradient of analyte is detected by colour reaction which can be measured in flow colorimetry. In FIA the flow rate, sample volume, temperature, residence time are all important in ascertaining the concentration of the anion. Thus FIA is continuous flow method. Large number of the samples can be analysed by it. The several anions have been analysed by FIA technique. The typical anions which are studied are Cl^-, CN^-, NO_3^-, SO_4^{2-}, F^- and S^{2-} in water.

14.4 CAPILLARY ION ELECTROPHORESIS: Experiment 14.2: Use of Capillary Electrophoresis

It offers multiple analytes in single step as in ion chromatography. It is rapid (~5 mts for total analysis). It can detect all anions by providing anionic fingerprints. Neutral organic, cations do not interfere in such anion analysis. A buffered solution absorbing anion and electroosmotic flow modifier is used with 75 μm with internal diameter. In silica capillary tube at 5 kV potential anions move. The sample is kept at cathodic end, while anions are separated on account of their difference in the mobility, which in turn can be detected as they displace the charge for change of anion leading to decrease in UV absorbance in the analyte of anions. The zinc is compared to background electrolyte. The detector polarity is reversed to provide positive mV response to data system. Time of migration vs concentration profile is plotted on graph or time of migration in peak area as compared to standard sample. The strongly interfering anion is fluoride. 5 mg/L formic acid can mask fluoride. The instrumental detection limit of the concentration is 0.1 mg/L (1 hr. sensitivity) however solution with high ionic strength furnish wrong results.

Reagents: Total organics concentration should be less than 50 μg/L. The chromate solution (100 m M) obtained by dissolving 23.41 gm $Na_2Cr_2O_4$ $4H_2O$ in 50 ml water are made to one litre. It has stability for 365 days. Electroosmotic flow modifier 100 m M. Dissolve 3.36 gm tetradeyltrimethyl ammonium bromide (TTAB) in 50 ml. of water. Buffer solution is obtained by dissolving 20.73 g of 2 [N-cyclohexyl-amino] ethane sulphate (CHES) in 500 ml H_2O made up to one litre. Calcium gluconate-0.43 gm in 1000 ml water. NaOH (100 m M) 8 gm M litre of NaOH; chromate electrolyte solution. The standard solution of anions (1000 mg/L).

Procedure: First condition the capillary, with zero UV detector. Apply 14 μA current keep analysis time 5 mts. Estimate migration time of each anion. The order of migration is

$$Cl^- > Br^- > NO_2^- > SO_3^{2-} > NO_3^- > F^- > PO_4^{3-}$$

First chloride peak is reference peak. A typical electropherogram is shown in Figs. 14.2 and 14.3.

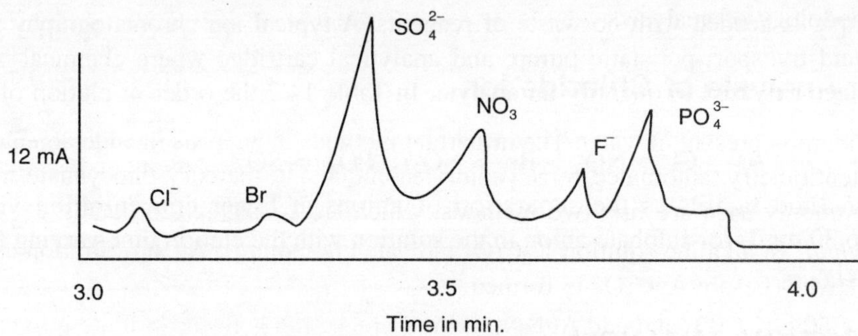

Fig. 14.2 Electrophoretic pattern

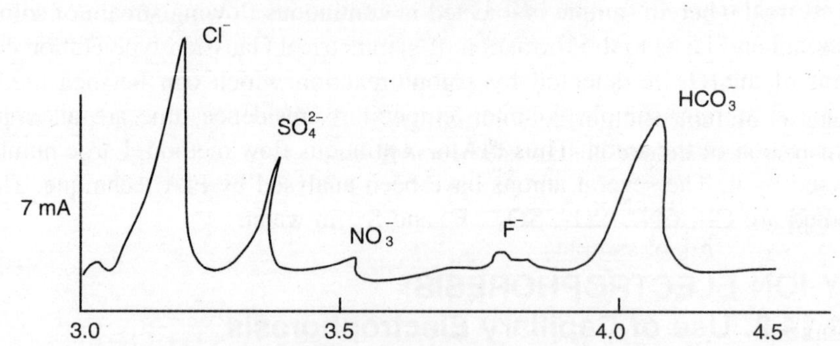

Fig. 14.3 Electrophoretic path for anions in water

Calculation: We use an expression C = A × F if C = analyte concentration. A = analyte from calibration curve, F = scale factor of dilution. The Table 14.3 gives comparison of capillary electrophoresis with analysis of anions by ion chromatogram.

The selective instrumental methods like ion chromatography, flow injection analysis and capillary anion electrophoresis for the characterisation of anions were considered. Each method has its own advantages as well as significant limitations. In subsequent sections specialised methods for the quantitative analysis of halide anions or anions from nitrogen oxy-anions of sulphur, and other common anions like carbonate-bicarbonate, cyanide are considered.

Table 14.3 Comparison of various methods for analysis of anions

Anion	*Source*	*Classical method*	*Ion chromatography*	*Capillary electrophoresis*
Cl^-	Potable water	5.5	5.1	5.0
		5.4	4.9	4.9
F^-	Potable water	1.2	0.9	0.9
		1.3	0.9	0.9

14.5 ANALYSIS OF HALIDES

This group of haloanions is composed of anions like fluoride, chloride, bromide and iodide. There is some kind of resemblance in their method of determination while in few places there is altogether sharp difference in the methodology of analysis.

Experiment 14.3: Analysis of Chloride Ion

This is one of the major anion present in water. The important methods of analysis include argentometry with silver nitrate, potentiometry, automated ferricyanide technique and mercury thiocyanate method. Out of these those commonly used are first two methods, which are considered in details.

Argentometric method: In alkaline solution K_2CrO_4 indicate end point of $AgNO_3$ titration as AgCl is precipitated quantitatively before $AgCrO_4$ is formed.

Reagents: Dissolve 50 g K_2CrO_4 and add $AgNO_3$, till red precipitate is formed, dilute it to one litre to get potassium chromate indicator solution. $AgNO_3$ (0.0141) dissolve 2.395 gm $AgNO_3$ in one litre (solution will have 1 ml = 500 μg Cl^-). NaCl (0.0141N) dissolve 824 gm NaCl in one litre water, $Al(OH)_3$ suspension: Dissolve: 125 gm Al_2SO_4 K_2SO_4. 24 H_2O in one litre; add 50 ml NH_4OH (con.) present ppt. NaOH (1N), H_2SO_4 (1N), water (30%), phenolphthalein as an indicator.

Procedure (A): Take an aliquot of chloride solution (pH ~ 7–10) and titrate it with $AgNO_3$ with K_2CrO_4 as indicator till persistent colour is stabilised for $AgCrO_4$.

$$\textit{Calculations:}\ \text{mg Cl/L} = \frac{(A-B)\times N\times 35.45}{\text{ml. of sample}}$$

where

A = ml of sample
B = ml of titrant
N = normality of $AgNO_3$

Procedure (B): The procedure A was for $AgNO_3$ method. Procedure B is for the mercuric nitrate ($Hg(NO_3)_2$) solution.

Reagents: NaCl (0.014 N), HNO_3 (1N), NaOH (0.1 N). Solution containing Cl^- when titrated with $Hg(NO_3)_2$ forms $HgCl_2$ at pH 3.0 where diphenylcarbazone is used as an indicator to note end point with formation of purple complex however Br^- and I^- interfere in titration. One should follow the following procedure. Take a sample solution, add 1 ml indicator which give green colour (pH ~ 2). Titrate it with 0.0141 N $Hg(NO_3)_2$ till purple colour is attained. If chloride concentration is high, some modification is required viz. Take sample add 0.5 ml indicator. Add 0.1 N HNO_3 till yellow and titrate it with $Hg(NO_3)_2$ till pink colour is retained. Indicator mixture consist of HNO_3 (4 ml) + 300 ml xylenol orange + 95% ethanol.

Calculations: $$\text{mg Cl}^-/\text{L} = \frac{(A-B)\times N\times 35.5}{\text{ml of sample}}$$

if A = ml of titration sample, B = ml of titrations for blank

N = normality of $Hg(NO_3)_2$

The potentiometry is used when coloured sample solution is present where in it is difficult to locate the end point. The end point is one where great change in potential is noted after addition of small volume of $AgNO_3$. Use electronic voltmeter with SCE or glass and Ag-Ag electrodes.

Potentiometric method. Reagents: 0.014 N NaCl, HNO_3, $AgNO_3$ (0.014 N), H_2SO_4 (1:1), H_2O_2(30%), NaOH (1N).

Procedure: Put NaCl in flask, add 2 ml HNO_3, immerse electrodes, add $AgNO_3$ as titrant till end point is reached (Note great change in reading).

Calculations: $$\text{mg of Cl}^-/\text{L} = \frac{(A-B)\times N\times 35.5}{\text{ml of sample}}$$

The symbols have same notations as in previous calculations. The automated ferricyanide method is used for industrial scale operation. A flow injection analysis with mercury thiocyanate can be also used. The catalytic reduction method and voltammetry are other options available for the analysis of iodide but these are less frequently used in routine work.

14.6 ANALYSIS OF FLUORIDE:
Experiment 14.4: Analysis of Fluoride in Water

There are several method for the analysis of fluoride however most acceptable methods are: ion selective electrode method, Colorimetry method, SPADNS complexon method or FIA can be also used.

Reagents: Standard fluoride solution, SPADNS (dissolve 958 mg of sodium 2-(parasulphophenylazo) 1–8 dihydroxy 3–6 naphthalene disulphonate) in 500 ml water. $ZrOCl_2$ $2H_2O$ (dissolve 133 mg salt in 25 ml water + 350 ml con HCl and dilute to 500 ml). Acid zirconyl SPADNS. Mix equal volumes of SPADNS with zirconyl solutions; sodium arsenite solution (dissolve 5 g Na_3AsO_3 in one litre water).

Procedure: Take an aliquot of solution containing fluoride ion. Add 5 ml of SPADNS and Zirconyl acid reagent or 10 ml of mixed acid Zirconyl-SPADNS reagent. After development of colour measure the absorbance at 570 nm.

Calculations: $\text{mg F}^-/\text{L} = \frac{A}{\text{ml of sample}}\times\frac{B}{C}$ if A = µg F^- from calibration curve, B = final volume of diluted solution, C = Volume of diluted sample for colour development.

When prepared 0 mgF^-/L standard is used calculation as follows:

$$\text{mg F}^-/\text{L} = \frac{A_0 - A_x}{A_0 - A_t} = \frac{A_2 - A_x}{A_2 - A_t}$$

A_0 = absorbance of prepared 0 mg F^-/L.

A_t = absorbance of 1 mg F^-/L standard

A_x = absorbance of prepared sample.

The complexon method uses Alizarine instead of SPADNS as the chromogenic reagent for colorimetry.

14.7 ANALYSIS OF BROMIDE:

Experiment 14.5: Analysis of Bromide in Water

All instrumental technique viz. IC, CE, FIA are applicable for its analysis however most preferred method is colorimetry with phenol red as chromogenic ligand. It is acting in the presence of chloromine T at pH 4.5 Cl^-, HCO_3^- in high concentration interfere the sensitivity is 0.1 mg Br^-/L.

Reagents: Acetate buffer pH 4.6 (dissolve 90 gm NaCl + 68 gm sodium acetate + 30 ml glacial acetic acid–all made to one litre with water. Phenol red indicator (dissolve 21 mg of indicator in 100 ml water). Chloramine T solution (dissolve 500 mg in 100 ml water) $Na_2S_2O_3$ (2M) (dissolve 49.6 gm in 100 ml water) stock solution of bromide (dissolve 744.6 mg KBr in one litre so 1 ml = 5μg Br^-).

Procedure: To an aliquot of solution containing bromide, add 2 ml buffer, 2ml of indicator and 0.5 ml chloramine T in total volume of 50 ml. Keep it aside for 20 mts then dechlorinate it by adding 0.5 ml $Na_2S_2O_3$. Measure the absorbance of resulting solution at 590 nm against reagent blank. From calibration curve prepared earlier with known concentration of bromide, findout the concentration of unknown sample.

Calculation: $\text{mgBr}^-/\text{L} = \text{mgBr}^-/\text{L (from Beer's law plot)} \times \text{dilution factor if L = litre.}$

Flow injection analysis is versatile method involving use of phenol red only.

14.8 ANALYSIS OF IODIDE:

Experiment 14.6: Analysis of Iodide in Water

For iodide colorimetry as well as leucocrystal violet amperiometry are best methods for analysis from sample of water. The former method is most popularly used for the analysis of iodide (50–6000 μg/L of iodide). The voltammetric method is also good. Analysis of iodide solution (dissolve 1.3081 gm kI in one litre water, 1 ml = 1 mg/L). Buffer solution (pH 3.8) is prepared by dissolving 192.2 gm citric acid in one litre NH_4OH (2N). (Add 130 ml concentrated NH_4OH in 700 ml water and make it to 10000 ml). Buffer solution (add 350 ml 2N NH_4OH in 670 ml citric acid). Also add 80 gm of ($NH_4H_2PO_4$ to above solution. Leucocrystal violet indicator, Add 1.5 gm of chormogenic ligand in 200 ml H_2O containing 3.2 ml conc. H_2SO_4. To 800 ml., add 2.5 gm $HgCl_2$ + lecuocrystal violet solution and adjust pH = 1.5, $KHSO_5$ is formed from 42.8% $KHSO_5$ ($KHSO_4$ + K_2SO_3), dissolve 1.5 gm in one litre water of $Na_2S_2O_3$.

Procedure: Take an aliquot of solution containing iodide. Add 1 ml citrate buffer and 0.5 ml $KHSO_5$ solution. Then add 1 ml chromogenic indicator (4 4′ 4″ methylidynetris (N–N′ dimethylamine) and dilute it to 100 ml. Record absorbance at 592 nm against reagent blank. From standard plot find out iodide in unknown sample use $Na_2S_2O_3$ (5 ml) if turbidity is present in the solution.

14.9 ANALYSIS OF AMMONICAL NITROGEN:

Experiment 14.7: Analysis of Ammonical Nitrogen

Apart from nitrogen, we consider ammonia (NH_3), Nitrite (NO_2), Nitrate (NO_3) and ammonical organic compounds. Ammonia is important environmental pollutant. Most popular method for analysis is titrimetric method involving Nesslersation. It is used for mercury which is another pollutant. Phenate is another good colorimetric method. Acetamide, urea, cyanides, interfere in both these methods. The method involves preliminary distillation step at pH ~ 9.5 to destroy cyanates and organic nitrogen compounds. Ion selective electrode is equally efficient for the analysis. Ion exchange can be also used to remove initial impurities including organic matter. Let us first consider here titrimetric method.

Reagents: Mixed indicator (dissolve 200 mg methyl red in 100% alcohol). Also dissolve 100 mg of methylene blue in 50 ml of ethanol. Boric acid solution (dissolve 20 gm H_3BO_3 in 1 litre water) H_2SO_4 (0.02 N). Prepare as usual from concentrated acid.

Procedure: Take an aliquot of 500 ml water, add 20 ml buffer, adjust pH ~ 9.5. (Remove chlorine by dechlorinating agent). Adjust pH = 7.0 add buffer (25 ml) to adjust solution to pH 9.5. Titrate it with standard H_2SO_4 (0.02 N) acid and calculate the concentration of ammonia in sample.

Calculations: $\text{mg } NH_3/L = \dfrac{(A - B) \times 280}{\text{ml of sample}}$ if A = volume of H_2SO_4 titre for sample solution, B = volume of H_2SO_4 titre for blank solution.

Other rapid methods for analysis are use of ion selective electrodes or one utilising ZSE with unknown addition. In phenate method, the blue coloured complex is formed with indophenol, hypochloride and phenol catalysed by sodium nitropruside, which can be colorimetrically measured.

Phenate method, Reagents: Phenol solution (mix 1 ml of phenol with 95% ethanol in 100 ml). Sodium nitropruside 0.5% (dissolve 50.5 gm reagent in 100 ml H_2O). Alkaline citrate (dissolve 200 gm trisodium citrate and 10 gm NaOH in total one litre of water. NaOCl (5%), oxidising solution (100 ml alkaline citrate with 25 ml NaOCl). The stock solution of ammonia is prepared as usual.

Procedure: To an aliquot of solution containing NH_3 add 1 ml phenol solution, 1 ml of sodium nitropruside and 2.5 ml of oxidising solution. Keep it for one hour at 22–27°C. Measure the absorbance of coloured sample solution at 640 nm against reagent blank.

Calculation: Use standard calibration curve to ascertain concentration of NH_3 in unknown sample solution. The phenate method can be automated. Flow injection analysis (FIA) is equally efficient technique involving the use of nitropruside, hypochlorite and phenolate ion.

14.10 ANALYSIS OF NITRITE (NO_2):

Experiment 14.8: Analysis of Nitrite in Water

Photometry is most suitable method for analysis of nitrite-nitrogen. Ion chromatograph, FIA are equally powerful methods of analysis. A red complex is formed at pH 2.0 by diazotised sulphanamide (*i.e.*, NEDA) N-(1 naphthyl) ethylene diamine dihydrochloride. Free chlorine or nitrogen trichloride interfere. Metals like Sb, Au, Bi, Fe, Pb and Hg promote precipitation in this media and cause interference.

Reagents: Nitrite free water, (distilled water can be used for dilution) colour reagent viz. NEDA to 800 ml water + 100 ml 85% H_3PO_4 + 10 gm sulphanilamide + 1 gm of NEDA.

Then make it to one litre with water. $Na_2C_2O_4$ (0.025 ml) dissolve 3.350 gm $Na_2C_2O_4$ in one litre water) $FeSO_4(NH_4)_2SO_4.6H_2O$–0.05 M (dissolve 19.60 salt in 20 ml conc. H_2SO_4) and make it to one litre. The stock solution of nitrite is made from $NaNO_2$ (Avoid moisture to prevent oxidation).

Procedure: Take an aliquot 50 ml of 0.01 M $KMnO_4$ + 5 ml conc. H_2SO_4 + 50 ml stock solution of nitrite. Shake and warm at 70°C. Discharge $KMnO_4$ colour by adding 10 ml 0.025 M $Na_2C_2O_4$. Titrate excess $Na_2C_2O_4$ with 0.01 M $KMnO_4$ to pink coloration.

Calculation: $NO_2^- \text{-N mg/L (A)} = \dfrac{(B \times C) - (D \times E) \times 7}{F}$

A = $mgNO_2^-$-N in stock, B = total ml of $KMnO_4$, C = normality of $KMnO_4$, D = total ml of reductant ($Na_2C_2O_4$), E = Normality of $Na_2C_2O_4$. F = ml of $NaNO_2$.

The Colorimetric method: Take nitrite solution (0.5 μg/ml). 0.01 M $KMnO_4$. To an aliquot of nitrile solution (pH 5–9); add 2 ml chromogenic reagent and mix. Allow it to settle and measure the absorbance of coloured solution at 543 nm. Adjust light path between 10–100 mm for nitrile concentration of 25–2 μg/L (if it is too low in concentration).

14.11 ANALYSIS OF NITRATE NO_3:

Experiment 14.9: Analysis of Nitrate in Water

The analysis is somewhat complicated as UV radiation at 220 nm is very useful for preliminary screening IC, CE, FIA are ideal methods of analysis. UV visible spectrophotometry is useful for screening when NO_3^- is (11 mg/L) measured at wavelength of 220 nm or 275 nm (provided if any interference is encountered). One can also use nitrate selective electrode. In the cadmium reduction method NO_3^- is reduced to NO_2^- in the presence of cadmium metal. NO_2^- resulted is diazotised and measured spectrophotometrically as described earlier with NEDA (0.01–1 mg/NO_3^- per litre). It can be measured as usual at 543 nm. We need reduction column like ion exchange column. The copper-cadmium granules in clean condition must be used.

Reagents: NH_4Cl EDTA (dissolve 13 gm NH_4Cl + 1.7 gm EDTA in 900 ml H_2O, HCl (6N), 2% $CuSO_4$ solution. The stock solution NO_3^- and NO_2^-. Put Cu-Cd granules in column, wash it with NH_4Cl-EDTA solution. The turbid solution should be rejected. Adjust pH to 7–9. A sample solution contains NH_4Cl. Then EDTA is passed through column at rate of 7 ml/mt. Take effluent add 2 ml colour forming reagent (NEDA) and after 2 hrs. standing measure absorbance at 543 nm against reagent blank. For unknown solution follow same procedure and find out NO_3^- concentration in unknown sample. This method can be used for automated cadmium reduction method. One can also use automated hydrazine reduction method. In FIA also one can use cadmium reduction method.

14.12 ANALYSIS OF ANIONS OF SULPHUR:

Experiment 14.10: Analysis of Sulphide in Water

There are in all three anions namely sulphide, sulphite and sulphate. Out of these first two have large number of methods including automation for their analysis however, sulphate still enjoys the classical gravimetric method of analysis as insoluble sulphate of barium or its nephelometric determination.

Sulphide: Threshold odour limit for H_2S is 0.25 μg/L. Three forms of sulphide are present.

There are total sulphide (e.g. H_2S), dissolved sulphides and unionised hydrogen sulphate. The reliable qualitative test is Sb_2S_3 to give yellow colour at 0.5 mg/L level. $Pb(CH_3COO)_2$ paper gives black colouration and is reliable test for sulphide. Iodometric titration and colorimetry with methylene blue complexation are two qualitative methods for the analysis. The standard solutions are prepared from $Na_2S \cdot 9H_2O$. Weight of sodium sulphide × 0.133 – gives amount of sulphide (S^2). A colorimetric method with methylene blue as chormogenic is mostly preferred.

Reagents: Amine-sulphuric acid solution (dissolve 27 gm N-N dimethyl-p-phenylene diamine in 50 ml. conc. H_2SO_4) + 20 ml water. Make up the solution to 100 ml mixture of reagent (dilute 25 ml amine-sulphuric acid solution with 975 ml (1:1) H_2SO_4. $FeCl_3$ (dissolve 100 gm in 40 ml water) H_2SO_4 (1:1). Diammonium hydrogen phosphate $(NH_4)_2$ HPO_4 in (800 ml by dissolving 400 gm). Methylene blue (I) solution (dissolve 1 gm in 1000 ml water). Standard solution from $Na_2S.9H_2O$ (1 ml = 8 mg/L). Methylene blue (II) solution. Dilute 100 ml ten times.

Procedure: Take an aliquot of solution containing sulphide, add 0.5 ml amine sulphuric acid and 0.13 ml $FeCl_3$. To other tube add 0.5 ml (1:1) H_2SO_4. 0.15 ml $FeCl_3$. Add 1.6 ml of $(NH_4)_2HPO_4$. The colour is developed in first tube. Then use it for colour comparison. Measure the absorbances of coloured solutions at 664 nm. Find out concentration of unknown sample by utilising standard calibration curve. Another reliable technique is the iodometric method. We use following procedure.

Reagents: HCl (6 N), Iodine solution (0.025 N) (dissolve 20–25 gm KI + 3.2 gm Iodine and dilute to one litre). You may standardise it against 0.025 N $Na_2S_2O_3$.

Procedure: Take known volume of standard stock solution (20 ml), add 2M HCl (6M). Back titrate with $Na_2S_2O_3$ with starch as an indicator with disappearance of blue colour at the end point of the titration.

$$\textit{Calculation:}\ \text{mg } S^{2-}/L = \frac{[(A \times B) - (C \times D)] \times 16{,}000}{\text{ml of sample}}$$

A = ml of iodine solution, B = normality of iodine solution, C = ml of $Na_2S_2O_3$, D = normality of $Na_2S_2O_3$. As small as 0.1 mg/L can be estimated from the sample solution. The ion selective electrode method as well as FIA with methylene blue as the chromogenic ligand methods are used on large scale.

14.13 ANALYSIS OF SULPHITE (SO_3^{2-}):

Experiment 14.11: Analysis of Sulphite

The iodometric titration method (limit 2 mg/L) and spectrophotometry with phenanthroline are two best methods for such analysis.

The Iodometry Reagents: H_2SO_4 (1:1) KI (0.002083 M) (dissolve 0.4458 gm KIO_3 + 4.35 gm KI + 3.10 mg $NaHCO_3$ in one litre water (1 ml ≡ 500 μg/LSO_3^{2-}). Sulphonic acid (NH_2SO_3H), EDTA (dissolve 2.5 gm in 100ml H_2O) use starch as indicator.

Procedure: Take a known volume of standard solution containing SO_3^{2-}. Then add 1 ml H_2SO_4 and 0.1 gm N H_2SO_3H crystal. Then add 50–100 ml EDTA solution. Add 1 ml starch then titrate against KI-KIO_3 mixture till blue colour disappears. Repeat the same procedure for the reagent blank analysis.

$$\textit{Calculations:}\ \text{mg } SO_3^{2-}/L = \frac{(A - B) \times M \times 6 \times 40{,}000}{\text{ml. of sample}}$$ where A = ml of titrant for sample,

B = ml. of titrant for blank, M = Molarity of KI-KIO_3.

Experiment 14.12: Spectrophotometry with Phenanthroline Method for SO_3 Analysis

Reagents: o-phenanthroline (0.03 M) (dissolve 5.95 gm of ligand in 100 ml ethanol) and dilute it to one litre. $(NH_4)_2SO_4\ Fe_2(SO_4)_3 \cdot 12H_2O$ (0.01 M) (dissolve 4.82 gm in one litre water containing 1 ml of H_2SO_4 con). Ammonium bifluoride (5%) (dissolve 25 gm NH_4HF_2 in 300 ml water. Potassium

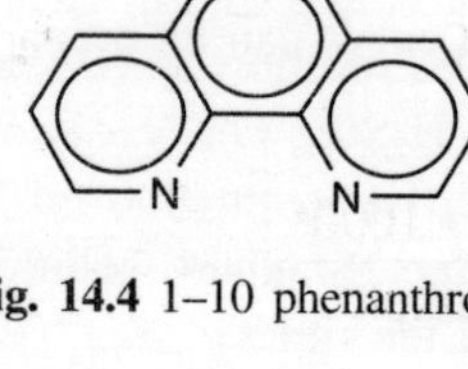

Fig. 14.4 1–10 phenanthroline

tetrachloromercurate (TCM), K_2HgCl_4 (0.04 M) dissolve 10.86 gm $HgCl_2$ + 5.96 gm KCl + 0.060 gm EDTA and dilute to litre. Adjust pH to 5.2. A saturated KI-KIO_3 solution (0.0125 M), HCl (1:1). Octyl alcohol, sulphamic acid 10% (dissolve 10 gm of acid in 100 ml H_2O), EDTA (2.3 g in 100 ml).

Procedure: An absorbing solution is obtained from 5 ml of chromogenic ligand + 0.5 ml of ferric ammonium sulphate solution + 25 ml water + 5 drops octyl alcohol + 1 ml of sulphamic acid—Dilute to 100 ml. Add 10 ml (1:1) H_2SO_4 to generate gas. After one hour, nitrogen purging mechanism, add 1 ml NH_4HF_2—dil. to 50 ml. Measure the absorbance of coloured complex at 510 nm (using 50 mm cell) against reagent blank treated in similar fashion. Construct Beer's law plot with standards.

Calculation: $$\text{mg } SO_3^{2-}/L = \frac{\mu g\ SO_3^{2-} \text{ from Beer's Law Plot}}{\text{ml. of sample}}$$

14.14 ANALYSIS OF SULPHATE (SO_4^{2-})

Amongst instruments ion chromatograph and capillary electrophoresis are most suitable method for the analysis of sulphate, however gravimetric method is most simple and reliable. The nephelometry can be also used for small concentration of SO_4^{2-} ion in the solution.

Experiment 14.13: Analysis of Sulphate in Water (Gravimetry)

The gravimetric method as $BaSO_4$ precipitate in HCl: Alkali metals interfere.

Reagents: Methyl red indicator (dissolve 100 mg in 100 ml water); HCl (1:1) $BaCl_2$ (dissolve 100 gm in 1000 ml H_2O) $AgNO_3$-HNO_3 mixture (dissolve 8.5 gm $AgNO_3$ and 0.5 ml HNO_3 (conc.) in total volume of 500 ml.

Procedure: Take an aliquot of solution containing sulphate anion. Make it to 150–250 ml. Adjust the pH to 4.5–5.0. Use indicator to check pH. Add 2 ml HCl and heat to boiling, slowly add $BaCl_2$ till precipitation is complete. Digest it at 80°–90°C overnight. Filter on ashless filter paper wash precipitate free of chloride ion. Transfer it in platinum crucible and ignite at 800°C for one hour. Then cool it and weigh.

Calculations: $$\text{mg } SO_4^{2-}/L = \frac{\text{mg } BaSO_4 \times 411.6}{\text{ml. of sample}}.$$

A method involving gravimetry with drying of residue is also fissible. The automated methylthymol blue method as well as FIA with same ligand is useful on industrial scale. Only nephelometric method is considered below for low concentration of sulphate is present.

Reagents: Buffer solution (A) (dissolve 30 gm $MgCl_2 \cdot 6H_2O$ + 5 gm CH_3COONa + 1 gm KNO_3 + 20 ml. CH_3COOH acid and make up to one liter with water.

Buffer solution (B) dissolve 30 g $MgCl_2 \cdot 6H_2O$ + 5gm CH_3COONa + 1 gm KNO_3 + 0.1 gm. Na_2SO_4 + 20 ml CH_3COOH all make to one liter with distilled water.

$BaCl_2$ and standard solution are prepared as described in the gravimetric method.

Procedure: Take sample solution containing sulphate anion. Add 20 ml of buffer solution + spoonful $BaCl_2$ crystals, stir for one minute. After stirring put solution in cuvettes and measure the absorbance or transmittance at 420 nm. The calibration curve can be prepared utilising standard solution. If necessary give correction to blank solution.

Calculations: $\text{mg } SO_4^{2-}/L = \dfrac{\text{mg } SO_4^{2-} \times 1000}{\text{ml. of sample}}$ (Use buffer solution A and B depending on level of SO_4^{2-} present in sample)

14.15 ANALYSIS OF CARBONATES AND BICARBONATES: Experiment 14.14: Analysis of CO_3^{2-} and HCO_3^{2-} in Water

The nomographic and titrimetric methods are used to estimate CO_2 from water. Titrimetric method involves normal titration with phenolphthalein or potentiometry analysis. Nomography is sensitive method and if pH is wrongly adjusted by 0.11 adds to error of 2–4 mg/L of CO_2 at pH of 7–7.3. However, use titrimetric method, CO_2 reacts with NaOH to form $NaHCO_3$ but Al, Cr, Cu, Fe interfere in analysis.

Reagents: CO_2 free water of pH ~ 6.0 and conductivity 22 μmho/cm. Potassium hydrogen phthalate solution 0.05 N (dissolve 15–20 gm salt in water. Use 10 gm and dilute to one litre. NaOH (0.1N) (0.02 N), H_2O_2 (30%). Bromophenol blue as indicator (pH ~ 3.7) (dissolve 100 mg in 100 ml H_2O) Meta-cresol, purple (pH ~ 8.3) (dissolve 100 mg is 100 ml H_2O) Phenolphthalein (pH ~ 8.3), $Na_2S_2O_3$ (dissolve 25 gm $Na_2S_2O_3 \cdot 5H_2O$–to one litre gives 1M).

Procedure: In presence of interefering ions take sample solution and add 5 ml 0.02N H_2SO_4 + 5 drops H_2O_2–cool to room temperature and titrate it with NaOH. Otherwise take sample; add 0.05 ml of 0.1 N $Na_2S_2O_3$ + 2–5 drops indicator and titrate. One can obtain end point by potentiometry in conventional manner.

Calculation: Acidity (as $CaCO_3$/L) $= \dfrac{(A \times B) - (C \times D) \times 50{,}000}{\text{ml of sample}}$

where A = ml of NaOH, B = normality of NaOH, C = ml of H_2SO_4, D = normality of H_2SO_4. Also $\text{mg } CO_2/L = \dfrac{A \times N \times 44{,}000}{\text{ml. of sample}}$ if A = ml of titrant, B = normality of NaOH. We can calculate various forms of alkalinity by calculation as shown below:

(1) Total CO_2 = mg total CO_2/L = A + 0.44 (2B + C), A = mg free CO_2, B = bicarbonate alkalinity, C = carbonate alkalinity

(2) Free $CO_2 = \text{mg } CO_2/L = 2 \times B \times 10^{(pH=10)}$

(3) Hydroxide alkalinity OH as mg $CaCO_3$/L = $5 \times 10^{(pH-10)}$

(4) Carbonate alkalinity = CO_3^{-2} as mg $CaCO_3$/L = $0.94 \times B \times 10^{(pH-10)}$

(5) Bicarbonate alkalinity HCO_3 as mg $CaCO_3$/L $= \dfrac{T - 5 \times 10^{(pH-10)}}{1 + 0.94 \times 10^{(pH-10)}}$

If T = total alkalinity mg $CaCO_3$/L.

14.16 ANALYSIS OF CYANIDE, CYANATES AND THIOCYANATES: Experiment 14.15: Analysis of Cyanide and Thiocyanate in Water

The cyanide may be present in water as simple cyanide or complex coordination compound, specially the alkali metal cyanide. The colorimetry and titrimetry are most dependable methods for their analysis. The total cyanide is analysed by titration preferably after distillation step for its analysis in waste water.

(a) Reagent: NaOH (dissolve 40 gm NaOH in one litre) $MgCl_2 6H_2O$ (dissolve 510 gm in one litre; H_2SO_4 (1:1) $PbCO_3$ powder, sulphanic acid (NH_2SO_3H).

Procedure: Take sample solution, add 10 ml NaOH, add 2 gm sulphamic acid, then add 50 ml (1:1) H_2SO_4 + 20 ml of $MgCl_2$. We get precipitate but it subsequently gets redissolved. Reflux for one hour. Carry out usual titration as described below.

Volumetric method: $AgNO_3$ is used as titrant to complex CN^- as $Ag(CN)_2^-$. Then Ag is detected by p-dimethylaminobezalrhodamine, 1mg/L sensitivity.

Reagents: Indicator solution: (dissolve 20 mg rhodamine in 100 ml acetone), $AgNO_3$ (dissolve 3.27 gm $AgNO_3$ in one litre, standardise again. NaCl with K_2CrO_4 as an indicator dilute 500 ml solution of $AgNO_3$, NaOH (dissolve 1.6 gm in 1000 ml).

Procedure: Take an aliquot of solution containing cyanide. Dilute with NaOH to 100 ml titrate it with standard $AgNO_3$ till colour changes from yellow to salmon blue. Blank is similarly titrated.

Calculations: $$\text{mg } CN^-/L = \frac{(A-B)\times 1000}{\text{ml. original soln}} \times \frac{20}{\text{ml. portion used}}$$

if A = ml of $AgNO_3$ stand for sample
B = ml $AgNO_3$ for blank

(b) Colorimetric analysis: One can use barbaturic acid-pyridine chloramine aschromogenia ligands.

Reagents: Chloromine T (dissolve 1 gm 100 ml water). Stock standard cyanide solution is obtained from (KCN + NaOH) (1.6 gm NaOH + 2.5 gm KCN diluted to 1000 ml. pyridine barbaturic acid reagent. Dissolve 15 gm barbaturic acid in 250 ml Water + 75 ml pyridine + 15 ml HCl (con). Dilute to known volume.

Acetate Buffer (410 gm CH_3COONa + Add CH_3COOH in glacial to pH 4.5).

Procedure: Take an aliquot containing cyanide anion, dilute it with NaOH (40 ml). It develops colour. Carry out same procedure for the blank also. Add 1 ml acetate buffer and 2 ml chloramine T. Then add 5 ml pyridine-barbaturic acid. Make it to the known volume. Measure the absorbance of coloured complex at 578 nm. Prepare a calibration curve and compute the amount of cyanide in the unknown solution from the standard calibration plot.

14.17 ANALYSIS OF CYANATES (CNO):

Experiment 14.16: Analysis of Cyanate in Water

It is unstable at low pH. It is hydrolysed to NH_3. Then amount of NH_3 is determined quantitatively in directly to ascertain (CNO) concentration.

Reagents: Dissolve 3.819 gm NH_4Cl in 1000 ml H_2O. Standard NH_4Cl solution NaOH (10 N), H_2SO_4 (1:1).

Procedure: Take a stock solution containing cyanate. Add 0.5 ml H_2SO_4 (1:11) to pH 2.0–2.5. Warm and cool. Add 1 ml 10 N NaOH (pH > 11.0). Record the potential after completion of the reaction. Calculations are shown below.

Calculations: mg NH_3–N from CNO = (A–B), A = mg NH_3-N/L in acidic hot sample. B = mg NH_3-N/L in untreated solution.

14.18 ANALYSIS OF THIOCYANATE (SCN^-):

Experiment 14.18: Analysis of Thiocyanate

It is colorimetrically analysed but Cr (VI), Fe (III), interfere.

Reagents: $Fe(NO_3)_3 8H_2O$ (404 gm in 800 ml H_2O + 80 ml HNO_3 con. — dil. to 1000 ml).

HNO_3 (0.1N). (Mix 6.4 ml con. HNO_3 to 800 ml and dilute to 1000 ml). The stock solution of SCN^- is obtained from KSCN (1.673 gm in 1000 ml H_2O), NaOH (4 gm/800 ml).

Fill column with ion exchange resin Ambertise XAD-7. Pour solution on column, add CH_3OH three times volume of 0.1 N NaOH. Take a solution containing thiocyanate. Adjust pH to 2.0. Measure the absorbance at 460 nm against the reagent blank prepared by similar procedure.

Calculation: It is calculated from following expression:

mg $SCN^-/L = (ma_1 + b) \times$ dilution factor where if a_1 = absorbance, m = slope of plot. The idea is to develop permanent colour with Fe^{3+} with free SCN^-.

In conclusion we can see that most commonly used method used for the analysis of anions from water are based upon spectrophotometric analysis involving the complexation reaction with suitable chromogenic reagent. With standard solutions Beer's law plot is constructed and from this calibration curve, the concentration of the unknown sample is evaluated. Apart from spectrophotometry, flow injection analysis was an extension of spectrophotometric analysis. The capillary electrophoresis as well as ion chromatography was less frequently used. However, there was growing preference for ion selective electrodes for rapid analysis in the industry.

ADDITIONAL READINGS

14.1 APHA, *Standard Methods for the Examination of Water and Waste Water*, 21st Ed. (2005).

14.2 I. Weiss, *Handbook of Ion Chromatography*, Johnson Dionex Corp, (1980).

14.3 W.A. Coakley, *Handbook of Automated Analysis*, Marcel & Dekker Publishers (1978).

14.4 P. Jandik and G. Bonn, *Capillary Electrophoresis of Small Molecules and Ions,* VCH Publishers (1993).

14.5 D.F. Boltz and J.A. Howell, Ed., *Colorimetric Determination of Nonmetals*, 2nd Ed., Wiley-Interscience (1978).

14.6 I.M. Kolthoff and N.H. Furman, *Potentiometric Titrations,* 2nd Ed., John Wiley & Co. (1931).

14.7 H.F. Walton, *Principles and Methods of Chemical Analysis*, Prentice-Hall (1964).

14.8 N.H. Furman Ed., *Standard Methods of Chemical Analysis*, 6th Ed., D. Van Nostrand & Co. (1962).

14.9 I.M. Kolthoff, E.J. Meehan, E.B. Sandell and S.Brucknestein, *Quantitative Chemical Analysis*, 4th Ed., MacMillan (1969).

14.10 W.F. Hillbrand, Lundell and Bright, *Applied Inorganic Analysis*, 2nd Ed., John Wiley (1953).

CHAPTER 15

Analysis of Organic Compounds in Water from Its Bulk Properties

15.1 INTRODUCTION

In comparison with metal pollutants or anions which are mostly inorganic in nature there are number of organic compounds present in the water. It is always possible to analyse every organic compound from the sample of water. However, it helps even if the group or kind of organics which are present in water are identified. The purpose of characterisation of organic compounds or some bulk properties are very useful for identification. Such bulk characteristics include dissolved oxygen (DO), biochemical or biological oxygen demand (BOD) or chemical oxygen demand (COD). The last method is very useful specially when organic compounds in water are non-degradable. Apart from these DO, BOD and COD one can also take the resort of the other properties such as total organic carbon (TOC), evaluation of ultimate BOD for 5-days, or respirometric BOD analysis which provides a valuable information regarding the nature of the organic compounds present in water. The analysis of TOC at high temperature or by persulphate oxidation technique provides a valuable information.

It is always beneficial to look in for the presence of the humic substances, oil and greases, phenols tannins and lignin in polluted water. Many components are UV absorbing which in turn mars the quality of potable water. In spite of the advancement in science of instrumentation, no reliable instrumental method has been in existence for analysis of DO and in fact Winkler's classical volumetric method or Alsterberg modified technique provides most useful and reliable method of the quantitative analysis of DO. In spite of the discovery of membrane electrodes and their use in DO analysis by potentiometry it had never provided a valuable and reproducible results for the presences of DO in water. Even common BOD_3 of three days was not preferred over ultimate BOD_5 of five days and also the respirometric BOD. For COD conventional redox titration was most reliable. The analysis of total organic carbon led to classical C, H, N, S analysis of the compounds. In fact it has been observed invariably that the dissolved oxygen (DO) in water gives fairly good idea of presence of organics in water. This is further confirmed by biochemical oxygen demand (BOD) and one can predict the existence of non-degradable organics like detergent, high molecular mass amines from the knowledge of chemical oxygen demand (COD). In spite of greater strides in instrumental analysis till date Winkler's method with iodometric titrations and G. Alsterberg's modification, when nitrites are present in water remains the most dependable volumetric method for the analysis of DO from sample of water. In fact many efforts were made to determine DO by ion selective electrodes but method is not that selective.

15.2 WINKLER'S IODOMETRIC TITRATION:
Experiment 15.1: Analysis of Dissolved Oxygen in Water

A divalent manganese gets easily oxidised to the hydroxide to higher valency. Further in the presence of iodine ion in acid media oxidises manganese (III), reverts to divalent manganese (II) with liberation of iodine equivalent to DO. Iodine is then titrated with standard $Na_2S_2O_3$. Alternatively liberated iodine can be analysed by spectrophotometer. The oxidising agents so also reducing agents interfere in such analysis. An azide modification by Alsterberg is valuable contribution in the litreature on DO analysis. It eliminates NO_2^- interferences. $KMnO_4$ can be used if Fe (II) is present. So also we can use H_3PO_4 for acidification. The sampling of water must be done with all care. The preservation of samples is also of importance. $MnSO_4$ is used to preserve it or H_2SO_4 acid if iodine demand is large. NaN_3 is used to eliminate NO_2^- from water provided its level is > 50 µg. The details of procedure are now considered in this experiment.

Reagents: $MnSO_4$ $2H_2O$ (dissolve 400 gm of salt in 1000 ml of distilled water). Alkali azide reagent (dissolve 500 gm NaOH (0.700 gm KOH) + 1.35 gm NaI in one litre water. Add 10 gm NaN_3 in 40 ml solution. The supersaturated samples dissolve in (10 gm NaN_3 in 500 ml) + 480 gm NaOH + 750 gm NaI + H_2SO_4 (con.). Starch iodide as indicator is used (dissolve 2 gm starch + 0.2 gm of salicylic acid in 100 ml water), $Na_2S_2O_3$ (dissolve 6.05 salt + 1.5 ml of 6 N NaOH + 0.1 ml solid NaOH and make it to one litre; standard potassium biiodate 0.002 M (dissolve 821.4 mg $KH(IO_3)_2$ in 1000 ml water. The standard solution is obtained from KI (2 gm/100) + 1 ml H_2SO_4 (6M) + 20 ml of iodate solution. Dilute it to 200 ml and standardise iodometrically by volumetric redox titration with starch as indicator.

Procedure: To sample of water, add 1 ml $MnSO_4$ + 1 ml of alkali-iodide azide reagent. Stopper the bottle and shake it briskly till the precipitate appears. Add 1 ml con H_2SO_4. Restopper and mix the solution well. Titrate 200 ml of original solution. For 2 ml of $MnSO_4$ and alkali-iodide azide reagent. Titrate with 0.02 M $Na_2S_2O_3$ solution to pale yellow colouration.

Calculations: For 200 ml sample 1 ml of 0.025 M Na_2SO_3 ≡ 1 mg DO/L. The sensitivity is 20 µg/L. One can give correction with $KMnO_4$ alum flocculation; $CuSO_4$-NH_2SO_2OH modification or use membrane electrode also.

Experiment 15.2 Modifications

If Fe (II) is present one must use $KMnO_4$ as oxidising agent. The relevant theory is presented in previous chapter on water pollution.

Reagents: 1 ml of KF + NaN_3 eliminates Fe (III) interference, but analysis of DO (modified) for the elimination of interference of sulphur compounds cannot be eliminated, $KMnO_4$ (6.3/L), $K_2C_2O_4$ (2/L), KF (400/L) are used for the modification of $KMnO_4$ method. The major portion of experiment is same as described earlier. While in the alum flocculation methods, when 10 ml of (10%) alum with 1–2 ml NH_4OH is added to the sample, the floc so formed removes interfering ions from the sample. $CuSO_4$ with sulphamic acid is also used as flocculating agent, to eliminate some interferences. Take (32 gm NH_2SO_2OH + 50 gm $CuSO_4$ $5H_2O$ made to 500 ml). The mixture is added to sample solution and clear supernatant liquid is used for titration (as azide modification) as described earlier in Expt. 15.1.

15.3 MEMBRANE ELECTRODES FOR ANALYSIS

They are being used for rapid analysis of dissolved oxygen. The membrane electrodes are also used in polarographic analysis. The Winklers method is unsuitable for field analysis. Therefore membrane electrode method was used. The membrane electrode is made of two solid metal electrodes with supporting electrolyte and is separated by thin membrane. In galvanic cell, external potential is applied to start reaction. Membrane permeable to oxygen like polyethylene or fluorocarbon are used in the cell. They facilitate analysis of DO *in situ.* For testing water with zero DO (obtained by using Na_2SO_3 + $CoCl_2$). it is used as 0.1 mg/L. DO can be thus analysed. For variation you should closely study Figs. 15.1 to 15.4 which show effect of temperature, stirring and salts.

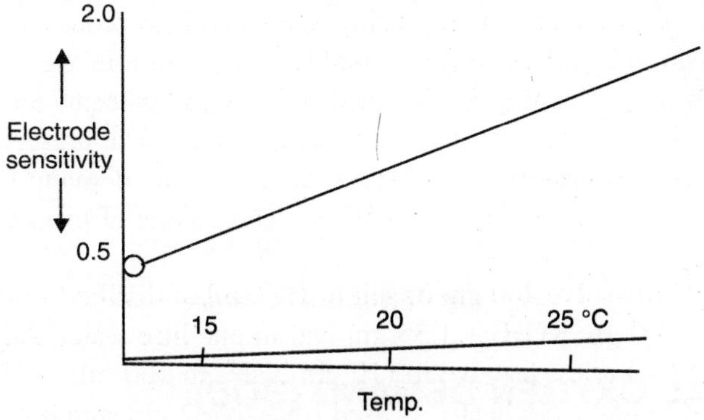

Fig. 15.1 Effect of temperature

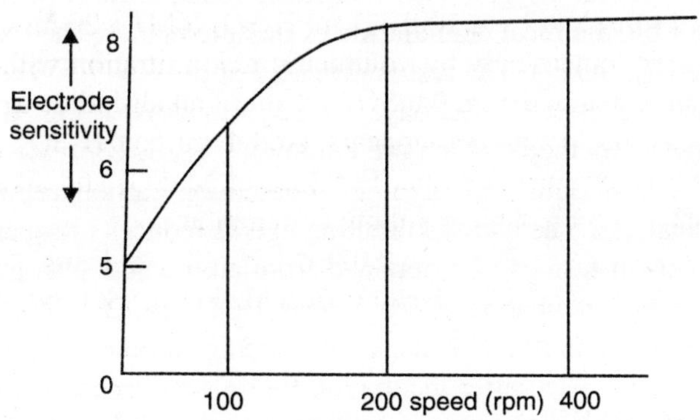

Fig. 15.2 Effect of stirring

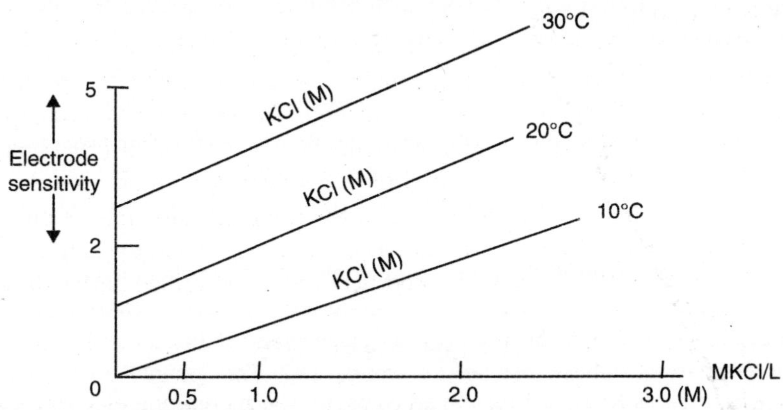

Fig. 15.3 Effect of salting out at various temperatures

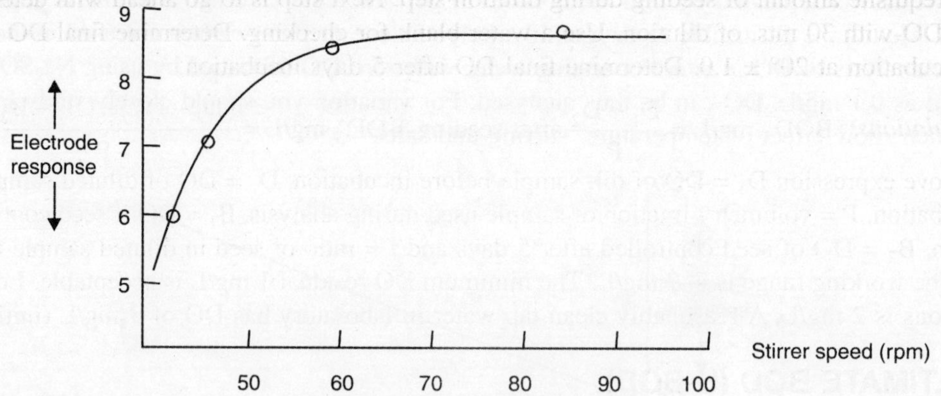

Fig. 15.4 Effect of stirring

15.4 BIOCHEMICAL OXYGEN DEMAND (BOD): Experiment 15.3: Analysis of BOD

This test measures the molecular oxygen used to oxidise reduced form of nitrogen. A test is specified with incubation period for biochemical degradation of organic matter. The oxygen is used to oxidise the inorganic matter (e.g. Fe (II), SO_3^{2-}). The test is provided at pH 6.5–7.5. The analysis considers 5 days BOD_5. The oxygen consumed after (60–90 days) *i.e.*, 2–3 months) is called as U-BOD, or continuous oxygen intake (respirometric method) which are commonly used. Ultimate BOD represents total oxygen needed for complete degradation of organics, containing carbonaceous and nitrogenous matter. NH_3 interferes in BOD analysis. The use of inhibiting agents reduces nitrogenous degradation influencing the value of BOD. This in turn can be obtained from ammonical nitrogen value. It is necessary to dilute sample solution before incubating it for 5 days BOD (BOD_5). The diluting water should not contain degradable materials. The BOD_5 consists of keeping sample in incubation for definite period. First DO is measured as usual. Then after incubation for definite interval say three days or five days it is again measured and difference between two reading is considered as the BOD_3. For experimentation one needs specially designed incubation bottles (~60 ml) with ground glass stoppers. An incubator controlling temperature within 20° ± 1°C (is thermostatically controlled) is used.

Reagents: Phosphate buffer (pH ~ 7.2) (Dissolve 8.5 gm KH_2PO_4 + 21.75 gm K_2HPO_4 + 33.4 gm Na_2HPO_4 + 1.7 gm NH_4Cl - dilute to one litre. $MgSO_4$ (dissolve 22.5 gm in one litre), $CaCl_2$ (dissolve 27.5 gm to one litre), $FeCl_3$ (dissolve 0.25 gm in one litre). Acid alkali solution each 1N (H_2SO_4 28 ml dilute to one litre). Alkali (40 gm NaOH made to 1000 ml with water) Na_2SO_3 (1.575 gm in one litre). Inhibitor (2 chloro 6-(trichloro methyl) pyridine, glucose-glutamic solution (150 mg Glucose + 150 mg Glutamic acid made to one litre with water), NH_4Cl (dissolve 1.15 gm to 500 ml, adjust pH 7.2 and dilute to one litre).

Procedure: To suitable volume of water, add 1 ml phosphate buffer and $MgSO_4$, $CaCl_2$, $FeCl_3$ solutions. Adjust temperature of mixture to 20°C. Then store water not more than 24 hrs. BOD test is bio assay. BOD can be tested with 150 mg each of glucose and glutamic acid. The seeds sometime are added to test BOD analysis. Check pH before testing DO. The sample containing alkalies, residual chlorine compounds, toxic substances, need a specialised treatment before incubation and DO analysis (initial). If nitrification inhibition is required add (TCMP) *i.e.*, 2-chloro-6 (tricholoromethyl) pyridine (3 mg) to BOD bottles prepare five sample solutions with varying degree of dilution.

Add requisite amount of seeding during dilution step. Next step is to go ahead with determination of initial DO-with 30 mts. of dilution. Use a water blank for checking. Determine final DO after 3 or 5 days incubation at 20° ± 1.0. Determine final DO after 5 days incubation.

Calculations: BOD_5 mg/L $= \dfrac{D_1 - D_2}{P}$ after seeding BOD_3 mg/L $= \dfrac{(D_1 - D_2) - (B_1 - B_2)f}{P}$

In above expression D_1 = DO of dil. sample before incubation, D_2 = DO of diluted sample after 5 days incubation, P = volumetry fraction of sample used during analysis, B_1 = DO if seed control before incubation, B_2 = DO of seed controlled after 5 days and f = ratio of seed in diluted sample to seed to control. The working range is 7–9 mg/L. The minimum DO residual 1 mg/L is acceptable. Lower limit of detections is 2 mg/L. A reasonably clean tap water in laboratory has DO of 9 mg/L (initially).

15.5 ULTIMATE BOD (U-BOD): Experiment 15.4: Analysis of (BOD_5) Five Days

It is an extension of BOD_5. A single sample is placed in incubator for indefinite period. DO is measured in stages. A plot of DO versus time gives U-BOD. If DO = 2 mg/L sample should be aerated. Glass serum bottle of 4–10 litres can be used. During experimentation do not use rubber or plastic stoppers as with time they themselves get degraded and spoiled. It is better to use 300 ml BOD bottles. The oxygen sensitive membrane electrode can be also used.

Procedure: Fill BOD bottles with water, measure DO, afterwhile incubate bottle for 2–8 days and periodically measure DO. Reaerate water if DO < 2 mg/L. See Table 15.1 for U-BOD.

Table 15.1 Comparative account of DO, BOD and COD for three months

S. No.	*Day(s)*	*DO (Direct)*	*N-BOD$_5$ (Normal)*	*COD*
1.	0	8.1	0	0
2.	3	5.6	0	2.5
3.	5	8.0	0	4.6
4.	10	8.0	0.46	8.9
5.	20	6.6	0.92	16.7
6.	50	8.0	0.92	24.0
7.	90	5.4	0.92	32

Calculations: BOD = (U-BOD $(1 - e^{-Kt})$
If BOD = oxygen measured at time, t_1 mg/L, t = time
k = first order uptake rate.

Table 15.2 Analysis of U-BOD (ultimate)

Estimated BOD (mg/L)	*Threshold BOD (mg/L)*	*% Diff.*
276	308–321	−10 −14
310	308–321	+1–3
303	308–321	−2–0

15.6 RESPIROMETRIC BOD ANALYSIS: Experiment 15.5: Analysis of Respirometric BOD

It gives direct information on oxygen used by micro-organisms from oxygen rich environment in closed container at fixed room temperature. Such methods help to assess extent of biodegradation taking place by action of chemical, or by toxicant, disinfection, pH adjustment or nutrient addition. There are four kinds of respirometers. The manometric respirators reveal oxygen intake leading to change in pressure, while volumetric respirator measures oxygen intake in increment changes. The third electrolyte respirator measures amount of oxygen generated by electrolysis and finally the last one direct input respirometer gives oxygen in sample. CO_2 is removed first. Liberation of any kind of gases (other than CO_2) causes interference. Minimum limits of analysis is 0.1 mg/L. The commercial respirometer are available in market.

Reagents: Phosphate buffer (pH 7.2), NH_4Cl (0.71 N), $CaCl_2$ (0.25 N), $MgSO_4$ (0.41N), $FeCl_3$ (0.018 N), KOH (6N), Acids (1N), Alkalies (1N), e.g. H_2SO_4, NaOH each (1N), Nitrification reagent as in BOD_5 analysis glucose-glutamic acid (15 gm/L), Na_2SO_3 (0.025 N). Trace elements: Mn, Zn, Mo, Fe in appropriate concentration, yeast extract 0.015%, Nutrient solution.

Procedure: Adjust solution to pH 7.0. Dechlorinate sample by addition of Na_2SO_3. Always use distilled water for sample dilution. Add nutrient and buffer solution to sample. Keep nutrient inhibition viz. TCMP may be used if necessary. Incubation temperature 20°C ± 1C.

Calculation: $C = [A - B\ (S_A/S_B)\ (1000/Ns)$

C = Corrected oxygen uptake of sample; A = measured oxygen by seeded sample, B = measured oxygen in sample of seed control, S_A = volume of seed in sample in A. S_B = volume of seed in sample B, N_s = volume of undiluted sample in A ml.

By using mixture of glucose/glutamic acid, instrument accuracy can be tested. However no standard is available to check accuracy of an instrument. In summary, we note BOD_5 *i.e.* oxygen consumed by microorganisms and inorganic materials for five days, while in U-BOD we measure the oxygen consumed after 60–90 days of incubation period and continuous uptake of oxygen (*i.e.* respirometric analysis) by sample of water. Thus U-BOD measures total oxygen needed to degrade entire organic carboceous matter. The oxidation of reduced form of nitrogen (e.g. NH_3), organic N_2 demand oxygen and is considered hindrance or the interference in BOD analysis. Therefore we use inhibitor during BOD analysis to eliminate these kinds of interferences.

15.7 CHEMICAL OXYGEN DEMAND (COD): Experiment 15.6: Analysis of COD

It is the amount of a specified oxidant that reacts with the sample under controlled experimental conditions. $K_2Cr_2O_7$ is used as the specific oxidant which gets reduced to Cr^{3+}. Both organic as well as inorganic compounds readily get oxidised in the presence of H_2SO_4. COD is measure of pollutants in waste water while BOD is measure of oxygen consuming micro-organisms while TOC represents total measurable organic carbon in the sample, while TOD (total oxygen demand) is a measure of amount of oxygen consumed by all elements in water. The COD > BOD because many organic compounds like detergents, high molecular mass amines, some pesticides do not undergo degradation unless vigorous oxidation conditions are employed. In COD, we employ open reflux method e.g. if COD is ~ 50 mg O_2/L. Then we expect interfering compounds are pyridine. Therefore Ag_2SO_4 is used as the catalyst to promote oxidation. The chloride ion (Cl^-) also interferes but in presence of Ag_2SO_4 it forms insoluble precipitate of silver chloride. Mercury sulphate instead of Ag_2SO_4 can also complex chloride ion. NH_3

is not oxidised. Sulphamic acid is used to eliminate interferences of nitrite anion. The sample is collected in glass bottles. Excess $K_2Cr_2O_7$ is used to oxidise organic matter. The unreacted $K_2Cr_2O_7$ is titrated with $FeSO_4$ $(NH_4)_2SO_4{\cdot}6H_2O$. Around 120 minutes refluxing (open) is necessary for completion of reaction. This is a open reflux technique.

Reagents: $K_2Cr_2O_7$ (0.04167 M) (dissolve 12.25 gm $K_2Cr_2O_7$ in 1000 ml water. Its concentration is 0.25 N H_2SO_4, (Add Ag_2SO_4 to con. H_2SO_4 15.5 gm/L). Ferrous indicator solution (dissolve 1.485 gm of 1–10 phenanthroline) and 695 mg $FeSO_4$ in 100 ml. $FeSO_4$ $(NH_4)_2SO_4.6H_2O$ (dissolve 98 gm salt + 20 ml. con. H_2SO_4 - dil. to one litre).

Dilute 25 ml $K_2Cr_2O_7$ to 100 ml, add 30 ml con. H_2SO_4 then titrate with 0.1 M ferrous ammonium sulphate (FAS) then

$$\text{Vol. of } 0.0417\ \text{M}\,K_2Cr_2O_7 = \frac{\text{Solution titrated, ml}}{\text{Vol. of FAS used in titrations}} \times 0.2500 \quad \text{if FAS = ferrous ammonium sulphate}$$

$HgSO_4$ solution, sulphonic acid, KHP – potassium hydrogen phthalate (dissolve 425 mg in 1000 ml). Fresh solutions should be used if possible during titration.

Procedure: (COD > 50 MgO_2/L). Add sample (50 ml) in reflux condenser. Add 1 gm of glass beads + 5 ml H_2SO_4–Cool–Add 25 ml 0.04167 M $K_2Cr_2O_7$ + 70 ml H_2SO_4 in flask and reflux for 2–3 hrs. with cooling water circulating in the condenser cool and dilute refluxed mixture. Titrate excess of $K_2Cr_2O_7$ against standard ferrous ammonium sulphate with ferrous ion as an indicator to show change in colour from blue green to reddish brown. The calculations are

$$\text{COD as mg } O_2/\text{L} = \frac{(A - B) \times M \times 8000}{\text{ml of sample}},$$

A = ml of ferrous ammonium sulphate for blank,
B = ml of ferrous ammonium sulphate used,
M = molarity of ferrous ammonium sulphate, 8000 = ml. eq. wt. of O_2 × 1000.

(a) Closed reflux method: Volatile organic compounds readily get oxidised in close chamber. COD (40–460 mgO_2/L) same reagents as used as in previous experiment are viz. $K_2Cr_2O_7$ (0.01667 M), H_2SO_4, Ferrous indicator, (0.10 M), ferrous ammonium sulphate, sulphamic acid, KH phthalate standard.

Procedure: As in previous experiment reflux mixture for 2hrs, and calculate COD as follows:

Calculations: $$\text{COD as mg } O_2/\text{L} = \frac{(A - B) \times M \times 8000}{\text{ml of sample}}$$

A = Ferrous ammonium sulphate ml used blank
B = Ferrous ammonium sulphate for sample
M = Molarity of titrant

Experiment 15.7: Analysis of COD by Spectrophotometry

The dichromate shows strong absorbance at 380–400 nm while Cr (III) absorbance at 604 nm wavelength and chromic acid at (λ nm 600 nm).

Reagents: 10.216 gm $K_2Cr_2O_7$ in 500 ml, H_2SO_4 (167 ml con), 33.3 gm $HgSO_4$ in one litre. All other reagents viz. H_2SO_4 KH sulphamic acid and potassium hydrogen phthalate are taken as in previous experiment.

Procedure: Measure suitable volume of sample and reagents. As usual digest for 2 hrs and cool the sample. Mix contents of reaction vessel in water and allow it to settle. Measure the absorbance of known volume of solution at 420 or 600 nm. At 600 nm use undigested blank as the reference solution. At 420 nm use reagent blank solution as the references solution. Absorbance between digested sample and digested blank reference is measured. Prepare calibration plot as usual with known concentrations of standard solutions containing oxygen.

Calculations: $$\text{COD as mg } O_2/L = \frac{\text{mg of } O_2 \text{ in final volume}}{\text{ml of sample}} \times 1000$$

15.8 TOTAL ORGANIC CARBON (TOC): Experiment 15.8: Analysis of TOC

This is the direct expression of total organic content. Although it cannot provide the same kind of information as is given by BOD or COD. TOC gives good idea for fixing the dose for the treatment of water or waste water. For potable water it is necessary to know TOC. It utilises high temperature catalysis with UV chemical oxidant in order to convert carbon or CO_2. Special kind of instruments are used to measure TOC as well as DOC (*i.e.*, dissolved organic carbon). Inorganic carbon is > TOC but at pH = 2.0, it gets converted to CO_2. One also separate TOC and inorganic carbon. The high temperature combustion method is also good. For high level of TOC, one can measure as low as 10 mg/L of TOC. There are three methods mostly used for analysis of TOC. They are:

(*a*) High temperature combustion method,

(*b*) Persulphate-UV or heated persulphate oxidation method, and

(*c*) Wet oxidation method

(*a*) *High Temperature Combustion Method:* Particles must be of small size. Cobalt oxide is used as oxidative catalyst $BaCrO_4$ is also used. CO_2 generated is measured by NDIR method spectrometrically.

This provides useful information on total organic carbon. The inorganic carbon can be removed by acidification and difference in weight is noted to know weight of organic carbon. Combustion temperature > 950°C. The sensitivity is 1 mg C/L.

Reagents: H_3PO_4 (or H_2SO_4) organic carbon (dissolve 2.125 gm anhydrous potassium biphthalate in one litre of water (1 ml = 1mg C/L). For inorganic carbon dissolve 4.4122 gm Na_2CO_3 + 3.497 gm $NaHCO_3$ made to one litre. Carrier gas is pure oxygen air for purging in the system.

Procedure: Good TOC analysers are commercially available in the market which can be used as per manufacturer's instruction.

(*b*) *Persulphate Oxidation Technique:* Organic carbon is oxidised to CO_2 by persulphate oxidation in the presence of UV radiation, CO_2 generated is measured by NDIR. Insufficient acidification will lead to incomplete release of CO_2 gas.

Reagents: Sodium peroxydisulphate (dissolve 100 gm in one litre)

Ammonium peroxydisulphate (dissolve 150 gm in one litre water)

Potassium peroxydisulphate 2% (dissolve 20 gm in one litre)

Procedure: Homogenised the sample (Filter sample if necessary). Transfer it in 30–50 ml in flask adjust pH = 2.0. Purge gas on it. The standard curve can be also prepared in similar manner.

(*c*) *Wet Oxidation Technique:* It is useful for water analysis. The sample is acidified (oxidised by persulphate in autoclave at 130°C). CO_2 generated is measured by NDIR. The sensitivity of method is 0.10 mg C/L.

Reagents: Same as described in Experiment 15.8A (*i*) method, H_3PO_4 (Add 3 ml acid of H_3PO_4 in one litre of water). $K_2S_2O_8$.

Procedure: Sample + 1.2 N H_3PO_4 homogenise sample + 2 gm $K_2S_2O_8$. Keep plane sealed sample blank and standards all in autoclave and digest for 4 hours at 120°C. One can study precision and accuracy of the method of analysis from following Table 15.3.

Table 15.3 Precision of TOC analysis

S. No.	*Nos. of samples*	*Mean mg TOC/L*	*Standard deviation ±%*
1.	9	2.2	5.9
2.	10	5.3	2.8
3.	10	9.9	1.1
4.	10	38.0	3.7

15.9 ANALYSIS OF OTHER ORGANIC COMPOUNDS

So far we have considered the most important parameters like DO, BOD, COD, TOC which throw some light on the presence of organic compounds in water. There are few impurities like humic substances, oil and greases; phenols, tannin and lignin, volatile acids. The UV absorption of organic components give some idea of the quality of waste water. Only one important method used for their characterisation from industrial effluents is described in following sections for humic acid, oil grease phenols and tanins.

Experiment 15.9: Analysis of Humic Substances in Water

Aquatic humic substances contain all organic compounds. They are analysed by column chromatography on diethyl ammonium ethyl cellulose (DEAE) by measuring DOC (*i.e.*, dissolved organic matter).

Reagents: Diethyl ammonium ethyl cellulose, HCl, (0.1 M, 0.5 N). NaOH (0.1N, 0.5 N), DOC standard KCl (0.01N), H_3PO_4 (con).

Procedure: DEAE column is obtained by putting it in column with 0.5 N HCl and adjust pH 4.0 for effluent. Then add 10 ml water, 1 gm DEAE mix and add to column (pH 6.0, 2 ml/mt flow rate). Add humic substance sample and eluate with 0.1 N NaOH. To the effluent lot add H_3PO_4 (pH 2) and remove CO_2. Determine volume and DOC of codified effluent.

Calculation: Calculation is as follows:

DOC = dissolved organic component, Humic substance mg DOC/L = $(A - B) \times C/D$ if A = average DOC of two sample of NaOH, B = average DOC of control NaOH, C = volume of eluate, D = volume of sample.

Experiment 15.10: Analysis of Oil, Grease, Phenols by Gravimetry

Oil and grease is obtained by partition gravimetric method: It is soluble in solvent. Exact composition cannot be ascertained. Dissolved oil and grease is extracted from water by suitable solvent. Emulsion may be formed. n-Hexane is used.

Reagents: HCl H_2SO_4 (1:1), n-Hexane, Methyl ter-butyl ether (MTBE), Na_2SO_4; solvent mixture.

Procedure: To sample add 1:1 HCl or H_2SO_4 till pH 2.0 is obtained. Then add n-hexane or solvent mixture (80% n-Hexane + 20% MTBE). Shake well. Allow it to settle and separate. Then separate organic phase, add 10 gm Na_2SO_4 to remove traces of moisture. Use centrifuge if two layers do not easily separate well. Distill solvent at 85°C. Cool and weigh it.

Calculations: mg oil grease/L = $\frac{(A-B)\times 1000}{\text{ml of sample}}$. A = total gain in weight which is calculated as residue in Hexane.

Phenols are hydroxy derivatives if B = amount of oil and grease in the sample. Benzene and condensate are nuclear. Direct photometric method is used for the analysis. Steam distilable phenol compounds react with 4-aminoantipyrine at pH 7.9 with $K_3Fe(CN)_6$ to form coloured dye which is measured at 500 nm. The sensitivity of method is 10 μg/100 ml solvent.

Reagents: Stock phenol solution; bromate-bromide (2.784 gm $KBrO_3$ in water + 10 gm KBr - made to one litre), HCl con., $Na_2S_2O_3$, (0.025 M), NH_4OH (0.5 N). Phosphate buffer (104.50, $KHPO_4$ and 72.3 KH_2PO_4 made up to litre, expected pH = 6.8), $K_3Fe(CN)_6$ (8 gm in 100 ml), $CHCl_3$, Na_2SO_4, KI-crystal.

Procedure: Put phenol solution (<0.5 mg) in flask. Add 2.5 ml 0.5 N NH_4OH (pH 7.9) + 1 ml of aminoantipyridine + 1 ml $K_3Fe(CN)_6$. Measure colour at 500 nm.

Calculations: mg phenol/L = $\frac{A}{B}\times 1000$ if A = mg phenol in standard solution and B = ml of original solution. Also use other formula mg phenol/L = $\frac{C\times D\times 1000}{E\times B}$, C = mg of standard phenol, D = absorbance of sample, E = absorbance of standard solution. B = ml of original solution.

Tannin and Lignin: They are determined by colorimetry. They contain water due to degradation of vegetables. They have chromatic hydroxyl group. They form blue coloured solution with tungstophosphoric acid (Folin) and Molybdophosphoric acid.

Experiment 15.11: Analysis of Tanin and Lignin

Reagents: Folin phenol (100 gm Na_2NO_4 + 25gm Na_2NaO_4) + 700 ml H_2O + 50 ml H_3PO_4 (85%) + 100 con HCl -boil-then add 150 gm Li_2SO_4 + 50 ml H_2O - cool Filter and store in bottle, carbonate tartrate reagent (200 gm Na_2CO_3 + 12 gm sodium tartrate + 750 ml H_2O dilute to one litre). The stock solution use tannic acid, tannin, lignin to prepare standard solution for making calibration curve.

Procedure: Take 50 ml aliquot, heat to 20°C + 1 ml Folin + 10 ml cabonate tartrate solution. Allow to stand 30 mts. to develop colour which in turn can be measured at 700 nm. From the standard calibration curve find out amount in unknown sample.

UV absorbing components: It is measured at 253.7 nm (or 254 nm) colloidal substance interfere. Oxidants also interfere. Fix pH below 4.0 or above pH 10.0.

Reagents: Clean water (no organics present), HCl (0.1 N), NaOH (0.1N), Phosphate buffer (pH 7.0) (dissolve 4.08 gm KH_2PO_4 and 284 gm Na_2HPO_4 (or K_2HPO_4) in 800 ml water—Dilute to one litre). Take appropriate sample volume (~50 ml). Use water blank and measure absorbance at 254 nm. Find out mean value of the absorbance from 2–3 samples.

$UV_A^{pH} = \left|\frac{\bar{A}}{B}\right| D$ if UV_A^{pH} mean UV absorbance, B = cell path, A = mean absorbance, D = dilution factor, D = find sample vol/initial sample volume.

In conclusion from the knowledge of DO, BOD, COD, TOC one can predict, what kind of organic compounds are present in water. If DO is high and BOD almost zero this states that no biodegradable matter and micro-organism are present in water however if COD is much higher than BOD it indicates samples have large concentration of non-degradable organic matter. The TOC does not throw direct light on organic compounds, such method involves several oxidation steps for analysis.

ADDITIONAL READINGS

15.1 M.D. Mackjnnon, *Measurement of Organic Carbon in Sea Water* in 'Marine Organic Chemistry, Evolution composition, Interactions' and the Chemistry of organic matter in sea water, Elsevier Scientific Publishers (1981).

15.2 APHA, *Standard Methods for Examination of Water and Waste Water*, 21st Ed. (2005).

15.3 R. Back, *Handbook of Decomposition Methods in Analytical Chemistry*, Blackie Glassgow, UK (1979).

15.4 D.C. Burreli, *Atomic Spectrometric Analysis of Heavy Metal, Pollutants in Water*, Ann Arbor, Science Pub. (1975).

15.5 K.E. Jarvis, A.L. Gray and R.S. Hook, *Inductively Coupled Plasma, Mass Spectrometry*, Blacki Academic Prof. Chap. Hall (1992).

15.6 K.J. Irgolic, Ed., *Frontiers of Bioinorganic Chemistry*, VCH Publishers, Germany (1986).

15.7 N.H. Furman, *Standard Methods of Chemical Analysis*, 6th Ed., D. Van Nostrand & Co. (1962).

15.8 F.B. Sandell, *Colorimetric Determination of Trace Metals*, 3rd Ed., Interscience, Publishers (1959).

15.9 S.M. Khopkar, *Environmental Pollution, Monitoring and Control*, New Age International Publishers (P) Ltd. New Delhi (2006).

15.10 S.M. Khopkar, *Basic Concepts of Analytical Chemistry*, 3rd Ed., New Age International Publishers (P) Ltd., New Delhi (2008).

CHAPTER 16

Analysis of Specific Organic Compounds in Water

16.1 INTRODUCTION

We have considered earlier bulk properties of the effluents. Such bulk properties include DO, BOD, COD and TOC. Although it was possible to characterise a class of compound, it had not thrown much light on the quality of individual organic compounds. For instance compounds like dissolved organic halogens, phenols, surfactants, tanins and lignins, trihalomethane compounds or humic acids, were identified. Although we could identify group of compounds with the help of chromatography (HPLC and GLC), capillary electrophoresis or flow injection analysis however it could not give definite information on the presence of individual organic compound. Some individual organic compounds are deleterious to human health and are toxic. Some efforts were successful in utilising the information from TOC and C, H, N, S analysis but it only gave information on the overall group of organic pollutants and not individual compound.

Some organic compounds like methane, 1, 2, dihalomethane or trihalomethanes, polychlorinated bisphenols (PCB), polynuclear aromatic hydrocarbons (PAH), and large group of the pesticides and herbicides are encountered in sample of water or waste water. Amongst these pesticides, we have carbamate pesticides, or organochloro pesticides and variety of herbicides. These compounds are certainly toxic and need definite identification.

The organic pollutants products are identified form TOC or C, H, N, S analysis. In most of the characterisation studies analytical absorption spectroscopy was invariably resorted because it was simple rapid and variety of chromogenic reagents were easily available for spectrophotometric analysis. This led to extensive use of flow injection analysis. However, in order to identify individual organic compounds we have to use GC-MS hyphenated method, solid phase microextraction, liquid-liquid extraction, coupled with gas chromatography, SE-MS *i.e.*, solvent extraction mass spectrometry, HPLC, etc. These techniques were not only rapid but also they provided definite information on the quantitative aspect at microgram concentration of the pollutants.

Several organic compounds are toxic and are encountered either in waste water or in atmospheric pollutants. Amongst these compounds, substituted methane or ethane, chlorinated organic compounds specially trihalomethanes, phenols, polychlorinated bisphenols, polynuclear aromatic hydrocarbons, carbamates. Organochlorine pesticides and herbicides are important compounds for which we will have to consider simple methods for their determination. In several cases such analysis include liquid-liquid extraction coupled with gas chromatography or gas chromatography coupled with mass spectrometry. These techniques are extensively used for the quantitative analysis. Except in analysis of methane no where conventional volumetry technique is useful. In few instance one can take a resort to high performance liquid chromatography (e.g. carbamates) or liquid column chromatography coupled with fluorescence measurements (e.g. Herbicides). In analysis of organic pollutants, sample collection (see Table 16.1) and its preservation is of great importance. The sample holding vials or tubes must not contain traces of organics and should avoid air bubbles. All samples must be collected favourably in duplicate. Appropriate preservative should be used for their storage in chilled atmosphere (4°C).

Table 16.1 Preservatives for volatile compounds (~2 weeks ability)

Component	*Chlorinated portion*	*Nonchlorinated portion*
Aromatics	6N HCl + reducing agent (e.g. Ascorbic acid)	HCl
Halocarbons	6N HCl + reducing agent (e.g. sodium sulphite)	HCl
EDE/DBC	None	None
THM	Reducing agent	None

16.2 GAS CHROMATOGRAPHIC SEPARATIONS

Most of the compounds need isolation as well as purification by solvent or gas extraction prior to their analysis.

1. *Gas chromatography:* Amongst chromatographic methods gas (GLC), chromatography is frequently utilised as several compounds are volatile. CH_4, N_2, Ar, He or H_2 are used as carrier gas. Methylene chloride, chloroform are common interfering compounds are encountered in such analysis. Sulphur is potential interfering element. It is eliminated by combining with Cu/Hg filings to form corresponding sulphide of metal.

The main source of interference in GLC are septum bleed, column bleed and ghost pecks (if excess compound is used). Amongst detectors viz. ECD, FID, TCD one prefers electrolytic conducting detectors (thermal) but ECD (electron capture) is also used for pesticide analysis; or analysis of drugs. It is sensitive for compounds having electronegative groups but it is insensitive towards functional groups. FID or flame ionisation detector is sensitive and preferred for analysis of organic carbon containing compounds. It has very good response, it has negligible response to impurities. It has large linear response, which is not affected much by variation in the flow rate or variation in temperature. It is reliable and it has low detector dead volume. Few limitations are that it has no response to noncombustible gases, it changes physical and chemical properties of the sample.

2. *Mass spectroscopy:* GC-MS is hyphenated method, Quadrupole model in mass spectrometry is best for analysis. m/e is studied for separation of components. From the fragmentation pattern, mass scan can be obtained. One can uses standard charts for the identification of the compound or pollutants.

3. *High performance liquid chromatography (HPLC):* It is applicable to nonvolatile compounds as well as thermally unstable compounds. Absorbance detector is best while the conductivity detector is used for the analysis of inorganic compounds. Photodiode array detector (PDAD) is most popular. It

directly measures absorbance of sample but it is not dependable at lower wavelength (viz. 200–230 nm). The posi column reactor (PCR) is place for derivatisation. Fluorescence detector is very useful as it is sensitive. We would take a very brief discussion on quality of volatile organic compound. This would follow the consideration of specific analysis of most important organic pollutants. The substituted methane, phenols, polychlorinated bisphenols (PCB), polyaromatic hydrocarbons, organochlorine pesticides and some herbicides like glyphosphate herbicides shall be considered in details. In most of these cases the methods used for quantitative analysis include extractive gas chromatography, or GC-MS or HPLC.

(4) *Closed loop stripping gas chromatography* is used for the analysis of broad spectrum organic pollutants. It is useful for those compounds which have the taste as well as odour. The organics are separated from water on activated carbon filter. The capillary column GLC is used for identification with FID detector. The compounds are detected at nanogram concentrations (*i.e.*, 10^{-9} gm/ml).

The high temperature stripping can be also carried out. Earthy musky smelling organic compounds, are best analysed by CLSA-GC-MS (*i.e.*, close loop stripping analysis–gas chromatography–mass spectrometry e.g. several pyrazines. One must have CLSA apparatus, filters, thermostatic bath, CS_2, acetone, chlorooctane, reagent styrene, lindane, etc. are analysed by this technique.

16.3 ANALYSIS OF VOLATILE ORGANIC COMPOUNDS

The hyphenated capillary GC are used for purgeable compounds. The calibration of the instrument is of paramount importance for accurate analysis. Large number of organic compounds can be analysed by this technique e.g. aromatic hydrocarbons, carbon tetrachloride, chloroform, chloro substituted aromatic hydrocarbons, naphthalene, vinylchloride, etc. In another method called as the purge and trap capillary column gas chromatography–mass spectrometry, methyl silicone coated packing should be preferably used. Programmed temperature model should be preferably used. Mass spectrometer is capable of scanning 35–300 amu mass level. The reagents used during such analysis are predominantly 2, 6 diphenylene, oxide polymer, methyl silicone, silica, methanol, HCl, vinyl chloride, ascorbic acid, but the standards must be calibrated before hand. The kind of organic compounds e.g. aromatic hydrocarbons, dichloromethanol ethane, styrene naphthalene, vinyl chloride, xylene, etc. can be analysed by these methods. The another technique is purge and trap capillary–column gas chromatography. It is principally used for analysis of purgeable hydrocarbons and organic aromatic compounds. The detector used in GC analysis is ECD and photoionization detectors. The kind of the organic compounds analysed by this technique includes halosubstituted hydrocarbons as methylene chloride, aromatic hydrocarbons; styrene naphthalene. The amounts ranging from 0.025–0.220 μg/L can be analysed with varying retention time. The standard deviation ranges from 0.007 to 0.150 while % recovery varies from 6.4–13.8. In following sections we shall consider individually organic pollutants and their method of analysis, which largely includes, GC, GC-MS, HPLC techniques. However in compounds like methane one can take resort to volumetric techniques. All GC methods need capillary columns, with purging gas, with ECD, FID as photoionisation detectors for quantitative analysis.

16.4 ANALYSIS OF METHANE: Experiment 16.1: Analysis of Methane (CH_4)

It is explosive gas which has limit of tolerance from 3.95–15% by volume. Two methods viz. combustibie gas indicators and volumetric method are used for its analysis.

(*a*) *The combustible gas indicator* and volumetric analysis are two important methods of analysis.

First method is for low concentration while second method is for high concentration (2–5 mg/L). The combustible gas indicator operates on principle of catalytic oxidation. The heat generated increases the resistance of filament. The deflection is shown in multimeter, which can be monitored by Wheatstone bridge provided if the change in the resistance is calibrated with % methane, but ethane interferes H_2S. H_2 also shows interference at lower pH. The purpose of analysis of methane is to assess the explosive hazards to the environment. 0.2 mg/L can be detected by this method. Figure 16.1 indicates the electrode circuit for the combustible model. R_3/R_4 resistance are adjusted R_2 is reference filament with air while R_1 is variable resistance depending on concentration of methane. A is micrometer which measures variation in current while V stands for voltmeter which is tuned to measure variation in resistance. As told change in resistance ($R_3 \rightarrow R_4$) is related to % methane in sample.

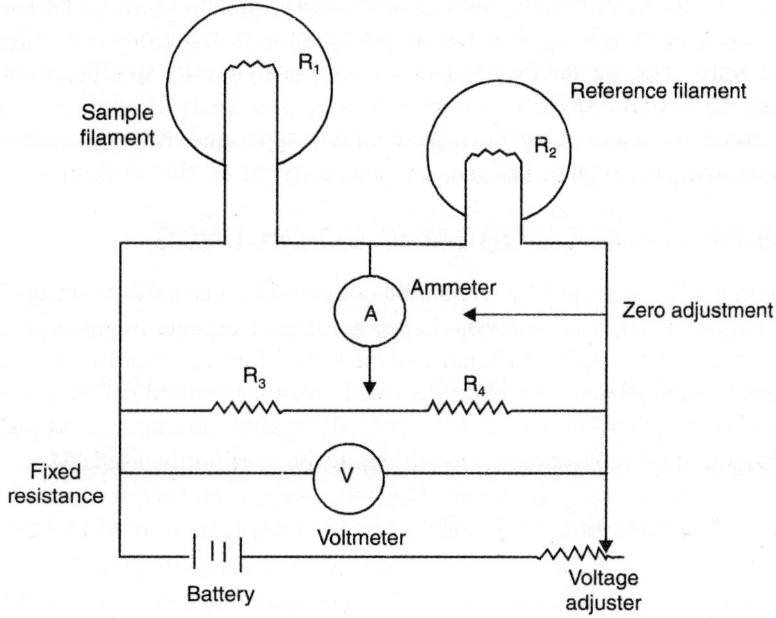

Fig. 16.1 Gas indicator

Reagents: NaOH pellets. It is very simple to operate.
Add NaOH if H_2S is present. The sample CH_4 is in chamber R_1 which is measured.

Calculations: Wt. of CH_4 (mg) $= P\left[\left(\frac{1.928V_g}{T+273}\right)+\left(\frac{890\,V}{H}\right)\right]$ wherein the symbols P = partial pressure of CH_4, T = temperature in °C, V_g = volume of gas, V = volume of liquid phase, H = Henry's constant. If room temperature is 20°C and atmosphere pressure is 100 K pascal. Then we have

$$\text{Weight of } CH_4 \text{ mg/L} = R_f\left(6.7\frac{V_0 - V_1}{V_1} + 0.24\right)$$

if R = scale reading, V_0 = total volume. The Henry constant (H) varies with temperature, V_1 = Volume of H_2O, f = instrumental factor. If atmospheric pressure is different f = 0.77 × 0.05 = 0.0385 is to be used.

(*b*) *Volumetric method:* CH_4 on mixing with O_2 converts to CO_2 and H_2 at platinum wire*.

---

* $CH_4 + O_2 \rightleftharpoons CO_2\uparrow + H_2\uparrow$

Low boiling hydrocarbons interfere. Excess of O_2 is passed in CH_4 and by differential absorption and volumetric changes CO_2 can be measured. One can easily calculate from stoichiometry % CH_4 from amount of CO_2 known by titration with mild alkali in usual acid-base titration.

16.5 ANALYSIS OF TRIHALOMETHANE: Experiment 16.2: Analysis of Trihalomethane

THM compounds are observed in water samples which are chlorinated e.g. chloroform is potential human carcinogen. The improper chlorination of water leads to formation of trihalomethanes. The extractive-gas chromatography is best method for analysis. Let us consider new method of liquid-liquid extraction gas chromatography. One can analyse chloroform, bromodichloro methane ($CHBrCl_2$), dibromochloromethane ($CHClBr_2$) and tribromoform ($CHBr_3$). The sample is extracted with pentane and is analysed by GLC with ECD as the detector. One needs only ten minutes for analysis. The actual procedure of analysis is as follows.

Reagents: Extractants are pentane, hexane, methylcyclohexane. Use methanol standard solution in water, 1–2 bromopropane is used as the internal standard.

Procedure: Extract carefully in two mts. pentane. Then centrifuge for two mts. Inject clean pentane and extract in GLC. We can calculate the amount of trihalomethane as

Calculations: $$\text{RF (response factor)} = \frac{\text{nominal amount of compound extd } (\mu g)}{\text{response (peak area into chromatogram)}}$$

$$\text{Amount of trihalomethane } \mu g/L = W_n = V_s \times C_s \quad V_s = \text{Volume of standard}$$

$$C_s = \text{Concentration of standard}$$

One can calculate total concentration of trihalomethanes by this procedure.

Table 16.2 Typical analysis of Trihalomethanes with GLC

S. No.	*Compound*	*Added (μg/L)*	*Amount Found (μg/L)*			*% recovery*	*% stand dev.*
1.	$CHCl_3$	20	18.7	19.5	19.7	95.6	–2.39
2.	$CHBrCl_2$	20	19.8	20.8	20.5	101.6	+2.10
3.	$CHBr_2Cl$	20	18.7	20.2	20.6	98.3	–3.58
4.	$CHBr_3$	20	17.4	19.2	19.8	93.5	–4.08

16.6 ANALYSIS OF PHENOLS: Experiment 16.3: Analysis of Phenols

Phenols have detrimental effect on health. They must be analysed in water. They are analysed by first liquid-liquid extraction. Gas chromatography can be used as hyphenated method viz. GC-MS must be used if the amount of phenol is too high. We have method for analysing total amount of phenol. Methylene chloride is an extractant. Then 2-propanol is used. FID is used as detector in GLC. The sensitivity varies from 0.31–16.0 μg/L. One can analyse chlorophenol, nitrophenol, methylphenol by this technique. All chemicals must be handled with at most care due to their toxicity e.g. pentafluorobenzyl bromide, 18C6 is injurious to health and therefore must be handled with maximum precaution.

Reagents: NaOH (dissolve 4 gm in 100 ml H_2O), H_2SO_4 1N (58 ml in 500 water), K_2CO_3,

Pentafluorobenzyl bromide, 18C6. Derivatisation reagent (1 mt pentafluorobenzyl bromide + 1 gm 18C6 made to 50 ml 2-propanol. The solvent used are acetone, methanol, methylene chloride, hexane, 2-propanol, toluene. Use silica gel and standard solution. Prepare first calibration curve with internal standard.

Procedure: Use two litre separatory funnel. Keep sample and solvent together and shake well for few minutes. Allow two layers to settle and separate. Pipet out organic phase and inject it in GC. All extraction must be carried out at pH 1–2. Use flame ionisation detector. Calibration could also be done with underivated sample of phenol. The ECD detector can be also used. The amount of phenol is found by calculations.

$$\text{Phenol } \mu g/L = \frac{(A)(V_i)}{(V_t)(V_s)}$$

if A = Amt. of sample injected; V_s = Volume of water, V_i = Volume of injected extracts, V_t = Volume of total extracts.

The analysis can be carried out with laboratory fortified sample. For high level of phenol GC-MS is used. During analysis of polychlorinated bisphenols which is commonly used in transformer can contaminate water. So liquid-liquid extraction gas chromatography method is used for analysis. Organochlorine pesticide can be simultaneously monitored for the purpose of analysis.

16.7 ANALYSIS OF POLYNUCLEAR AROMATIC HYDROCARBONS: Experiment 16.4: Analysis of Polynuclear Aromatic Hydrocarbons (PAH)

These are the by-products of combustion of oil and petrol. HPLC, GC, GC-MS are preferably used for their quantitative analysis. The sample is extracted with methylene chloride and injected into HPLC instrument provided, with UV and fluorescence detector. The compounds like anthracene, phenanthrene, chrysene and their substituted derivatives are encountered in water. The detection limit varies from 0.013–2.3 μg/L.

Reagents: $Na_2S_2O_3$ cyclohexane, methanol, methylene chloride, acetonitrile. Silica gel (100–200). The stock solution is used for construction of calibration curve.

Procedure: Put sample in separating funnel and extract it with appropriate extractant. Add 1–10 ml extract with 4 ml-cyclohexane on liquid column with silica gel as the stationary phase. The sample can be concentrated with 4 ml of acetonitrile. Inject suitable volume of extract either in HPLC or GLC. Prepare as usual calibration curve. Determine the concentration of every compound from calibration plots. For unknown sample repeat same process and find out from graph its concentration.

Amongst PAH those compounds analysed by HPLC and GLC are listed in Table 16.3.

Table 16.3 Conditions of analysis

Detail of analysis	*(max/min)*
Sensitive HPLC	0.016–2.3 μg/K
Retention time (GC)	4.5–38.6 mts
Test concentration	10–100 μg/L
Limit of analysis	1.3–43.2 μg/L
Precison	0.21–0.39 μg/L
Column capacitance (HPLC)/GLC	12.6–28.7

16.8 ANALYSIS OF CARBAMATE PESTICIDES: Experiment 16.5: Analysis of Carbamate Pesticides

This analysis of pesticidal group comprises compound like N-methycarbmoyloxime and N-methylcarbamates. These compounds are heat sensitive and labile. Therefore GLC cannot be used for their analysis but HPLC technique can be extended for their characterisation. It is followed by post column hydrolysis (pH 10). Reverse phase HPLC works well. The hydrolysis products are fluorinating compounds like OPA & MERC. High chlorine compounds show interference. The commercially obtained carbamate pesticides include Baygon, oxamyl, carbamyl, aldicarb, methonyl, carbofuran, aldicarb sulphoxide, aldicarb sulphone and 3 hydroxycarbofuran.

Reagents: Methanol NaOH (0.05 N), 2 Mercaptoethanol (Mix 10 ml ethanol with 10 ml of acetonitrile; $Na_2B_4 10H_2O$ (dissolve 19.1 gm in one litre) $CH_2ClCOOH$ (2.5 M). The stock solution pesticides OPA reagent (dissolves 0.10 gm phthaldehyde in 10 ml of methanol, add 1 ml borate solution. It is stable for minimum of 72 hours *i.e.* 3 days for $Na_2S_3O_3$.

Procedure: Carry out HPLC operations and use internal standard (BOMC).

Calculation: $R_f = \dfrac{A_s \times C_i}{A_i \times C_s}$ A_s = peak area, A_i = peak area of internal standard solution.

C_i = concentration of internal standard; C_s = concentration of standard sample

Concentration of carbamate μg/L (internal stand) $C = \left[\dfrac{A_s \times C_i}{A_i \times R_f}\right]$ if A_s = response of sample and

for external standard $C = \left[\dfrac{A_s \times C_i}{A_s}\right]$

The drinking water contains 0.99–1.03 μg/L of some carbamates.

The average recovery by HPLC method ranges from 47%–108%.

16.9 ANALYSIS OF ORGANOCHLORO PESTICIDES: Experiment 16.6: Analysis of Organochloro Pesticide

Their origin traces back to discharge from agricultural industries. Many compounds are bioaccumulative and also carcinogenic in nature. The important members of these series of pesticides are: Lindane, heptachlor, aldrin, dialdrin, captan, endrine, DDT, BHC, endosulphone. These pesticides are extracted in suitable solvent, evaporated and injected into GLC column for further analysis. In GLC the detector used is ECD. For need of purification column chromatography is used to control purity. Two column procedure is also preferred. PCB interfere in analysis. So also phthalate esters interfere during analysis. The sensitivity of method (10 μg/L) which is quite satisfactory.

Reagents: Hexane, petroleum ether, diethyl ether, ethyl acetate, methylene chloride, magnesia-silica gel (60–90 mesh), Na_2SO_4, Glass wool, carrier gas (N_2, Ar + CH_4).

Procedure: Inject suitable volume of pesticide in GLC column and (~1 μL). Follow same technique for the analysis of standard as well as the unknown sample. After GLC, extract pesticide in diethyl ether (13%) or methylene chloride. Again subject it to gas chromatography train for better separation and purification.

Calculation: D is dilution factor in decimal. The concentration of the pesticide ion is calculated as

$$\mu g/L = \frac{A \times B \times C \times D}{E \times F \times G}.$$

The significance of various symbols is as follows:

A = μg standard pesticide, B = peak height of sample, C = extraction volume μL, D = dilution factor, E = peak height of standard, F = volume of extract injected, G = volume of sample extracted.

16.10 ANALYSIS OF ACIDIC AND GLYPHOSATE HERBICIDES: Experiment 16.7: Analysis of Herbicides

The phenoxy acid herbicides and chlorinated phenoxy acid herbicides are important members. The extractive gas chromatography is used for analysis of herbicides with FCD while glyphostal-N (phosphomethyl) glycine is another important as detector. Herbicide encountered in drinking water is (Round up and Rodeo as the commercial agricultural product in) (LD_{50} = 1570 mg/kg). For the analysis of GLYPH and AMPA liquid-column chromatography is most useful. Among acidic herbicides we have 2,2 dichloropropanoic acid (dal pan), 3, 6 dichloro 2-methoxy benzoic acid (dicamba), 2,4, (dichlorophenoxy) acetic acid, (2,4, D) pentachlorophenol (PClP), 2, 4, 5, trichlorophenoxy propionic acid (2, 4, 5 TP/Silvex), Pidoram, dinoseb, bentazon and (2,4,5, T) *i.e.*, 2, 4, 5 (trichlorophenoxy), acetic acid. Under glyphosate herbicides as told before we have GLYPH and AMPA which are relatively toxic and are in use.

The herbicide is extracted in solvent like (MtBE) *i.e.*, 3-methyl tertiary butyl ether at low pH. The product is esterified with diazomethane. The resulting methyl ethyl derivative is chromatographed with ECD detector in GLC instrument. Most of the compounds are carcinogenic and must be handled with utmost care. The following reagents are used during analysis of acidic herbicides:

Reagent: $Na_2S_2O_3$, Herbicide stock solution in MtBE, Internal Standard (1, 2, 3 trichloropropane), surrogate (2, 3, 5, 6 tetrafluorobenzoic acid). Prepare a calibration curve using standard in MtBe on GLC.

Procedure: Carry out microextraction for 30 ml sample, add 9 gm Na_2SO_4 and shake + 20% NaOH (pH ~ 12.0). Cool for 1 hr + 3 ml MtBE. Then add 1.5 ml con. H_2SO_4 + 3 gm $CuSO_4$ + 5 gm Na_2SO_4. Shake vials on centrifuge. One can proceed then with separation, concentrations and derivatisation. This is followed by gas chromatographic separation with suitable carrier gas. One can also use capillary column. One can calculate response factor from which concentration of unknown is found out.

Calculations: $R_F = \dfrac{(A_n)(C_i)}{(A_i)(C_n)}$ if A_1 = area of herbicide, C_i = concentration of internal standard compound of herbicide, We can calculate concentration using R_F = value as

$$C = \text{mg/L herbicide } \mu g/L = \frac{(A_n)(i_s)}{(A_i)(R_F)(V)}$$

where C = con. of compound, A_n = area of herbicide. i_s = Internal standard amount, A_1 = Area of internal stand, V = volume of water extracted. See important herbicides in Table 16.4.

Table 16.4 Analysis of typical acidic herbicides and their physical properties

S. No.	*Herbicides popular name*	*Mol. wt.*	*M.p. °C*	*Det level μg/L*	*Total %*	*% recovery (mean)*
1.	Bentzon	240.3	140	0.01	76	84
2.	Dicamba (2, 4, D)	221.4	140	0.02	82	84
3.	Pentachlorophenol (PCP)	226	190	0.02	68	107
4.	2,4,T TP (Silvex)	269	177	0.01	102	107
5.	2, 4, 5 T	255	15	0.012	76	104
6.	Dinoseb	240	42	0.04	85	90
7.	Picloram	241	220	0.01	82	80
8.	Dalpan	142	14	0.1	96	114
9.	TCP (Internal standard)	147	90	–	–	–
10.	TFBA (Internal standard)	194	150	–	–	–
11.	2 4 D (Internal standard)	221	116	0.1	105	75

16.11 ANALYSIS OF GLYPHOSATE HERBICIDES: Experiment 16.8: Analysis of Phosate Herbicides

Two products viz GLYPH and AMPA are separated by anion/cation exchange chromatography and are measured by post column derivatisation. The detection limit for two above compounds is respectively 25 and 0.5 μg/L. Analysis is carried as per following procedure.

Reagents: H_3PO_4, H_2SO_4, HCl, CH_3OH, KH_2PO_4, EDTA, NaCl, NaOH, $Ca(OCl_2)$, G-phthaldehyde, 2-mercaptoethanol, H_3BO_3, KOH, Fluorogenic agent, Glyphosate and AMPA fortified standards. Oxidising agent (dissolve 0.5 gm $Ca(OCl)_2$. in 500 ml water + 1.7 gm KH_2PO_4 + 11.6 NaCl + 0.4 gm NaOH–Dilute.

Procedure: Pass solution and oxidant at rate of 0.5 ml/mt in HPLC. Prepare standard plots with known standards. Fortify 10 ml sample with 0.1 ml EDTA and inject it in HPLC. Filter if suspended matter is present.

In summary we can generalise that most powerful methods for the analysis of individual organic compound is HPLC and GLC. However liquid-liquid extraction is fruitfully utilised during analysis along with GLC. Further for quantitative analysis GLC is coupled with Mass Spectrometry. Variety of the detectors are used in GLC but TCD and FID appear to be most useful in analysis. The solid-phase microextraction is new technique and it has come to forefront only in recent years. The purge and trap capillary column gas chromatography is used for the analysis of volatile organic compounds while classical volumetric method is useful for the analysis of methane.

The combination of instrumental methods has facilitated the analysis of variety of organic pollutants. Thus not only they include pesticides and herbicides but also common toxic materials. The various compounds included are methane, trihalomethane, halosubstituted acetic acid and aldehydes. Heptanol, Aldrine, PCB compounds, heptachlor, 4-4′ DDT, pyrene, chrysene, carbaryl oxamyl, Baygon (propoxur), several kinds of pesticides such as lindane, aldrin, dialdrin, PCB-1260, 2, 4 D, 2, 4, 5-T, 2, 4, DB, etc. Many of these compounds are deadly toxic and in many western countries they are banned (e.g. DDT) and not allowed to be used in agriculture practice.

In analysis of water these compounds do not occur very frequently. Only when there is interaction of potable or industrial water with waste water or sewage the problem arises. Further the analysis of all these compounds should be done only when exigencies arise.

ADDITIONAL READINGS

16.1 APHA, *Standard Methods for the Examination of Water and Waste Water,* 22nd Ed., (2005).

16.2 L.H. Keith, *Principles of Environmental Testing*, Am. Chem. Soc. (1988).

16.3 L.M. Dennis and M.L. Nichols, *Gas Analysis*, MacMillan (1929).

16.4 J.S. Heldane and J.L. Graham, *Methods of Air Analysis*, Charles Griffin & Co., London (1935).

16.5 W.I. Budde and J.W. Eiehlberger, *Organic Analysis Using Gas Chromatography, Mass Spectrometry,* Ann Arbor Pub. (1979).

16.6 Alceon Corf, *An Overview of Available Information on the Toxicity of Drinking Water, Disinfectants and Their By-products*, Cambridge (1993).

16.7 N.M. Mellair and B.J. Bonelli, *Basic Chromatography*, Consolidated Printing, Berkely (1969).

16.8 N.V. Steeve, *Handbook of Laboratory Safety*, Chemical Rubber Co., Cleveland, Ohio (1977).

16.9 S.M. Khopkar, *Environmental Pollution, Monitoring and Control*, New Age International Publishers (P) Ltd., New Delhi (2006).

16.10 S.M. Khopkar, *Basic Concepts of Analytical Chemistry*, 3rd Ed., New Age International Publishers (P) Ltd., New Delhi (2008).

CHAPTER 17

Microbiological Examination of the Water

17.1 INTRODUCTION

Microbiological examination of the water is equally important as is chemical characterisation of water. However the periodicity at which such examination is carried out is much different. Many parameters in water are analysed once in 24 hrs. few in once in week and some once in month. Similarly in microbiological examination, analysis of BOD and COD is the first signal it gives to carry forward microbiological examination. This is because the presence of bacteria or micro-organisms in water would enhance the value of BOD. Further the substances which are non-degradable yield COD which will be still higher in its magnitude.

The microbiological examination of water can be viewed from three angles. The first being presence of coliform, faecal coliform, streptococci bacteria present in water. This also includes iron and sulphur bacteria like crenthroxine. The second category includes pathogenic bacteria which spoil the quality of specially potable water. This class includes actinomycetes also. While the third category is viruses. Such viruses include those causing cholera, enteric fever, fungi, kind of all organism leading to ill effect on human health.

As regards frequency of analysis ordinarily we conduct examination of first category of organism regularly per day. These include coliform, streptococci and analogous organism. While analysis of pathogens is carried out only if there is direct or indirect threat to quality of water and the human health. Finally the viruses are analysed in special circumstances when there is an emergency or when there is widespread propagation and spreading of diseases like cholera. The kind of experimental methodology used for the immediate characterisation of water is quite simple and dependable. A good microscope for getting plate count, a steady incubator for multiplication of organism at controlled temperature and in few occasion the photometer are adequate.

So far we have considered physical as well as the chemical examination of water. Similarly good amount of the information can be obtained from microbiological examination. In addition to the presence of viruses and pathogens which are injurious to health, there are other kinds of the micro-organism like coliform, faecal coliform, streptococci which throw light on the quality of water. The viruses or pathogens presence is special case when enteric disease spreads. The characterisation of water with plate count on coliform or streptococci gives fairly good amount and adequate reliable information to know whether water is fit for drinking purposes by human beings. Therefore in this chapter we would consider in greater details kind of special equipment used, determination of plate count, preparation of the culture media, evaluation of total microbial count, use of fermentation phenomena and membrane filters for the characterisation with passing reference to pathogens, viruses and fungi present in the sample of water.

17.2 LABORATORY APPARATUS AND EQUIPMENT USED

It is necessary to first get ourselves conversant with the kind of apparatus used or special equipment needed for the characterisation of the quality of water. Apart from the common glass apparatus which is generally used in chemical laboratory, we need special kind of equipment. Thus first we need good incubator. It controls the temperature at fixed point throughout the experimentation. One cannot equate refrigerator with incubator as in case of referigerator. the temperature goes on varying according to materials stored in the refrigerator. We need a dependable thermometer. The commonly used water baths are also required during operation. There is also need for hot air sterilising ovens. Autoclaves provide uniform temperature (121°C). A good steriliser with automatic controls and one which is capable of carrying out complete sterilisation cycle is required. A gas like burshane or CNG, ethylene oxide can be used for heating, we need optical counting equipment consisting of pore and spread plates or the membrane filter suitable to carry out colony counting. A binocular microscope with magnification of 10 to 15 X is generally used. A pH meter is also needed. So also common chemical balance, pipettes, graduated cylinders, media preparation utensils are required.

The plastic containers of specific made can be used provided they are very carefully sterilised. For plate count, petri dishes are needed. We need sometimes pressurised petri dishes. The membrane filters are held by seamless stainless steel holders. A special kind of fermentation tubes or vials are easily available. The sample bottles must be absolutely clean and properly sterilised. The wide mouth bottles are preferable during quantitative analysis.

17.3 STERILISATION AND MEDIA PREPARATION

The glassware must be properly cleaned before sterilisation. Pipettes made of copper are not suitable but one made out of stainless steel are ideal for the purpose. The 60 minutes sterilisation must be carried around at 170°C. Autoclaves can be used for dependable sterilisation. The next important operation is the preparation of culture media. Dehydrated media is at a times used. The culture media should not be stale and more than one week old but can be stored longer. Avoid sunlight exposure. Fermentation tubes must be stored at 25°C. The pH of solution used indicate quality of culture media. The change in pH should not be more than 0.1–0.2 units. The media should be sterilised at least for two hours. Only the distilled water should be used for the preparation of the media. Water should not contain excess chlorine generally used for disinfection. The dehydrated media should always be used. Commercially prepared media are available for fast work. The phosphate buffer of pH 7.2 needs careful preparation. Peptone water is obtained from 10% solution at pH 6.8 (See Table 17.2 to 17.5).

17.4 SAMPLE COLLECTION, ITS PRESERVATION AND STORAGE

The samples must be carefully collected in sterilised container latter being properly washed before use. A removal of chlorine can be done if required with sodium thiosulphate solution. Better collect sample and provide aseptic environment for storage. A special care should be taken while sampling the raw or surface water. While sampling in boat, upstream collection is preferred. The special sampling apparatus is also available in the market. The average size of the sample should have capacity of at least 100 ml in volume. The microbiological examination should start as soon as sampling is completed. For potable water preservation at < 10°C is necessary. One should hold sample at <10°C and complete entire analysis within 8 hrs. As regards culture media there are specification to suit each kind of sample and end use of results so obtained in the normal chemical analysis. Therefore microbiological examination is carried out accordingly.

17.5 RAPID CHARACTERISATION TECHNIQUES

This approach is sometimes necessary and very useful e.g. Faecal coliform can be detected within seven hours. A media specially prepared from peptone, yeast, lactose, mannitol, NaCl, sodium lauryl sulphate, bromocresol phenol red and agar is needed in definite proportion. The media after preparation is used for detection. One must filter sample through membrane filter, containing M-7HFC agar medium and incubate it to 45°C. You can observe yellow faecal coliform colony in microscope.

Table 17.1 Rapid detection techniques

S. No.	*Microbial groups*	*Technique*	*Time of operation*	*Sensitivity (cells/ml)*
1.	Microflora (Non-specific)	Fluorescence	One hr	100,000
		Photometry	0.02 hr	100,000
2.	Faecal coliform	Electrochemical	1–7 hrs	1,000,000
		GLC	9–12 hrs	> 50
		Photometry	8–20 hrs	10,000,000
3.	Gram negative bacteria	Fluorescence	2–3 hrs	28–30
		Limulus Test	2 hrs	500–3000

The special rapid techniques are summarised in Table 17.1. With the exception of spectrophotometry other methods are of no use for routine analysis. The bioluminescence test is viable for microbial count; as assay is fast it needs only 60 minutes. The radiometric detection is useful for faecal coliforma counting.

17.6 DETECTION OF COLIPHAGE: Experiment 17.1: Detection of Coliphage in Water

Coliforma and coliphage have direct relation as ultimately coliphage gets converted to coliform to indicate quality of water.

Apparatus and Reagents: Hot culture *Escherichia coli*: Media of two kinds viz. Tryptin on (TSA) obtained from Tryptone (15 gm), Soytone (5 gm), NaCl (5 gm), Agar (15 gm) and one litre water pH = 7.3 is adjusted. Other media is Tryptic assay (TSB) obtained from Tryptone (17 gm), soytone (3 gm), Dextrose (2.5 gm), NaCl (5 gm), K_2HPO_4 (2.5 gm), all in one litre water with pH 7.3. Third reagent is modified tryptic assay (MTSA). To TSB, add NH_4NO_3, $Sr(NO_3)_2$ and agar in ratio of 1.6:0.21:1.5 pH maintained at 7.3 glycerine (10% in trypticase), 2, 3, 5 triphenyl tetrazolium chloride (1% w/v) in C_2H_5OH. Add MTSA of 4.5°C to increase plaque visibility.

Procedure: Freeze hoze preparation. Inoculate *E. coli* from stock agar in 10 ml of TSB and 10% glycerine. Each tube is inoculated in 25 minutes TSB + 10% glycerine at 35°C. Measure the absorbance with reagent blank in absence of *E. coli*. The assay procedure is as follows. Good for > 5 coliphage/100 ml. Dilute if necessary. Thaw tube of frozen host *E. coli* at 44°C on waterbath. Add 1 ml of host *E. coli* culture and 0.08 ml TPTZ to MTSA. Mix thoroughly and pour it in petri dish and cover it by agar gel. Incubate it at 35°C. Count plaque after 4–6 hrs. of incubation.

Results and Calculations: Count plaques on each plate and record. Obtain same result on five plate and add up the result of all five plates. Based on coliphage counts estimate total and faecal coliform numbers expressed as:

Total coliform: $\therefore$ $\log y = 0.627 (\log x) + 1.864$ where in x = Total coliform/100 ml

y = coliphages/100 ml while faecal coliform are calculated as follows:

$\log y = 0.805 (\log x) + 0.895$ if n = faecal coliform/100 ml; y = coliphage/100 ml.

Thus coliform, faecal coliform and streptococci are indicator bacteria to state about the quality of water or waste water. The test for the stressed organism is conducted to see how many micro-organisms are killed and how many are incapable of further growth by maring in turn quality of water.

Experiment 17.2: To Ascertain Coliform and Faecal Coliform in Water

Reagents: Two layer agar (in absence of glucose) with 2 hr incubation at 35°C followed by 1.5 hr incubation. Pre-enriched phenol red lactose broth incubated at 35°C. Modify high temperature by preincubation of M-FC culture for 22 hrs. Eliminate rosolic acid from M-FC and incubate at 44°C for 1 hr faecal coliform colonies are blue with verification of stressed bacteria. During total coliform count the addition of sodium sulphite improves detection of coliform bacteria. The modification of media and process decreases selectivity and differentiation of faecal coliform colonies. The recovery of stressed faecal streptococci with membrane filtration using the broth medium faecal streptococci are comparable with fermentation test.

17.7 HETEROTROPHIC PLATE COUNT (HPC):

Experiment 17.3: To Ascertain Plate Count in Water

The heterotrophic plate count (HPC) is also called standard plate count. It is a process to ascertain number of live heterotrophic bacteria present in water. Colonies originate from pairs, chains, cluster cell all of which are in colony forming units. For assessing this plate count there are three methods of which spread plates however is best being pour plate count.

As told pour plate method, spread plate method and membrane filter methods are known methods. The first method can be tested on 0.1–2ml sample of water. The second method does not promote heat shock but small volume of sample can be processed. The last method facilitates testing large volumes of water with no production of heat shock, but it is expensive due to use of membrane filter. It has small display area. There are number of samples with date in separate components.

Procedure: (a) Plate Count Agar: Use it for first two methods. The media is Tryptone (5 gm), yeast (2.5 gm), glucose (1 gm), agar (15 gm) made with one litre water. Adjust pH to 7.0, autoclave at 121°C for 15 mts. Next the HPCA agar made from peptone (20 gm), geltin (25 gm), glycerol (100 ml). Agar 15 gm made with one litre water R2A agar made from yeast (0.5 gm), protose (0.5 gm), casaminic acid (0.5 gm), glucose (0.5 gm), starch (0.5 gm), K_2HPO_4 (0.3 gm), $MgSO_4$ (0.05 gm), sodium pyruvate 0.3 gm, agar 15 gm and one litre of water. Adjust pH 7.2. Finally NWRI agar (HPCA) is made from peptone (3 gm), casein (0.5 gm), K_2HPO_4 (0.2 gm), $MgSO_4$ (0.05 gm), $FeCl_3$ (0.001 gm). Agar 15 gm and water one litre. Incubate at 38°C for 48 hrs. Control humidity during incubation. The counting and recording to pour and surface plate method be done after incubation. Use counting apparatus, pipette out 2 ml of sample of water. The bacterial count per ml is evaluated.

$$\text{Colony forming unit/ml} = \frac{\text{colonies counted}}{\text{actual vol. of sample dish, ml}} = \text{Bacterial count per ml}$$

The term colony forming units (CFU) is description of method, so temperature and time is specified to complete heterotrophic plate count (CFU). Avoid inaccuracies in counting.

(*b*) *Pour Plate Method:* Select proper dilution of the sample (~10^{-2} dil). Pipette out required quantity of sample in sterile media. Pour 10–12 ml fraction of liquified medium at 40°C in dish incubate and count as was done earlier.

This is not very common method for getting plate count. The next popular method is:

(*c*) *Spread Plate Method:* This is the most commonly used method.

Apparatus and Reagent: Glass rods, pipettes, Turn table incubator.

Procedure: Agar is used as media to get best results. Pour 15 ml of medium in sterile petri dish. Allow agar to solidify then use predried plates. Pipette out 0.5 ml of sample on turn table. Let inoculum be absorbed into medium before incubation. Pipette out (0.1–0.5 ml) sample on surface of predried agar plate and proceed for counting as described earlier.

(*d*) *Membrane Filter Method:* Agar is used as the media, 5 ml of it is put in dish. Allow it to solidify at room temperature. Take fixed volume of the sample, filter fixed volume through sterile 47 mm (0.45 μm) membrane filter under partial vacuum. Incubate at 35°C for 2 days. Take out plate and proceed for plate count as described in earlier sections.

So far we have considered four methods under heterotrophic plate count. They provide rough estimate of total number of viable bacteria signifying quality of water. Such plate count is useful in assessing efficiency of different treatment process and it facilitates in plant control test. Such counting is valuable for ascertaining quality of the finished water in distribution centre as indicator of microbial regrowth and sediment built up. In order to ascertain (rapidly) quality of water, the assessment of plate count, faecal coliforms and faecal streptococci is good enough. The consideration of stressed organism is equally important. If indicator bacteria are stressed or injured in water, they are capable of growing and formation of colony. A good portion (~10–90%) may be detected. These injured bacteria continue to pose potential health risk in potable water. A separate procedure is followed for recovery of injured total coliform by using membrane method in order to improve the quality of the water.

Table 17.2 Time and temperature of sterilisation of techniques

S. No.	*Technique*	*Total time at 120°C*	*S. No.*	*Technique*	*Total time at 120°C*
1.	Membrane filter	10 mts.	4.	Covered membranes filter peak	15 mts.
2.	Carbohydrate medium	12–15 mts.	5.	Buffered solution	15 mts.
3.	Contaminated broth	30 mts.	6.	Water for rinsing	30 mts.

It will be of interest at this stage to consider behaviour of media, sterilisation, weight loss. The Table 17.2 shows time required for sterilisation by various techniques. One needs large time if there is a kind of contamination. The culture media which are generally prepared do not have life span of more than 1/2 months e.g. the membrane filter broth gets spoiled within 96 hours only as indicated in Table 17.3. The average weight used by various kinds of media is shown in Table 17.4. It will be noted that one needs large weight (> 20 gm) if we use gelatin or peptone as media, while excellent work on culturing can be easily carried out if as small as 0.5 gm is taken like casaminic acid, soluble starch and polypeptone. Finally Table 17.5 indicates weight loss which is related to weight of plate. As temperature rises, weight loss is significant.

Table 17.3 Life span of cultured media

S. No.	*Type of media*	*Lifetime in weeks*	*S. No.*	*Type of media*	*Time in weeks*
1.	Membrane filter broth	1/2 weeks	4.	Large agar volume	12 weeks
2.	Membrane filter agar	2 weeks	5.	Pound plates of agar	2 weeks
3.	Agar broth (loose) packing	2 weeks	6.	Agar broth (tight packing)	2 weeks

Table 17.4 Kind of media and quantity used during analysis

S. No.	*Kind of media*	*Amount (gms)*	*S. No.*	*Kind of media*	*Amount (gms)*
1.	Tryptone	5	8.	Peptone	20
2.	Agar	15	9.	Gelatin	25
3.	Glucose	1.0	10.	Glycerol	10 ml
4.	Yeast extract	2.5	11.	Polypeptone	0.5
5.	Casaminic acid	0.5	12.	Soluble starch	0.5
6.	K_2HPO_4	0.3	13.	$MgSO_4.7H_2O$	0.05
7.	$FeCl_3$	0.001	14.	Sodium pyruvate	0.5

Table 17.5 Weight loss of water with temperature by plates (agar)

S. No.	*Temp. °C*	*Weight (1 gm)*	*Wt. (3 gm)*	*Wt. (4 gm)*	*Wt. (2 gm)*	*Wt. (3 gm)*	*Wt. (4 gm)*
		(With lid inverted)			*(Inverted lid removed)*		
1.	24	32	90	125	7	10.5	14
2.	48	6	18	24	1.3	1.9	2.8
3.	60	4	12	16	–	–	–
		For closed plate			For open plate		

Apart from the methods described earlier it is also possible to know the plate count by epifluorescence technique as well as by utilising P-17 strain and NOX method. The outline of these two techniques is presented in subsequent sections of this chapter.

17.8 TOTAL MICROBIAL COUNT FOR PLATES:
Experiment 17.4: To Ascertain Microbiological Plate Count

There are two prominent methods in order to assess microbial plate count. They are *Epifluorescence* microscopic method and the other are called as strain NOX method. We would consider both in details from experimentation point view of analysis. The plate count obtained by these methods is much larger than one obtained by heterotrophic plate count because errors can occur in selectivity of growth media. The cell clumping has very gradual growth rates. The former method provides direct plate count. The details of experiment are as follows:

(a) Apparatus reagents: Microscope, filters, blender, counter, membrane filter, syringe. Phosphate buffer of pH 7.2 (dissolve 13.6 gm K_2HPO_4 in one litre water). Then fix above 5% glutaraldehyde in phosphate buffer, fluorochrome 0.1% acridine orange in phosphate buffer, Immersion oil.

Procedure: Take known volume of sample of water (~9 ml). Add 1 ml fixative at 4°C. Dilute sample, mix in blender and dilute with phosphate buffer. Put 1 ml of sampled water on non-fluorescent polycarbonate filter on cellulose membrane. Then add by syringe 1 ml of fluorochrome and 3 ml of phosphate buffer. Filter under vacuum and wash it with buffer, remove polycarbonate, filter and air dry for 5 mts. Put filter on drop of immersion oil on microscope slide. Examine ten selected fields to measure 10–50 cells.

Calculations: Total cell/ml = (average cell/square) × (square × filter) × (dil. factor/sample vol. ml). The second method is also called as Pseudomonas fluorescence strain P-17. Spirillum strain NOX method. Turbid water strongly interferes.

(b) Alternate method: Apparatus reagents: Incubator, water bath, pipette, flask, sodium acetate solution (400 mg/litre) (dissolve 2.267 CH_3COONa in one litre of water). Autoclave sample, $Na_2S_2O_3$ dissolve 30 gm in one litre water). Buffer water; $Na_2S_2O_8$ 10% (dissolve 100 gm in one litre). Mineral salt solution (dissolve 171 mg K_2HPO_4,707 mg NH_4Cl, 1.44 gm KNO_3 in one litre water). Culture of strains P-17 and NOX.

Procedure: Prepare stock inoculum from P-17 and NOX by transfering growth from slant culture, on agar neutralised chlorinated water sample. Add 125 ml of acetic acid solution. Make available count of culture. Inoculate with 500 mg colony forming unit.

$$\text{Volume of inoculum} = = \frac{(500\ \text{CFU/ml}) \times (40\text{ml/vial})}{\text{CFU/ml. stock solution}}$$ Yield must be constant throughout.

Calculation: Average vial count .... results for 3ds, we calculate AOC

$$\mu\text{g/ml AOC} = \frac{(\text{mean P-17 CFU/ml})(2.44 \times 10^{-7}\ \text{acetate-C/CFU})}{+\ (\text{mean NOX CFU/ml})(3.45 \times 10^{-7}\ \mu\text{g oxalate-C/CFU})(1000\ \text{ml/L})}$$

17.9 COLIFORM AND FERMENTATION:
Experiment 17.5: To Study Fermentation Count

The coliform group consists of several kinds of bacteria belonging to enterobacteriaceae group. They are anaerobic gram-negative, nonspore forming rod shaped bacteria which can fement the lactose to liberate gas within two days at 308°K. Their population is reported as most probable number or MPN count. The counting decides the kind of treatment plan. We therefore consider standard fermentation method; those involving bacterial density, detect presence/absence of coliform, faecal coliform procedure, *Escherichia coli* test.

The Fermentation Technique: Lauryl tryptose broth is used. The reagents used are tryptose (20 gm), Lactose (5 gm), K_2HPO_4 (3 gm), KH_2PO_4 (3 gm), NaCl (5 gm), sodium lauryl sulphate (100 mg) and one litre water. Add all above components to water and adjust its pH to 6.8–0.01 gm/L with bromocresol purple. First determine whether there is acid production.

(*a*) *Procedure:* Take 7–8 fermentation tubes and add 5, 10, 20 ml of sample. Mix the test portion. Incubate them at 35°C for one day. The generation of gas in tubes is positive test for coliform. Use the culture media containing peptone (10 gm), lactose (10 gm), oxgail (20 gm), brilliant green (0.010 gm) and one litre of water. Adjust pH to 7.2 and sterilise the mixture and incubate it for 24 hrs. The completed test performance may be necessary. The following reagents are used. Nutrient agar containing peptone (5 gm) beef extract (3 gm), Agar (15 gm) and one litre water which is sterilised. The gram stain reagent is obtained from $(NH_4)_2C_2O_4$ (0.8 gm), Crystal violet (2 gm in ethanol). The preparation of gram stain reagent also includes following steps viz. grind iodine (1 gm) with 2 gm KI. Safranine dye (2.5 gm) acetone + ethanol (95% pure).

(*b*) *Procedure:* First streak the plates then incubate plates for 24 hrs at 35°C. Pick a coliform colony. Incubate broth tubes. In gram stain technique, the formation of acid gas within 48 hrs indicates presence of gram negative bacteria present in the sample of water.

Bacteria Density Determination: The following procedure describes the evaluation of MPN count. It is calculated from following expression:

$$\text{MPN value known} \times \frac{10}{\text{Volume tested}}$$

$$= \text{MPN/100 ml} = \frac{\text{Nos. of positive tubes} \times 100}{[\text{ml sample negative tubes} \times \text{ml sample in all tubes}]^{1/2}}$$

$$\text{MPN/100 ml} = (\text{Table MPN/100 ml}) \times 10/V \text{ if } V = \text{Vol. of sample}$$

Table 17.6 MPN count analysis

S.No.	*Sample*	*Dilution of 1 ml*	*Combination (+ve)*	*MPN count noted*	*Remarks*
1.	a	5/5	5–2–0	5000	High count
2.	b	5/5	5–4–2	2200	Medium count
3.	c	0/5	0–1–0	20	Very low count
4.	d	5/5	5–3–2	1400	Average count
5.	e	5/5	5–3–2	1400	Average count

Experiment 17.6: To Ascertain Presence of Faecal Coliform

It is to be tested at elevated temperature. One can find out their presence by the procedure described outlined in following section.

Reagents: Tryptose (20 gm), Lactose (5 gm), Bile salt mixture, (1.5 gm), K_2HPO_4 (4 gm), KH_2PO_4 (1.5 gm), NaCl (5 gm) all mixed up in one litre of water. Adjust pH to 6.9 and sterilise it and use close tubes with plastic caps.

Procedure: After incubation for 48 hrs at 35°C, segregate tubes which show the generation of the gas, then shake such fermentation tubes with 3–5 mm loop to transfer growth. Then incubate the inoculated EC broth for as 24 hrs at 44°C. Put EC tubes in water bath for 1/2 hr. The gas production with growth in EC broth culture in 24 hrs is complete with positive coliform reaction. Calculate MPN from nos. of positive EC broth tubes. A direct test is also possible.

Experiment 17.7: To Ascertain *E. coli* Count

Escheria coli Population: This belongs to coliform group only indicating faecal contamination of water. They are called *E. coli* also. They are Enteric pathogens as indicated earlier.

Reagents: Trypticase (20 gm), lactose (5 gm), bile salts (1.5 gm), K_2HPO_4 (4 gm); KH_2PO_4 (1.5 gm), NaCl (5 gm), 4 methyl umbelliferyl-β-n-glucuronide (0.05 gm) with one litre water. Adjust the pH to 6.0 after sterilisation. No inverted tube is required. Close tubes with plastic caps.

Procedure: After 48 hrs incubation, note whether the gas has been generated. Shake those tubes and transfer growth in bottle. Incubate inoculated tubes on water bath for 24 hrs. Examine all tubes showing growth for fluorescence at 360 nm. The presence of blue fluorescence indicates presence of *E. coli* in sample. From multiple tubes calculate MPN count of *E. coli* present by using formula indicated in Section 17.8.

17.10 COLIFORM BY MEMBRANE FILTRATION

This provides reproducible method and can be used for study of large number of sample volumes. Membrane fillrate is beneficial in monitoring the quality of potable water. Turbid water cannot be tested by membrane filteration technique. The coliform group is defined as those anaerobic gram negative, nonspore red shaped bacteria developed in 24 hrs at 35°C in lactose. The toxic substances reduces their number. In comparison to multiple tube fermentation method, membrane filtration is very efficient. This technique has few limitations viz. turbid water cannot be tested by this technique. With purified culture media, they produce cytochromic oxidase and positive β-galactosidase test reactions. Pink, blue, yellow, white or the colourless colonies with no shade are classified as Non-coliforms. Members filtration technique can test saline water but not wastewater as latter contain toxic substances. The volume of water tested is 100 ml. The small volume is good for water with high densitiy. The coliform population is properly counted by this method.

Experiment 17.8: Analysis of Coliform by Filtration Method

(a) Procedure for total coliform with membrane filter: The glassware used consist of sample bottles, dilution bottles, pipettes and cylinders, culture media, vessel and dishes membrane filter, forceps, incubator and microscope membrane, filter must hold all coliform.

Reagents: Yeast (1.2 gm), Trypticase (3.7 gm), Thiotone (3.7 gm), Tryptose (7.5 gm), Lactose (9.4 gm), K_2HPO_4 (1 gm), NaCl (3.7 gm), Sodium desooxycholate (0.1 gm), Sodium lauryl sulphate (0.05 gm), Na_2SO_3 (1.6 gm), Fuchsin (0.8 gm), Agar (1.5 gm), water (1 litre). Adjust the final pH to 7.2. Avoid exposure to sunlight. The M-endo medium contains in addition to above components polypeptone (10 gm), thiopeptone (5 gm). Take care to prepare agar solution. Prepare broth in same manner or agar by heating and cooling and storing in sterile container. Those of the bacteria which produce red colony within 24 hrs at 85°C on the endotype (*i.e.* second media) confirms their presence. For testing, use 100 ml sample of potable water.

Procedure: The sample is diluted by sterile water. Use sterile filtration unit. Incubate it at 35°C for 24 hrs. The suggested volume of sample for membrane filteration work for potable water is 100 ml and for well water it is 50 ml while river water (10 ml) and sewage (0.1 ml) needs small volume. To count the membrane filter colonies use low power (10–15 x) microscope. Do not count more than 200 colonies.

$$\textit{Calculation:} \quad \frac{\text{Total coliform}}{100\text{ ml sample}} = \frac{\text{Counted coliform} \times 100}{\text{ml of sample filtered}}$$

$$\%\text{ verified coliforms} = \frac{\text{Nos. of verified colonies}}{\text{Total nos. of colonies}} \times 100$$

(*b*) *Membrane filter and faecal coliform:* Membrane filter technique can be used for ascertaining them. The temperature of incubator is 45°C with excess of lactose in the sample.

Reagents: Tryptose (10 gm), Polypeptone (5 gm), Yeast (3 gm), NaCl (5 gm), Lactose (12.5 gm). Bile salt (1.5 gm), Aniline blue (0.1 gm), Agar (15 gm), water one litre in sterile bottles. No autoclave is used. Final pH should be 7.4. Use M-FC medium with tight fitting dishes.

Procedure: The counting procedure for faecal coliform by membrane filter method provides blue colony in said medium. Then count colonies with low power binocular microscope.

$$\textit{Calculations:} \text{ Faecal coliform/gm} = \frac{\text{counted coliform}}{(\text{dilution}) \times \%\text{ dry solids}} \text{ with denomination in decimal form.}$$

17.11 ENZYME SUBSTRATE COLIFORM TEST

A hydrolysable substrate is used. Both the *E.coli* and coliform can be counted. In this method, total coliform group is defined as all those bacteria's having enzyme β-D-galactosidase cleavage on the coloured substrate. *E. coli* are defined as bacteria providing total positive total coliform response. They release fluorogenase coloured ortho-nitrophenyl-β-D galactopyranoside (ONPG) as is used to detect β-D-galactosidase enzyme hydrolyses substrates.

Escherichia coli (*E. coli*) is detected by fluorogenic substrate as 4-methylumbelliferyl-β-D-glucuronide (MUG). It gives fluorescent product at 366 nm. The test is used for potable water. The coloured water is not given for enzyme substrate test. We get readymade formations. One must avoid exposure to sunlight. There are multitube and multiwell procedures. With ONPG no colour is shown but CPRG shows yellow format. The colour change for various media is different e.g. with ONPG-MUG-yellow colour for coliform and blue colour for *E. coli* is observed while for CPRG-MUG we have red and blue (*E. coli*) fluorescence for *E. coli* positive.

17.12 DIFFERENTIATION OF COLIFORM BACTERIA

This mode helps to detect the presence of *E. coli*. Each form of coliform cannot ferment lactose. Every species cannot react uniformly in media. Some kind of water do represent the presences of Enterobacter, Klebsiella citrobacter and Escherichia organism. Pure cultural media is necessary for identification. The oxidase positive gram negative nonspore forming rods are not coliform bacteria. Coliform in general is a facultative anaerobic gram negative nonspore forming rods which ferment lactose with liberation of gas at 35°C in two day in membrane filter. It gives dark red rods. The commercially available media and reagents provide a procedure for differentiation. The test includes lactose fermentation, ONPG hydrolysis, Indian test, methyl red test, etc. The indole is produced by metabolism of tryptophan. A dark red colour in amylalcohol indicates presence of coliform. In methyl red test, after the incubation for two days, it gives red colour of MR negative and yellow colour of MR positive. The citrate test indicates the ability of coliform to use citrate as the source of the carbon which gives blue colour with sodium citrate in the presence of $MgSO_4$, $(NH_4)_2$ HPO_4 and K_2HPO_4 and NaCl with bromothymol blue. In oxidase test we use trypton (15 gm), soy tone (5 gm), NaCl (5 gm), Agar (15 gm) and one litre water at pH 7.5. Tetramethyl p-phenylene diamine dihydrochloride (1%) in water indicates dark purple colour on glass rod showing presence of positive oxidase.

17.13 FAECAL STREPTOCOCCUS AND ENTEROCOCCUS GROUPS

This group comprises of number of species. It is found in gastrointestinal tract of animals and in human faeces. The FC/FS *i.e.*, faecal coliform/faecal streptococcus is not always reliable to trace origin of the pollution. The enterococcus group cultivates in 6.5% NaCl buffered at pH 9.6 at 10 and 40°C in an incubator.

Table 17.7 Weight loss for agar plates with varying temperature

Temperature °C	*Weight in gms. (3 gms original)*	*Weight in gms (4 gms original)*
24	95	125
35	51	67
50	18	24
60	12	16

Table 17.8 Sample volume for total and faecal coliform test

S. No.	*Kind of water*	*Total coliform (ml)*	*For faecal coliform (ml)*	*Remarks*
1.	Potable water	100 ml	50 ml	Low faecal form
2.	Well water	50 ml	50 ml	Equal coli and faecal form
3.	River water	1 ml	100 ml	High faecal form
4.	Sewage	0.1–0.00001	0.001	Very low broth form

Table 17.9 Accuracy of analysis for membrane filter coliform (100 ml sample of water)

S. No.	*Coliform colonies at 9.5%*	*Confidence limit at 99%*	*Confidence limit*	*Remarks*
1.	0	0	3.7	Good limit
2.	5	1.6	11.7	Reasonable
3.	10	4.7	18.4	Satisfactory
4.	15	8.4	24.8	Average
5.	20	12.2	30.8	High limit
6.	25	16.1	40.0	Very high limit

One can use multiple test tube (MT) or membrane filtration (MF) method for differentiation. In multiple test tube we use polypeptone (15 gm), beef extracts (5 gm), glucose (7.5 gm), NaCl (7.5 gm),

NaN_3 (0.2 gm) and one litre of sterile water at pH = 7.2. This is azide dextrose broth. On exposure for 24–48 hrs at 35°C, we get black brown colonies with brown halos. This confirms the presence of enterococcus group. In membrane filter test, one uses m-enterococcus agar for streptococcus. We use Tryptose (20 gm), yeast (5 gm), glucose (2 gm), K_2HPO_4 (4 gm), NaN_3 (0.4 gm), 2, 3, 5 triphenyl tetrazolium chloride (100 mg), Agar (10 gm) and one litre sterilised water. It is Autoclaved at 120°C for 15 mts (pH 6.6). Filter it in sterile membrane. Then incubate for two days at 40°C count pink red entercoccus on filter paper using fluorescent lamp and a magnifying lens.

Table 17.10 Sample volume needed for total coliform test

S. No.	*Quality of water*	*Volume of filtered water range*	*S. No.*	*Quality of water*	*Volume to be utilised*
1.	Potable water	100 ml	4.	Water supply	0.1–10 ml
2.	Well water	10–100 ml	5.	River water	1–0.001 ml
3.	Lake water	10–100 ml	6.	Treated sewage	0.1–0.001 ml

17.14 IRON AND SULPHUR BACTERIA

Their presence is headache and can cause deposition of iron (III) and sulphur in potable or industrial water. These are single cell organisms. They cause fouling and plug walls of containers with red colour to water, bad odour and offensive taste, frothing and enhancement in turbidity. These bacteria procure energy from the process of oxidation of Fe (II) to Fe (III). They can be microscopically identified especially the sediments indicate their presence and can be dissolved in 1M hydrochloric acid. The floating or hanging slide method can be used for their characterisation with the help of photographs of the species. The sulphur organism are of single cell and oxidise organic sulphur to SO_4^{2-}. They grow anaerobically. They are in four kinds as green purple bacteria, colourless filaments, large colourless nonfilamentous form and last oxidising/reducing bacteria but first variety is most common. Purple form occurs only if H_2S is present in water. The colourless form occurs if water contains O_2 and H_2S. The nonfilament type colourless species occurs if decaying algae is present. Crenothrix polysphere are in reality iron bacteria.

Experiment 17.9: To Ascertain Population of Iron and Sulphur Bacteria

Only two varieties can be easily isolated. A large variety of media are used for culturing them. With 1.5% agar, we use casitone (5 gm) glycerol (10 ml), yeast (1 gm) and one litre sterile water. The isolation medium is made out of glucose (0.15 gm) (NH_4) SO_4 (500 mg), Ca $(NO_3)_2$ (10 mg), K_2PO_4 (50 mg) $MgSO_4$ (50 mg), KCl (100 mg), agar (10 gm), cyanocobalamin (10 mg), thiamine (40 mg) water one litre. For iron bacteria for maintenance we use sucrose (1 gm) casitone (75 mg), yeast (25 mg), trypticase (25 mg), agar (10 gm), the vitamins as cyanocobalamin (10 mg), thiamine (40 mg) and one litre sterlised water. The iron oxidising medium is possible from $(NH_4)_2SO_4$ (3 gm), KCl (100 gm), K_2HPO_4 (500 mg), $MgSO_4$ (500 mg), $Ca(NO_3)_2$ (10 mg), H_2SO_4 (1 ml), and H_2O. The ferrous sulphide agar is composed of not only FeS (500 ml), but also Na_2S(15 gm), $FeSO_4$ $(NH_4)_2SO_4$ $6H_2O$ (70 gm), Agar (30 gm/L). The sulphur reducing medium contains sodium lactate (3.5 gm), Beef extract (1 gm), peptone (2 gm), $MgSO_4$ (2gm), Na_2SO_4 (1.5 gm), K_2HPO_4 (0.3 gm), Ferrous salt (0.10 gm), $CaCl_2$ (0.1 gm), sodium ascorbate (0.1 gm) and one litre of water at pH = 7.5. The thiosulphate oxidising medium has $Na_2S_2O_3$ (10 gm), K_2HPO_4 (2 gm), $MgSO_4$ (0.1), $CaCl_2$ (0.1 gm), $(NH_4)_2SO_4$ (0.1 gm), $FeCl_3$ (0.02 gm) with one litre of sterile water. The sulphur media consists of S (10 gm) K_2HPO_4 (3 gm), $MgSO_4$ (0.5 gm), $(NH_4)_2SO_4$ (10.3 gm), $CaCl_2$ (0.25 gm), $FeCl_3$ (0.02 gm) and one litre of sterile water at pH 4.8.

The accompanying trace elements along with iron and sulphur are Zn, Mn, Cu, Co, Mo. Let us consider procedure for characterisation of iron bacteria. The media should be put in autoclave (50 ml sample) and inoculate for 24 hrs at 25°C using dissecting type microscope locate filament type bacteria. The carbon tetrachloride medium is good for detection.

The sulphur reducing bacteria are also microscopically examined in anaerobic conditions. The tubes get blackened in presence of sulphur bacteria. Media for enumeration or isolation of various species of sulphur are commercially available.

17.15 ACTINOMYCETES:
Experiment 17.10: To Count Actinomyces

Earth musky odour, affects the quality of water. Such odour cannot be easily removed. The volatile metabolite produce bad smell. A gasomin and 2 methyl isobornet have been isolated which are responsible for generation of odour. Also the filament blue green algae produce odour. They cause poor waste water treatment. They have fungal type of morphology. They are actually not fungi but filament branching bacteria. The water must be analysed promptly at 10°C. The plate counting is quite simple and rapid surface colonies are identified. One should use diluted sample.

Reagents: Consist of starch (10 gm), Casein (0.5 gm), KNO_3 (2 gm), NaCl (2 gm), K_2HPO_4 (2 gm), $MgSO_4$ (0.05 gm), $CaCO_3$ (0.02 gm), $FeSO_4$ (0.01 gm), Agar (15 gm) and one litre water. pH adjustment is not required for double layer plates. Prepare 3 plate for each dilution. Transfer starch agar to solidify in dish. At 45°C add liquified starch agar. Pipette inoculated sample on hardened agar. 30–500 colonies are microscopically counted. They are reported per millimeter of water or gram of soil. They are shiny, soft weak and regular.

17.16 PATHOGENIC BACTERIA

The pathogens cause human disease. Therefore they must be removed from potable water if present. They enter directly in body or through skin inhalation, ingestion, aspiration in human beings. They can enter through mouth, ears and eyes also. In the nature of salmonella, shigella, *E. coli,* Enterobacterial fungi is considered in some details. The indicator organisms like coliform, faecal coliform, *E. coli* contents give some idea about presence of pathogens in water. Even faecal streptococci, is also indicator organism. The salmonella characterisation is of significance. They are present at low concentration in surface waters and can be detected by the use of swap of specific dimensions, however such analysis must be done within 6 hours. The diatomaceous earth is used for their detection. For large volume samples one has to use borosilicate glass microfibres bonded with epoxy resin. Membrane filter technique is most reliable for their characterisation. It is preserved in peptone water. The quantitative analysis can be carried out through M-endo, MF culture. Cystine broth inhibits gram positive and non-pathogenic bacteria. The selective F-broth allows recovery of all salmonella. Tetrathionate broth promotes growth of salmonella. Brilliant green dye is used for detection. The media must suppress coliforma during detection of salmonella. Bismuth sulphite agar is also used after 24 hrs incubation. They have predominantly green (black) colour. Many biochemical reactions share characteristics with salmonella. The preliminary screening, biochemical reaction and fermentation reaction can also confirm their presence in water. One can resort to immuno-fluorescence procedure for detection of salmonella. The salmonella density in water provides information on quantitative contents of water. Conventional estimation of MPN count gives idea about their population. Amongst them the important pathogenic bacteria is shigella.

Shigella: is acute gastrointestinal disease of human beings caused by four species as group A, B, C and D forms. It affects intestinal mucous membrane to cause dysentery, fever and diarrhoea. Inadequate chlorination raises their population in water as they are sensitive to chlorination at normal levels. They survive 4 days in river water while in exceptional cases they survive for 21 days in nonculturable state. The quantitative method for that estimation are not available. Membrane filtration and centrifugation can be used for their characterisation.

17.17 ESCHERICHIA COLI

Escherichia coli (*E. coli*) is normal inhabitants of human digestive tract which causes diarrhoeal diseases in humans. They are present in four groups as enterotoxigenic, enterohaemorrhagic, enteroinvasive and enterpathogenic forms. They can be grouped on basis of their serology. Water borne outbreaks are caused by the presence of *E. coli*. The lauryl tryptose broth (LTB) is used for MPN count. They are not influenced by elevated temperature. Biochemical kits are used for their quantification. No standard method for assessment is available. The coliforma method involving lactose fermentation are not suitable for their detection. High temperature and use of sodium lauryl sulphate for MPN analysis are also not suitable. They must be isolated by biochemical kits. Apple juice contains such organisms. Faecal coliform membrane filter method can be used for their assessment. The pathogenic potential of *E. coli* isolate can be assessed by testing its virulent properties e.g. Agglutination kit can be used for such purpose.

17.18 VIBRIO CHOLERAE

Vibrio cholerae is cholera causing bacteria leading to severe diarrhoea. Bengal Sero group (0.139) produces epidemic. Since their amount is low, enrichment or concentration is necessary. The samples are enriched in peptone alkaline broth for 6–8 hrs at 35°C in incubator. Sucrose fermentation gives yellow colonies. Several tests are used to identify them *V. cholera* like gram negative rod, glucose fermentation, growth in broth, etc. By employing different test they can be classified in different groups. Some important tests include indole, luminescence, lysine decarboxylase, fermentation and enzyme production. Fluorescent antibody techniques can be also used, nucleic acid probe is rarely used for their identification.

17.19 ENTERIC VIRUSES

They are contaminating water through excretion of the faeces or urine of infected persons. They affect gastrointestinal tract of human. They are transmitted by faecal oral route. They are found in sewage, waste water and potable water. In temperate climate they multiply faster. Hepatitis A is such virus which grows faster in cold climate. They are not normal flora in intestinal tract. Very few virus units can produce infection. Water virologist should test their presence in potable water. The collection, concentration and identification are important stages of their analysis. From small sample volumes they are concentrated by adsorption to elution from microporous filter. The electropositive filters facilitates their concentration in water, however many times microprocessor gets blocked and clogged. The process of adsorption and elution is most efficient in the process of concentration. The small as well as large volumes of samples can be handled in the process of adsorption and elution. The virus concentration can be also accomplished with $Al(OH)_3$ adsorption and precipitation. Polyethyleneglycol (PEG) hydro extraction is ultrafiltration process in cellulose dialysis. It is possible to isolate viruses from SPM in water and sewage.

17.20 FUNGI

They are ubiquitously distributed achlorophyllous. They are found at places where non-living organism are absent. Fungi do contain yeasts, and filamentous or moulds. They are parasitic in nature. Fungi are useful for ascertaining the quality of water. They are aerobic only few are of anaerobic metabolism kind. They occur in aqueous environment. They survive in the water treatment: tastes and odours in water are associated with prokaryotic organisms. The chemical coagulation and disinfection was effective for elimination of fungi. They also occur in recreational water like swimming pools, ponds, etc. Their growth in water is fast and several growth paths have been recognised. The excessive organic load promotes growth of fungi. A single cell (spore) can give rise to fungal growth. There are two techniques used for that assessment of their presence. The pour plate and spread plate techniques are utilised for their characterisation and assessment. The membrane filtration technique can be also used for identification. It furnishes information on colony forming unit. Zoospore fungi is a special class of fungi normally encountered in polluted water.

17.21 PATHOGENIC PROTOZOA

They are found as intestinal protozoa in the potable water. They cause digestive disturbances. Their microscopic test can help to ascertain their presence in water. They had caused havoc in USA in diseases called Gialdialamblia in year 1980. Faeces of man and the animal causes severe contamination of the water. Newer methods for analysis or evolving year by year in this field. Apart from *Vibrio cholerae* other injurious species are liptospiral and yersinia anteroplitica but their discussion is out of scope for our studies.

ADDITIONAL READINGS

17.1 APHA, *Standard Methods for the Examination of Water and Waste Water*, 21st Ed. (2005).

17.2 T.A. Ratlia, *The Laboratory Quality Assurance System*, Van Nostrand Reinhold, New York (1990).

17.3 R.S. Breed and W.D. Dotterer, *The Number of Colonies Allowable on the Satisfactory Agar Plates*, Tech Bull 53, New York, Agricultural Exp. Station (1933).

17.4 APHA, *Standard Method for the Examination of Dairy Products*, 16th Ed., Washington (1993).

17.5 B.J. Dutka, *Membrane Filtration Applications Techniques and Problems*, Marcel Dekker (1981).

17.6 V.B.D. Sherman, *A Guide to the Identification of Generas of Bacteria*, Williams & Wilkins, USA (1967).

17.7 P. Gerhards, *Manual Methods for General Bacteriology* (1981).

17.8 N.R. Krieg and J.G. Holt, Ed. Bergrey's *Manual of Systematic Bacteriology*, Vol. 1, Williams & Wilkins, Baltimore (1984).

17.9 E.H. Lennette, A. Ballows, W.J. Hanster and H.J. Shadomy, Ed., *Manual of Clinical Microbiology*, 5th Ed. Am. Soc. Microbiol. (1985).

17.10 S.T. Cowen and K.J. Steel, *Manual for the Identification of Medical Bacteria*, Cambridge, UK (1965).

CHAPTER 18

Analysis of Industrial Effluents

18.1 INTRODUCTION

Water pollution is serious problem, air pollution is still worst. However, pollution caused by waste water (also termed as Industrial effluents) is equally troublesome problem with greater dimension. The water which is used in chemical processing, materials management, steel and foundry as well as pharmaceutical and fine chemical industries after use is disposed off as the waste water. In real sense this water is really not waste but can be further processed and reused for various purposes as cooling, washing, cleaning and agricultural purposes after some treatment or processing.

The quality of waste water depends upon the source. For instance if waste water is emerging from food and beverage industry it will have high BOD with good amounts of proteins and carbohydrates. While water emerging from service industries like tanneries (*i.e.*, leather industry) would contain high concentration of chromium which is used for hide treatment. Waste water emerging from pharmaceutical industry would contain variety of drugs and analogous products thrown as waste. Finally waste water emerging from steel and materials industry will have large proportion of metal pollutants. All these effluents have high BOD, large COD and heavy TOC.

To ascertain the quality of waste water we need to go ahead with its characterisation. The task is somewhat complicated as it is mixture of different kinds of effluents. We will have to ascertain from its bulk properties, the magnitude of BOD and COD. Then specific analysis can be carried out only after the characterisation of physical properties such as measurement of density, viscosity, conductivity, turbidance, colour, odour and temperature. Only after such characterisation, one can proceed with specific analysis of organic pollutants with gas chromatography, UV, visible, IR, NMR spectral methods and finally TOC analysis. If one is interested in knowing contents of inorganic pollutants, one has to resort to conventional absorption spectroscopy, including AAS, or ICP-AES technique, neutron activation and voltammetry.

In final analysis one has to ascertain kind of anions like PO_4^{3-}, SO_4^{2-}, X^- (halides) as well as dissolved gases like chlorine and oxygen. The ISI *i.e.* Indian Standard Institute has provided standards for tolerance limits of pollutants from various sources such as distilleries, food, dyes, fertilisers, steel and paper products industry.

Alternatively water after use is disposed off into the sea, or a stream. Such water which is used for domestic or industrial purposes is considered useless from the point of further use and is therefore disposed off. Such disposable water is called as waste water, indicating thereby that it is waste as far as its consumption is concerned. Out of the total quantity water consumed by human being, more than 50% of it is consumed for industrial activity; and only a small proportion is used for drinking purposes. The water after use emerging out of industry is better termed as industrial effluent, while water which emerges out after domestic consumption largely in form of sewage is called as domestic effluent. It is not proper to term this water as waste water, because it is always possible to recover useful material from it and reuse it by recycling for some other purposes. The pollutants associated with industrial effluents are organic matter, inorganic dissolved solids, fertilising materials; thermal constituents in the form of heat, suspended solids and micro-organism and pathogens. Organic pollutants decrease limit of dissolved oxygen, impart bad odour and colour to the effluent. In water contaminated with marine environment, there would be substantial rise in its salinity. If there is a high concentration of phosphate and nitrate, growth of aquatic plants will rise. Thermal pollution of industrial effluents causes deoxygenation and causes serious ecological problems. Food processing industry effluent causes viruses and bacterial pollution which are health hazards.

18.2 OBJECTIVE OF ANALYSIS

The main purpose of the analysis of the industrial effluents is to evaluate appropriate methods of treatment of water, with the aim to reuse or dispose, ascertain quality of water and aim at recovery of valuable products from waste effluents. The purpose of analysis is to assess the material balance for the process to permit evaluation of efficiency of treatment. In order to ascertain the above objectives, it will be necessary to analyse various parameters, which would throw light on quality of waste water. The choice of methods for such analysis should aim at furnishing reliable information, give accurate and sensitive results. The frequency of analysis is also important. The method of sampling and transportation would also decide the actual choice of particular method of analysis. The measurement can be carried out *in situ,* or may be based on automated techniques. The *in situ* analysis involves on the spot field analysis, where we may not always get representative sample. Instrumental methods for *in situ* analysis are limited; however, automated methods furnish valuable information on physical, chemical and biological characteristic of the industrial effluents. The flow colorimetry methods are best suited for automated analysis. When close control of quality of water is needed, then regular monitoring is essential. Environmental analytical chemist should select his approach for analysis after careful diagnosis of the total problem. Such a diagnosis is dependent on the objective of analysis, speed and accuracy sought, interferences, environmental conditions of sampling and actual measurement of various parameters of industrial effluents.

18.3 KINDS OF MEASUREMENTS

In comparison to potable analysis, the analysis of waste water is complicated; the reason being one encounters in water all kind of impurities and deleterious material which are dumped into stream by different industries. Further, the waste water from different industries, mixes together, before merging into the sea or disposal point, leading to formation of very complex mixtures. Such water has a bad colour, offensive odour, turbidity, bad taste; high or too low pH, have dissolved solids, organic materials in the form of pesticides; and metal pollutants like As, Ba, Cd, Cr, Cu, Pb, Fe, Mn, Se, Ag or Zn and anions like F^-, NO_3^-, PO_4^{3-}, SO_4^{2-}, CN^-, etc. These constituents would certainly affect quality of aquatic life, promote corrosion, indicate hardness and they have a tendency of formation of foam. The intensive measurements are carried out to characterise the quality of water. As told before, *in situ* analysis give immediate idea about the quality of water. Instrumental methods for *in situ* analysis involve measurement of temperature, turbidity, pH, dissolved oxygen, conductivity as they furnish rapid results. The automated methods are somewhat different, as they involve wet methods of analysis. A typical set-up for automated analysis comprises of sample turn-table, proportionating pump to draw effluent sample, a suitable mixing coil, an efficient separator for segregation of solids and liquids; thermal set-up and Online analyser comprising of refractometer or colorimeter, or conductometer or any suitable devise coupled with the recorder. The analytical data so obtained can be stored in computer and retrieved when required. Such kind of retrieval system does exist for surveillance of rivers and streams in Canada and America at US-EPA.

18.4 QUALITY OF INDUSTRIAL EFFLUENTS

It is difficult to classify industrial effluents on the basis of its composition, due to the diversity of modern industrial process and practices. It may be possible to preferably classify waste water on the basis of its contents, or source from which it emerges. For instance, effluent from food industry contains too much of organic matter, while effluent from textile industry contains detergents, while one from plating industry contains cyanide and metal pollutants. Such classification therefore is based upon the analytical procedures useful for specific industrial situation.

(a) Effluent from Chemical Industries

The waste water emerging from chemical industries, such as heavy chemicals and fine chemical factory or explosive factories, fertiliser plants and pesticide industry belongs to this category. The waste water is coloured with strong offensive odour, is acidic (pH 2.6); power plant waste is highly acidic (pH 1.2) while waste water from pesticide industry contains abnormally high concentration of organic phosphorus, which is toxic to flora and fauna in water. Phosphatic industry waste contains tallow and slimes, suspended matter with acidic character (pH 1.5–2.0). All waste have high BOD. The non-degradable matter increases BOD due to presence of soaps, detergents and fatty acids.

(b) Waste Water from Food, Beverage and Drug Industry

The food and beverage factories are growing in metropolitan cities like Mumbai, Kolkata, Delhi and Chennai. However, in western India we have a large network of pharmaceutical and drug industries. The effluents from such sources have a very high BOD, suspended organic solids, dissolved organic matter; so also effluent from a canned food, dairy products, meat poultry preparation, contains high level of proteins, carbohydrates, amino acids, alkali metals and phosphates. Rice processing industry waste contains a high level of starch. The distillery waste which is fermented contains a high level of nitrogen, with pH 10.8 and BOD greater than 430 mg/litre. Liquid effluents from fish processing industry has a strong odour, pickel waste is badly coloured. Table 18.1 gives characteristics of distillery waste. It has a high concentration of nitrogen, chloride, sulphates with very high value for BOD.

(c) Effluents from Textile and Apparel Industries

Table 18.2 presents characteristics of waste water from textile industries at different stages of processing. The quality of waste water from various textile operations is given in terms of pH, BOD and solid contents, while Table 18.3 gives quality of water not only from textile industry but also from tanneries, steel, paper and rayon industries. A close examination of Table 18.3 will indicate that the average pH of waste water in all cases exceeds 8.0. The BOD is low for paper industry, while COD is high for apparel or textile industry, the rayon waste contains high level of zinc, while paper industry waste water contains high levels of sulphate; finally leather industry has high level of the chlorides, alkalies and metals like Cr, as $K_2Cr_2O_7$ which is used for its treatment as hide seeking.

Table 18.1 Characteristics of typical distillery waste water

S. No.	*Parameter*	*Level*	*Remarks*
1.	Colour	Brown-black	Varies
2.	Odour	Offensive	Bad smell
3.	Temperature	89–96°C	Warm
4.	pH	4.0–4.6	Acidic
5.	BOD	3500–4500 mg/litre	High value
6.	COD	3500–9300 mg/litre	Average
7.	Total solids	5200–8600 mg/litre	High
8.	Suspended solids	2000–14000 mg/litre	Medium
9.	Total nitrogen	1000–1500 ppm	Very high
10.	Sodium	120–250 ppm	Normal
11.	Potassium	800–12000 ppm	Average
12.	Calcium	300–800 ppm	Average
13.	Iron	30–80 ppm	High
14.	Sulphate	1500–4500 ppm	High
15.	Chloride	4500–8000 ppm	Very high

Table 18.2 Significant features of textile industries waste during processing

S. No.	*Stages of processing*	*Special features*
1.	Desizing	High BOD, pH, total solids, alkalinity and high temperature
2.	Bleaching	High BOD, pH and suspended solids, chlorine
3.	Mercerising	Low BOD and solids, high alkalinity
4.	Dyeing	High BOD, neutral or alkaline pH, unreacted dyes
5.	Printing	Colouring matter, acidic effluent, suspended matter
6.	Finishing	Waxes, starch product, solids, low pH

Table 18.3 Characteristics of industrial effluents from miscellaneous industries

S. No.	*Parameter*	*Textile industry*	*Viscose rayon industry*	*Leather industry*	*Paper facıory*	*Steel foundry*
1.	pH	7.0–11.8	8.6–9.4	8.0	9.0	8.5
2.	BOD	100–900	–	1700	200	10,000
3.	COD	250–1800	50–200	3500	800	25,000
4.	Sulphate (mg/litre)	–	1500–7000		20–800	
5.	Chloride (mg/litre)	–	–	4600	200	–
6.	Zinc (mg/litre)	–	6–20	–	200–250	–

The industrial effluent from textile industry in particular has high BOD, suspended solid and high temperature. They have a bad colour due to dyeing and bleaching operation and is alkaline in nature.

The waste water from woollen mills contains soaps, oils, detergents; while ammonia, coppersalt, acetic-acid is found in silk mill waste. The leather industry waste contains high percentage of proteins, precipitated lime, chromium in the form of $K_2Cr_2O_7$, sodium chloride, ammoniacal nitrogen and suspended solids.

(d) Waste Water from Material Industry

Such industries compose of paper pulp, steel, electroplating, foundry, smithy, rubber and glass industries. Table 18.4 summarises kinds of pollutants liberated, by such material industries. The paper mill waste has colloidal colouring matter, sulphate non-degradable lignin sulphonates, waxes, dyes and bleaching agents. While steel industry waste is acidic due to presence of high mineral contents, cyanides and alkalis; the pickel waste contains zinc, chromium, iron, and rubber industries waste contains latex coagulated rubber, suspended solids and has a pH of 3.2–7.9.

Table 18.4 Characteristics of waste from material industries

S. No.	*Source*	*Kind of pollutants present*
1.	Leather industry	Sulphides, organic, lime and chromates and colour. Alkaline pH, large BOD/COD, dissolved solids
2.	Textiles and apparel	Alkaline pH, sodium, organic matters, dissolved and suspended solids and colouring dyes.
3.	Milk industry and food products	High BOD, colloidal solids, slow rate of biodegradation, offensive odours, dissolved and suspended solids.
4.	Slaughter houses or abattoir	Very high BOD, bad odours, proteins and toxic solvents.
5.	Drugs and pharmaceuticals	High or low pH, total solids, toxic organics, metal pollutants, organic compounds.
6.	Electroplating bath	Acidic and toxic with metal cyanides
7.	Paper pulp factories	Strong intense colour, sulphide of sodium, alkaline earths, lignin sulphonate, black liquor

(e) Effluent from Energy and Power Industry

We mainly consider effluent emerging from thermal as well as nuclear power plants. Such effluents contain silica, sulphates, fluorides, iron, aluminium, manganese, calcium, and indium, etc. Nuclear waste contains radioactive waste. Table 18.5 indicates analysis of effluent from power related industries. It contains high level of several metals.

(f) Effluents from Service Industries

It includes laundry and photographic process industries. The waste from laundry contains soap and detergents and is alkaline in nature, while waste from photographic industry is composed of developers hypo, silver bromide spent solutions, fixing inorganic reducing agents. They contain large amount of suspended solids with turbidity of 300–400 JU. They also contain non-degradable organic matter. The study of physical and chemical properties of waste gives an idea about its source, e.g. Table 18.6 gives concentration of metals in Thane Creek in Mumbai. The Thane Creek is a confluence of industrial waste from alkali, rayon, heavy chemical and petrochemical industries. Level of trace elements in the creek is shown. It will give an idea about the extent of pollution at several stations, which were located on both sides of the creek (Table 18.6).

Table 18.5 Analysis of effluents from typical power industries

Metals	*Sample No. (in μg)*			
	A	*B*	*C*	*D*
Pb	150	2640	–	750
Cu	5000	–	2000	500
Al	500	1250	–	500
V	500	425	450	925
Fe	–	2500	–	2000
Zn	400	3750	–	4000
Hg	2650	250	462	625

Table 18.6 Concentration of metals from Thane Creek in Mumbai

S. No.	*Station at*	*Concentration (μg/litre)*		
		Pb	*Zn*	*Mn*
1.	Sewree	33.0	21.2	6.8
2.	Mahul A	24.1	93.6	12.7
3.	Mahul B	84.0	79.2	3.8
4.	Between Trombay Jetty and Mahul	24.0	51.2	5.2
5.	Trombay Jetty	18.2	64.8	6.7
6.	Near Vashi bridge	19.2	110.0	13.8
7.	In between Thane and Vashi bridge	13.8	110.0	3.0
8.	Thane Town	15.2	86.5	6.8
9.	Vashi A	30.5	70.5	3.6
10.	Vashi B	21.0	106.8	6.8

Table 18.7 gives tolerance limit of industrial effluent as per ISI specification No. 2490. It shows interesting features, viz. suspended solids do not exceed 500 mg/litre, while dissolved solids are within limit of 2000 mg/litre; ammonical nitrogen is 50 mg/litre, pH is 5.5, sulphate and chloride contents are less than 1000 mg/litre. The metals like Pb, Hg are present in small concentration, while high level of zinc are observed in the sample of effluents.

18.5 SAMPLING AND ANALYSIS OF EFFLUENTS

A sample must be randomly drawn and should be a representative of entire contents. The total number of collected grab samples must satisfy the statistical requirements of the sampling programme. Human error can be eliminated by the use of manual as well as an automated sampling device. The determination of gases like O_2, CO_2, Cl_2, CH_4, or H_2S should be preferably done *in situ* analysis. At the same time temperature and pH of effluent must be noted on site. The effluent analysis comprises of three components, viz. the physical methods for characterisation; the analysis of organic pollutants, involving non-specific and specific analysis; the analysis of metals largely involving instrumental methods of analysis of

Table 18.7 Tolerance limits for industrial effluents (IS: 2490 Part I, 1981) (mgs/Litre)

S. No.	*Parameter*	*Surface discharge*	*Effluent (public sewers)*	*Effluent (soil irrigation)*
1.	Suspended solids	100	600	200
2.	Dissolved solids	2100	2100	2100
3.	Oil and grease	10	20	10
4.	Ammoniacal nitrogen	50	50	50
5.	BOD	30	350	100
6.	Temperature, °C	40	45	38
7.	pH	5.5–9	6.5–9	5.5
8.	Chloride	1000	1000	600
9.	Sulphate	1000	1000	1000
10.	Lead	0.1	1	–
11.	Mercury	0.01	0.01	–
12.	Zinc	5	15	–

anions and dissolved gases. We would first consider various physical methods of analysis and examine what parameters can be analysed. The details of methods of analysis are given in earlier chapters.

18.6 PHYSICAL METHODS OF CHARACTERISATION

Such methods of analysis depends upon the purpose of analysis, nature of constituents present, and the nature of analytical procedure to be adopted. The important physical properties which are generally accounted for waste water are: density and viscosity; temperature; conductivity; turbidity; particulate matter and dissolved solids; oil and grease; colour, radioactivity and odour.

(a) Density and Viscosity

The determination of density is of utmost importance for characterisation of sea brines, to indicate dissolved materials. Such determination is carried out at 20°C by finding weight of exact volume of waste water. Secondly viscosity determination provides direct information on the resistance of liquid to flow, if water contains solid sludge. This information will be very useful while dealing with oil pollution. Pure water has viscosity of 1.009 cps at room temperature. The usefulness of such analysis is demonstrated by determination of hydrological parameters of Ulhas river presented in Table 18.8 to trace the source of impurities.

(b) Temperature

When thermal pollution is present, the water from cooling towers or venturiscrubbers is warm. This information throws light on existence of oxygen and flora and fauna in waste water. It also gives an idea about biochemical activity in effluent. It is of great relevance while assessing consequences of thermal pollution on quality of water.

(c) Electrical Conductance

The measurement of electrical conductivity of effluent gives information about ionic contents of waste water. The quality of surface or ocean water can be quickly ascertained for salinity by simple measurement

Table 18.8 Hydrological parameters of Ulhas river effluents at twelve monitoring stations in Mumbai

Station No.	*pH*	*Temp. °C*	*NO_2–N mg/litre*	*PO_4^{3-} mg/litre*	*SO_4^{2-}*	*Alkalinity (phenolphthalein)*	*Total alkalinity*
1.	8.0	30.0	0.9	0.09	2.83	6.9	71
2.	7.8	31.5	0.4	0.09	2.94	6.9	70
3.	7.8	32.0	0.7	0.08	1.01	6.8	72
4.	7.6	32.0	2.0	0.05	2.15	7.3	74
5.	7.5	31.8	2.1	0.08	2.29	7.8	79
6.	7.6	32.5	1.8	1.1	2.42	7.9	80
7.	7.7	32.0	1.1	0.05	0.7	6.8	72
8.	8.0	32.0	0.8	0.05	0.9	6.7	73
9.	7.5	31.0	8.4	0.06	2.1	8.0	78
10.	7.6	32.5	7.2	0.06	2.2	7.8	81
11.	7.8	33.0	7.3	0.05	2.9	7.2	72
12.	7.8	33.0	5.8	0.05	2.9	8.1	81

of electrical conductivity. Such measurement should be carried out at a constant temperature. Such analysis is also useful in chemical oceanography to ascertain quality of waste water. The conductivity meter such as Wheatstone bridge, is useful for quick determinations.

(d) Turbidance

It is a measure of light scattering characteristics of water. Such turbidity is ascribed to the presence of colloidal and suspended particulate matter in waste water. Jackson Turbidity test is based on measuring the length of light path traversed through the solution, at which if viewed flame of standard candle disappears. If effluent is highly turbid, it should be diluted with turbidity free water. If turbidity is > 1000 JU, Jackson candle method should be used, but if it is < 25 NTU, a nephelometer is best for the purpose for the measurement.

(e) Volatile and Dissolved Solids

This information can be obtained by filtration and weighing as is done in gravimetric analysis. The total suspended matter can be removed by ultrafiltration, while dissolved matter can be weighed after evaporation of known volume of the effluent. This test is good for control of plant operation in water. Such analysis was carried out at Thane Creek estuary, the results of which are, presented in Table 18.9. It has furnished very valuable information on source of contamination.

(f) Oil and Immiscible Liquids

These are removed by extraction, when the effluent is emerging from an oil refinery. The information on oil, grease and immiscible liquids throws light on the quality of the effluents. Benzene, chloroform, carbontetrachloride, diethylether is used for the purpose of such extraction. The infrared spectroscopy can be easily utilised for the gross determination of volatile and non-volatile oily matter in the effluent.

(g) Colour

The colour of the effluent may be its true colour or an apparent colour. As described earlier platinate units are used for expressing intensity of such colour. It is also expressed in Hazen units. Majority of

Table 18.9 Hydrological parameters of Thane Creek effluent at twelve monitoring stations in Mumbai (μg/L)

Station No.	*pH*	*Temp. °C*	NO_2^-	PO_4^{3-}	SO_4^{2-}	*Phenolphthalein alkanity*	*Total alkanity*
1.	7.6	31	3.75	0.087	2.87	7	80
2.	7.8	31	6.0	0.11	2.91	8	88
3.	8.1	37	7.25	0.26	2.78	8	92
4.	8.1	34	9.4	0.21	2.81	8	94
5.	8.0	33.5	6.2	0.14	2.84	10	98
6.	7.4	31.5	2.9	0.09	2.89	8.2	79
7.	7.6	32	2.65	0.21	2.74	10.1	82
8.	7.5	32.5	3.1	0.16	2.69	9.1	79.8
9.	7.5	32	4.0	0.13	3.19	10	81
10.	7.6	33	3.5	0.10	2.94	9	80
11.	7.8	34	2.85	0.12	2.72	9.5	81
12.	7.8	35	5.15	0.13	3.0	8.1	81

the industrial effluents are coloured. A comparison is made in Nessler tube with standard colour solutions made from potassium hexachloroplatinate tinted with small amounts of cobalt chloride. A 1 mg/litre of Pt gives a standard unit of colour. A solution containing 500 mg/litre of Pt has 500 units. The samples with more than 70 units of colour are diluted first and then compared with the standard solutions. It is also possible to measure colour by tristimulus colorimeter.

(h) Odour Measurements

It is a measure of physiological response to olfactory nerves. Odour is present due to volatile organic matter and biological materials such as algae and actinomycetes. Odour is expressed in term of (TON), threshold odour number. TON represents the greatest dilution of sample that still persists, perceptile residual odour. Odour depends upon temperature, hence all measurements should be carried out at fixed temperature, say at 25, 40 or 60°C. When the effluent emerges from food, beverage industry or canning and fish processing plant, it is imperative to measure odour intensity.

(i) Radioactivity

It can be measured only if the effluent originates from a nuclear reactor. Waste water from such sources are in a state of nuclear instability and emits ionising radiation. The quality of water around nuclear reactor is hazardous.

18.7 ANALYSIS OF ORGANIC POLLUTANTS

The details of such analysis are provided in earlier chapters. Here we consider only principles.The presence of organic matter has damaging influence on quality of water. To know the exact nature of organic pollutants, it is essential to carry out non-specific and specific analysis of the effluent. The non-specific methods involve evaluation of BOD, COD (*i.e.* chemical oxygen demand), total organic carbon (TOC), while specific analysis involves characterisation and analysis involving UV, visible, IR, NMR, fluorescence, GC-MS determination. We would first consider non-specific analysis. It would account for quantities of oxidisable organic material present in the effluent. BOD is initially employed for measurement of susceptibility of waste to biological oxidation. On the contrary, the measurement of concentration of organic matter oxidisable by dichromate reflux method is ascertained by chemical oxygen demand.

(a) BOD of Industrial Effluents

As described earlier the biochemical oxygen demand test is a bioassay method which involves measurement of quantities of oxygen consumed during biological oxidation of organic waste matter under controlled conditions. BOD is defined basically as the amount of oxygen required by bacteria, while stabilising decomposable organic matter under aerobic conditions. Thus BOD test is used to ascertain pollution strength of waste in terms of oxygen required, if discharged into natural water sources which is acidic in nature. This is a bioassay test for which toxic substances in water must be absent to promote bacterial growth. It is a wet oxidation procedure. The rate of such oxidation depends upon population of number of microorganisms and temperature. Inorganic nitrogen and nitrifying bacteria would reduce BOD. The addition of methylene blue or allyl thiourea (ATU) inhibits nitrifying bacteria. Thus the amount of oxygen required is fixed, BOD gives idea of gross concentration of organic matter and oxygen consuming tendency of waste water. BOD is useful for determination of amount of oxygen likely to be consumed by waste upon discharge in receiving water. The main limitations of 5-day BOD test is the need of patience to wait for five days to get reliable information.

Since some organic matter is resistant to biochemical oxidation, the IDDO, *i.e.* immediate determination of dissolved oxygen demand was devised. This is 15-minute test to ascertain the rate of oxidation of ferrous to ferric or sulphite to sulphate or sulphide to sulphate. In order to avoid photosynthetic bacterial degradation, the test is carried out in dark. Toxic matter retard biological activity. Aromatic hydrocarbons are resistant to biological activity. As a rule aromatic compounds undergo decomposition, but aliphatic compounds like ester do not undergo degradation. Similarly, macromolecules or biologically inactive and inert matter do not decompose, irrespective of the condition one employs. In such a case one has to take resort to determination of COD.

(b) Chemical Oxygen Demand

The experimental details of determination are provided in earlier chapter. This is widely used as a means of measuring pollution strength of domestic and industrial effluents. It allows measurement of total amount of oxygen required for oxidation to CO_2. It is based on the premise that all organic compounds can be oxidised by strong oxidising agents in acidic media. The COD values are greated than BOD values. COD is too large, if great amounts of biologically resistant organic matter is present, e.g. lignin. COD unfortunately cannot differentiate between biologically oxidisable and biologically inert organic matter. The rate of oxidation is not precisely understood in COD analysis. The outstanding advantage of COD in comparison to BOD is short time required for determination. In three hours all valuable information can be obtained. A variety of oxidising agents like ceric sulphate, potassium iodate are used, but potassium dichromate is best for the oxidation. Other useful agent are $KMnO_4$, $K_2S_2O_8$, $HClO_4$, HIO_3 and HNO_3. In order for $K_2Cr_2O_7$ to oxidise organic matter, the solution is made acidic in 50% sulphuric acid, under anhydrous condition. Reflux condensers are used to prevent loss of volatile organic compounds. Low molecular weight fatty acids do not get easily oxidised, unless a catalyst like silver salt is added, e.g. pyridine and aromatic hydrocarbons are not easily oxidised. There is interference caused by chloride and Fe(II). The former is eliminated by addition of mercury salt, which precipitates chloride as HgCl, e.g. $Hg^{2+} + 2Cl^- = HgCl_2$ ($K = 1.7 \times 10^{13}$). Sulphonic acid is added to oxidise nitrites to nitrates. The chloride interfere as: $6Cl^- + Cr_2O_7^{2-} + 14H^+ \rightarrow 3Cl_2 + 2Cr^{3+} + 7H_2O$, hence mercuric sulphate is added to destroy excess chloride. Organic nitrogen reduces dichromate, but slow addition of acid gives good results, specially when volatile matter is present.

The measure of 80% theoretical oxygen need, or 90% of oxygen requirement with silver catalyst can be assessed by COD. At a time a ratio of COD/BOD is used in data interpretation. During analysis, 0.125 N $K_2Cr_2O_7$ is used, but 0.25 N $K_2Cr_2O_7$ gives best results. An excess of oxidising agent must be present in COD analysis. Hence excess is added and actual amount reduced is determined. Ferrous ammonium sulphate is used for back titration, e.g.

$$6Fe^{2+} + Cr_2O_7^{2-} + 14H^+ \rightarrow 6Fe^{3+} + 2Cr^{3+} + 7H_2O$$

Ferroin (ferrous 1.10 phenanthroline sulphate) is used as an indicator; with sharp colour change. The COD test is very useful for analysis of industrial waste. Within short times results are obtained. COD gives an idea of toxic conditions and the presence of biologically resistant organic compounds in the effluent.

(c) Total Organic Carbon (TOC) Analysis

This is fact involves analysis of carbon, hydrogen, nitrogen, sulphur, phosphorus and oxygen contents of the effluent. Such analysis supplements BOD or COD determination for the estimation of pollution potential of industrial waste. Nitrogen is estimated volumetrically by Kjeldahl's method, phosphorus is estimated gravimetrically as yellow ammonium phosphomolybdate, while sulphur is gravimetrically estimated, as sulphate and carbon as usual is determined by Carius method. For details refer to earlier chapters.

(d) Specific Analysis of Organic Pollutants

As discussed in earlier chapters on air pollution, we can assess the presence of organic pollutant by spectral methods. The problem is complex, as we have to handle a mixture of organic compounds from the effluent. The compounds are further present in very small concentration. Biochemical degradation or chemical transformation complicates the analysis. The nature of waste is complex and variable. The components are present in diluted form. It should be concentrated and separated before spectral characterisation. Such concentration is done by evaporation or distillation or by cryogenic sampling. They can also be isolated by the process of solvent extraction, selective precipitation, fractional crystallisation, selective adsorption or separation involving paper, thin layer or gas chromatography. Organic compounds are identified by detection of the functional groups by use of chromogenic ligands from characteristic colour reaction. UV, IR, NMR are used for further confirmation, e.g. thermal pollution can be detected by IR measurements. Phenolic compounds present are recognised by UV spectroscopy due to bathochromic blue shift to longer wavelength in spectrophotometric measurements. The phenolic compounds are characterised at nanogram level by mass spectrometry. Carbohydrates are detected and determined by spectropolarimetry and phosphorimetry. GLC and fluorimetry helps us to analyse major compounds like phenols, pesticides and surfactants. Pesticides are found in agricultural waste, while surfactants and detergents are encountered in textile and domestic effluent. They cause foaming. Biodegradable surfactants are limited in use. They are less harmful and hence environmentally friendly. Biodegradation is a measure of susceptibility of organic compound to microbial metabolism, which may proceed either in aerobic or anaerobic environment. This process is identified by type and number of micro-organism, structure of organic compound and other environmentally conducive conditions. Sewage micro-organisms are used for biodegradibility measurement. NMR/ESR is used for confirming the structure if one has any doubt.

18.8 ANALYSIS OF METAL POLLUTANTS IN EFFLUENTS

The experimental details are described under AAS, ICP-AES, capillary electrophoresis methods in earlier chapters. Several industrial effluents from alloy industry, electoplating baths, refineries, fertiliser complex, tanneries, electronic industry consists of a large concentration of metal pollutants. Out of 104 elements discovered only 17 are toxic and out of these five to six are encountered in chemical effluents. Table 18.10 indicates presence of metal pollutant from industrial effluents. The main metals encountered are mercury, cadmium, chromium and lead. The metals were analysed by spectral methods as well as by ring-oven technique. Interestingly, the results match well with one another. The spectrophotometric methods have maximum applications in these types of analysis due to their interference. Sometimes it is necessary to undertake separation and concentration of these ions before analysis due to interference. Ion exchange or solvent extraction is commonly used. The common methods used for such determination include analytical, molecular and atomic absorption spectroscopy; emission spectroscopy, activation analysis and electroanalytical methods.

Table 18.10 Metal pollutants from typical industrial effluents

Metal pollutant	*Amount found µg/10 ml*	*Average µg/ml*	*By ring-oven method µg/litre*
Hg	0.10	0.01	10.0
	0.25	0.025	25.0
Cd	0.10	0.01	10.0
	0.30	0.033	33.0
Cr	0.10	0.01	10.0
	0.10	0.01	10.0
Pb	0.50	0.05	55.0
	0.60	0.06	60.0

(a) Absorption Spectroscopy

These methods mainly cover molecular and atomic absorption spectroscopy. In former Beer's law, ($A = abc$) is used. A suitable chromogenic ligand is complexed with metal pollutant to form coloured complex. The sensitivity of analysis is enhanced by controlling various factors such as pH, λ_{max}; reagent or metal concentration. The most popular technique is Atomic Absorption Spectroscopy (AAS). The principal merit of AAS is that, no separation of interfering ions is necessary, multiple samples for multi-element analysis can be analysed at nanogram levels. This technique has made dramatic change in metal pollutant analysis. Only limitation of AAS is, metals like mercury are to be analysed in vacuum by cold vapour AAS, while metals like Sn, Sb, Se, Te are analysed with hydride generation. Although flame AAS gives rapid results, graphite furnace gives reproducible results.

(b) Emission Spectroscopy

This technique comprises of (AES) which includes mainly flame emission and inductively coupled plasma atomic emission spectroscopy (ICP-AES). Molecular emission spectroscopy covers fluorescence spectroscopy. The flame photometry is largely used for analysis of alkali and alkaline metals, which are commonly present in sea brine. ICP-AES technique is an extension of AES with argon plasma torch. The advantage is argon flame ensures complete atomisation permitting multi-element analysis and possibility of rapid determination. In molecular luminescence methods viz. in fluorescence spectroscopy metals are analysed with Morine(Be), dibenzoylmethane (U), alizarin (Al), 8-hydroxyquinoline (Zn). Tables 18.11 and 18.12 are presented results of analysis of metal pollutants with AAS from Ulhas river effluent and Thane Creek estuary. The important metals like Mn, Fe, Cu, Zn, Cd and Pb were analysed by AAS. Their important source is effluent emerging from surrounding

Table 18.11 AAS determination of elements from Thane Creek estuary in Mumbai

Station number	*Concentration (μg/litre)*					
	Mn	*Fe*	*Cu*	*Zn*	*Cd*	*Pb*
1.	3.40	54.00	10.60	82.80	1.50	19.2
2.	6.30	45.00	13.80	73.80	1.40	12.7
3.	2.60	60.30	10.80	38.40	1.10	13.8
4.	12.10	44.00	10.10	69.40	1.30	16.2
5.	17.50	47.00	10.80	45.60	1.60	13.5
6.	6.80	52.00	14.20	69.00	1.70	10.5
7.	3.80	73.60	9.20	70.50	1.20	17.2
8.	2.50	80.40	7.80	80.20	1.40	17.8
9.	3.60	64.20	10.90	70.50	1.30	27.2
10.	6.80	72.40	18.80	106.80	1.00	17.8
11.	2.40	58.80	10.40	108.90	1.10	21.8
12.	13.60	147.20	10.10	52.00	1.50	17.8

The highest level of metal is underlined in Table 18.11 and 18.12.

Table 18.12 AAS determination of elements from Ulhas river effluent in Mumbai

Station number	*Concentration in μg/litre*					
	Mn	*Fe*	*Cu*	*Zn*	*Cd*	*Pb*
1.	3.6	50.2	11.4	53.5	1.4	19.0
2.	3.9	37.5	13.2	73.5	1.4	16.6
3.	3.5	32.5	7.2	32.0	1.1	14.4
4.	4.4	29.1	24.0	35.0	1.9	15.8
5.	2.7	32.5	12.4	29.5	1.4	13.7
6.	2.6	37.5	6.8	37.2	1.4	14.1
7.	8.0	31.5	8.5	84.0	1.9	52.0
8.	10.2	39.5	12.3	45.6	1.7	55.0
9.	8.2	42.6	16.4	110.0	1.4	15.3
10.	7.8	41.5	8.2	97.5	1.3	11.2
11.	7.8	30.4	5.8	104.2	1.2	12.5
12.	5.4	40.5	6.2	96.4	1.4	11.4
13.	6.7	38.5	4.5	82.5	1.5	10.8

factories, *e.g.*, pesticides (Cu), rayon (Zn), nuclear reactors (Cd), auto-exhaust emission (Pb), chloroalkali industry factories, e.g. (Hg), and foundry or smithy (Mn), etc. The monitoring was carried out from various stations located on both sides of the river (Table 18.6). The metals were concentrated on Chelex-100 resin and eluted with dilute hydrochloric acid and were determined by AAS. The results are interesting. A work on chemical speciation with ion chromatography can also indicate interesting information on quality of effluent.

(c) Electroanalytical Analysis Including Voltammetry

In Polarographic analysis, Ilkovic equation is used, which states that $i_d = 607\ n\ D^{1/2}\ m^{2/3}\ t^{1/6}\ C$, where i_d = diffusion current, D = ficks diffusion coefficient, m = rate of flow of mercury, t = time in seconds between two successive drops and C = concentration. If we keep *D, m, t, n,* constant, then we have $i_d \propto C$. It follows a linear relationship with concentration. The distinct advantage of polarographic method is use of half-wave potential for identification of metal pollutant. Amongst other electrometric methods, conductance measurement and ion selective analyser gives valuable information. The latter is used for analysis of anions. The experimental details are provided in earlier chapters.

18.9 ANALYSIS OF ANIONS AND DISSOLVED GASES

Most common anions which are present in effluent are SiO_2, S^{2-}, SO_4^{2-}, SO_3^{2-}, F^- and CN^-. These anions are concentrated by anion exchange chromatography. The fluoride is analysed by zirconium alizarine spectrophotometry method. Sulphides occur in tannery waste, septic tanks, oil refinery, tannery and viscose rayon waste. It is detected with lead acetate paper, and is determined by complexation with p-aminodimethyl-aniline hydrochloride, to produce methylene blue colour dye, which is measured spectrophotometrically. The sulphites are found in paper and pulp industries waste, while cyanide is encountered in effluents from electroplating. It is determined by interaction with ferroin at a pH 9.2–9.7. The blue complex so formed is extracted in chloroform and determined spectrophotometrically. Chlorides are determined by Volhard's or Mohr's method, involving precipitation titrations.

Amongst dissolved gases we have to primarily consider dissolved oxygen and chlorine. The dissolved oxygen is determined by Winkler's method, with azide modification as nitrate interfere in iodometric titration. The reaction of interference and elimination of interference are given as follows:

$$2NO_2^- + 2I^- + 4H^+ \rightarrow I_2 + N_2O_2 + 2H_2O$$

$$2N_2O_2 + O_2 + 2H_2O \rightarrow 4NO_2^- + 4H^+$$

$$NaN_3 + H^+ \rightarrow HN_3 + Na^+ \text{ (Elimination of } NO_2^-\text{)}$$

$$N_3H + NO_2^- + H^+ \rightarrow N_2 + N_2O + H_2O$$

This modification is termed as Winkler's Ashtelberg's method. The method for analysis of dissolved chlorine is same as O-tolutine arsenite spectrophotometry. The best method for analysis however will be ion analyser fitted with suitable ion selective electrodes.

Table 18.13 gives Indian Standards on tolerance limits of metals from different effluents.

There are several parts dealing with waste from distillery tanning, strawboard, electroplating, dye-stuff, coke oven, phosphate and nitrogen in fertiliser industry. The cane sugar diary, paper and tanneries are also covered. This should prove as the useful guideline for analysis, as well as for selecting method of treatment for the effluents.

Table 18.13 Indian standards (ISI) on tolerance limits of different effluents

	Tolerance limits for effluents discharged into surface waters	*Serial No. of prescribed standards*
1.	Part I: General limits	2490–1974 (Part I)
2.	Part II: Distillery industry	2490–1974 (Part II)
3.	Part III: Tanning industry	2490–1974 (Part III)
4.	Part IV: Strawboard industry	2490–1974 (Part IV)
5.	Part V: Electroplating industry	2490–1974 (Part V)
6.	Part VI: Dyestuff and dye manufacturing industry	2490–1976 (Part VI)
7.	Part VII: Coke ovens	2490–1976 (Part VII)
8.	Part VIII: Phosphatic fertiliser industry	2490–1976 (Part VIII)
9.	Part IX: Nitrogenous fertiliser industry	2490–1977 (Part IX)
10.	Tolerance limits for effluents discharged into marine areas	7968–1976
11.	Effluents discharged into sewers	3306–1974
12.	Effluents discharged on land irrigation	3307–1977
13.	Inland surface waters subject to pollution	2296–1974
14.	Sewage effluents discharged into inland surface waters	4764–1973
15.	Criteria for controlling pollution of marine coastal areas	7967–1976
16.	Distillery effluents	8032–1976
17.	Effluents of cane sugar industry	4903–1968
18.	Effluents of dairy industry	8682–1977
19.	Effluents of electroplating	7453–1974
20.	Effluents of pulp, paper and board	5061–1978 (Part I)
21.	Effluents of tanning industry	5183–1977
22.	Steel plant effluents	8073–1976

ADDITIONAL READINGS

18.1 G.M. Gair, J.C. Gayer and D.A. Okun, *Water and Waste Water Engineering*, Vol. 1 and 2, John Wiley & Sons (1968).

18.2 R.A. Horn, *The Chemistry of Our Environment*, John Wiley & Sons, New York (1978).

18.3 C.N. Sawyer and P.J. McCarty, *Chemistry for Environmental Engineering*, 3rd Ed., McGraw-Hill and Co. (2001).

18.4 K.H. Mancy, *Instrumental Analysis for Water Pollution Control*, Ann Arbor Science Publ., Ann Arbor, Michigan (1971).

18.5 D. Purves, *Trace Element Contamination of the Environment,* Elsevier, Netherlands (1977).

18.6 H.J.M. Bowen, *Environmental Chemistry of the Elements*, Academic Press, London (1979).

18.7 K.H. Mancy and W.J. Weber, *Analysis of Industrial Waste Water,* Wiley Inc. (1971)

18.8 S.M. Khopkar, *Environmental Pollution, Monitoring and Control*, New Age International Publishers (P) Ltd., New Delhi (2001).

18.9 S.M. Khopkar, *Basic Concept of Analytical Chem.,* 3rd Ed., New Age International Publishers (P) Ltd., New Delhi (2008).

18.10 S.K. Schoo and S.M. Khopkar, *Ind. J. Env. Protection*, **5**, 25 (1985); *Ind. Env. J.* **15**, 37 (1986). Environmental Management and Planning 291 (1988), *Ind. J. Envi. Prot.* **12** (1992).

18.11 S. Cherion, M.N. Gandhi and S.M. Khopkar, *Ind. J. Env. Prot.* **12** (1992).

18.12 S.N. Nimbhorkar and S.M. Khopkar, *Ind. J. Tech.* **24**, 658 (1985).

18.13 S.M. Khopkar, *Ind. J. Env. Prot.* **7**, 268 (1987).

CHAPTER 19

Domestic Effluents and Their Analysis

19.1 INTRODUCTION

India is country where several of its towns have beautiful lakes providing a recreation facilities for people, unfortunately some towns in India have unplanned growth and very few like Chandigarh or Bhubaneswar which are well planned cities with various sectors providing all basic facilities. These planned towns have well planned sewage system which maintains cleaner atmosphere. Unfortunately in some cities like Mumbai, Bangalore, Kolkata have exhibited a tremendous growth of population with virtually providing with no systematic plan for disposal of sewage. In several regions, towns sewage is merrily discharged into river or brooks. Although such sewage gets diluted thereby reducing harmful effects on human beings it is not always rendered completely harmless. Even in metropolitan towns, new colonies or dwellings are coming up with absolutely no systematic planning for discharge of waste. In last two decades, the solid waste has created greater menacing problem in cities and the biomedical waste disposal is real challenge to an environmental workers.

The problem of sewage/sludge disposal if properly handled can lead to better utilisation of effluent. In future there is bound to be growing challenge of conservation of waste water. As solution to this problem several industrial houses and establishments in cities have started recycling and treating the waste water and reutilising for washing, farming and cooling purposes. Even the solid waste generated, is treated with vermicules and used as the most effective fertiliser for soil. Several farms in Prisons and rail siding have embarked on the experiment of recycling and reuse of solid/liquid waste as the manure.

To facilitate future planning, we must understand what is the composition of domestic effluent? How such effluent can be treated to render it safe for handling with aerobic and anaerobic treatment of sewage, or tertiary treatment of the domestic effluent? The handling of such effluents poses a serious problem due to its offensive odour and hazardous effect on health of workers. The analysis or characterisation of the effluents is greater challenge to environmental workers. All such aspects dealing completely with sewage, sludge and sullage are therefore discussed at greater length in this chapter.

In the present chapter, it is envisaged to consider domestic effluents emerging out from houses, towns and municipalities. This waste water not only flows from toilets but also emerges out of kitchen after its use for washing, bathing, cooling and other household activities. This was termed as sewage, which largely contains human faeces and urine while water used for bathing, washing and other household activities containing soaps, detergents, solids. The latter is termed as sullage. As the industries grew, the number of small-scale industrial operations have also risen. In the absence of planned programme for disposal of industrial waste water, small-scale industry operators think it fit to let off waste water into sewage disposal channels. Therefore the modern day sewage in reality consists of sewage, sullage and small-scale industries waste water. In view of this complex nature, it is thought worthwhile to call sewage and sullage as the domestic effluent.

19.2 WHAT IS DOMESTIC EFFLUENT?

As told in preceding section, domestic effluent is composed of sewage, sullage and waste water from man's activities in environment. On an average a human being discards 135–270 gram of faeces and 1–1.5 litre of urine per day. If we consider its composition, a faeces contains about 66–80% moisture, 88–97% organic matter, 5–7% nitrogen; 3–5% phosphorus; 1.0–2.5% potassium, 44–55% carbon and 4.5% of calcium. The urine is composed of 96% moisture, the rest 4% constitutes 85% organic matter; 19% nitrogen; 5% phosphorus; 4.5% potassium; 17% carbon and approximately 6% of calcium. Since it contains pathogenic organism, it is extremely dangerous for health. Sewage has offensive odour and bad colour. Stale sewage has dark colour.

If one looks at the overall composition of domestic effluent, it contains 99.9% water and 0.1% solids; out of this 0.1% solids it contains 30% inorganic matter and 70% organic matter. The inorganic solid contains grit, metals and salts, while the major portion of organic solids contain proteins, fats and carbohydrates. Sullage consists of chemical detergents, soaps, fats, grease which are used for washing utensils in kitchen, clothes and milk products (Table 19.1). The strength of sewage gives an idea about its quality. Such strength is expressed in terms of 5-day BOD and COD. One has to consider COD,

Table 19.1 Nature of typical domestic effluent

S. No.	*Parameter*	*Values in mg/litre*
1.	pH	5.5 to 9.0
2.	Total suspended solids	600
3.	Temperature	< 45°C
4.	BOD (5-day)	500
5.	Oil and grease	100
6.	Phenolic compounds	5
7.	Cyanides	2.0
8.	Radioactive materials	10^{-7}
9.	Chromium	2.0
10.	Copper	3.0
11.	Lead	1.0
12.	Nickel	2.0
13.	Zinc	15.0
14.	Ammoniacal nitrogen	–
15.	T.D.S.	2100
16.	Sulphate	1000
17.	Chlorides	600
18.	Boron	2.0
19.	Sodium	60

because lot of non-degradable materials like soaps, detergents, esters and flavours are present in domestic effluent. On an average a sewage with 200 mg/litre BOD < 400 mg/litre or COD is termed as weak sewage, while with 500 mg/litre of BOD or 1000 mg/litre COD it is termed as strong sewage and one with > 750 mg/litre of BOD or > 1500 mg/litre of COD it is termed as extremely harmful sewage (Table 19.2). If we consider the average composition of sewage, it contains 5-day BOD of 280 mg/litre, 402 mg/litre of suspended solids, 1000 mg/litre of dissolved solids, 200 mg/litre of chloride and 25 mg/litre of ammoniacal nitrogen. With rise in temperature, the 5-day BOD value also increases.

Table 19.2 Comparison of 5-day BOD/COD contents of effluents

Domestic effluent contents	*COD mg/litre*	*5-BOD mg/litre*	*BOD in % of COD*
Ethanol	2080	1820	87.4
Acetone	2170	1680	76.5
Acetic acid	1070	860	80.0
Phenol	2380	1100	46.0
Glucose	1010	600	59.3

19.3 TREATMENT OF DOMESTIC EFFLUENT

The principal objective of treating sewage of domestic effluent is two-folds to reduce toxicity and prevent the possibility of spread of diseases by pathogens and to eliminate pollution of surface and ground water. One of the ideal ways to prevent such pollution of surface waters is to make beneficial end use after pretreatment, e.g. with protein production, irrigation of soil, growth of algae and fish. The pipes carrying sewage are called sewers. When one considers the treatment techniques, one has to know what is the objective of such a treatment? The main aim is first to remove inorganic solids like grit by the process of sedimentation. This is followed by degradation or disintegration of organic matter into easily assimilable form, with liberation of methane by biological treatment and finally refining the effluent so as to make it fit for reuse for the agricultural purpose or similar home applications after disinfection. We therefore classify treatment in the following category as: primary, secondary and tertiary treatment.

Table 19.3 Nature of domestic effluent before and after its primary treatment

Content	*Percentage* *Before*	*After*	*Remarks*
Moisture	93.99	96.4	Increase
Ashless substance	77.50	81.4	Increase
Ash	22.50	38.9	Increase
N_2	3.42	3.10	Decrease
Fats	13.47	9.30	Decrease
Carbohydrate	9.08	7.20	Decrease
Hemicellulose	8.90	7.30	Decrease

It is worthwhile at this stage to consider quality of sewage and sludge before and after primary treatment as shown in Table 19.3 and note increase or decrease of parameter.

It will be seen that although moisture contents rises by 3%, at the same time fats, carbohydrates and cellulose contents are substantially lowered, thereby making it less harmful to human beings. The main objective of any treatment procedure is to remove coarse particles, extract pollutants, absorbs or distill or neutralise particles if possible, undertake disinfection by chlorination to reduce pathogens and have biological treatment to promote fermentation and liberate ion of methane. It would be worthwhile to consider briefly the various methods of treatment, so as to design proper analytical methods for analysis.

19.4 PRIMARY TREATMENT OF DOMESTIC EFFLUENT

This operation mainly involves collection of sewage and its initial treatment. Such treatment should consider the criteria from the view point of health, reuse, ecological balance, public nuisance, cultural and operational aspects and cost-benefit analysis. No one system satisfies all the above criteria. The physical process mainly involves screening, removal of grit and organic suspended solids by sedimentation, removal of large floating objects. Grit, also called detritus, is heavy inorganic fraction of sewage solids, which settles fast in sedimentation process. Such grit is used for land fill or disposed off by burial. Sedimentation is the gravitational separation of suspension into its components solid and liquid phases. A high degree of clarification is aimed at removal of solids. This reduces load on secondary biological treatment and simplifies treatment in tertiary methods.

19.5 SECONDARY TREATMENT

As such, out of 0.1% of solids in sewage, 70% is organic matter consisting of fats, carbohydrates and proteins, which are responsible for generation of bad odour and growth of disease producing micro-organism; and bacteria, virus, algae and protozoa. Bacteria, algae and protozoa promote biological degradation. Bacteria are non-photosynthetic single-cell organism. The oxidation of organic matter provides with food. They can tolerate acidic and alkaline conditions and withstand high temperature. The disease carrying bacteria are cholera, dysentery, typhoid, fever and diarrhoea, and are infective. The bacteria useful in aerobic treatment of sewage are rod shaped, facultative and mesophilic. Viruses are peculiar microbes, which do not use organic or inorganic compounds from sewage for their growth, they reproduce by invading almost all cells, whose reproductive process. They redirect to manufacture more virus particles. Algae are mostly multicellular photosynthetic organisms; CO_2 is used by them to reproduce. Their growth is stimulated by phosphates and nitrates from water. They cause what is called as nutrient enrichment or 'Eutrophication' of water and extensive algal growth in the form of algal bloom. Finally, protozoa are unicellular proteins are ancestral to animal kingdom. They are responsible for purification of sewage, e.g. in their presence COD is lowered from 198 to 124 mg/litre, while organic nitrogen is reduced from 14 to 7 mg/litre, while suspended solids diminish by 86 to 26 mg/litre. The intestinal parasites include worms, flukes, and are not necessarily present in every sewage.

(a) Aerobic Treatment

The biological treatment depends upon factors such as susceptibility of organic substances for biochemical oxidation, concentration of pollutant within optimum level, pH of biological media, absences of surface active substances and presence of micro nutrients like nitrogen, phosphorus, potassium, carbon. Sewage treatment involves segregation of solid and liquid phases and their further treatment in aerobic and anaerobic conditions. The aerobic treatment gives activated sludge, while anaerobic treatment gives septic and digested sludge. Both processes involve adsorption of pollutants on activated sludge, and mineralisation of pollutants with micro-organism. The first process is rapid (10–15 minutes) while the second process is lengthy (more than 60 minutes).

In aerobic process aeration tanks, biological films, biological ponds, or irrigation fields are used for treatment. In trickling filters (biological filters spread with micro-organism) sewage is spread uniformly at short intervals. This process is capable reducing BOD of sewage from 200 to 15 mg/litre. The micro-organisms like green algae, blue-green algae, diatoms, protozoa are present respectively in numbers as 226×10^9, 140×10^9, 24×10^9 and 66×10^9. The important changes occurring during treatment include decrease in concentration of pollutant or its dilution and absorption of pollutant on activated sludge, decrease in amount of organic matter and conversion of ammoniacal as well as nitrite nitrogen to nitrate nitrogen, *i.e.* nitrification of sewage and sludge. For trickling filters overload with organic matter, filtration surface becomes clogged and efficiency gradually decreases. One gram of activated sludge contains 1×10^{12} bacteria with surface area of 1200 metre^2 in biocenosis of activated sludge. It has a good absorbing power. Protozoa promote accumulation of nitrogen in media, to enhance quality of fertiliser of sludge.

(b) Anaerobic Treatment

These are mainly used to treat sludge (*i.e.* solid mass in sewage). The principal goal is to change the physical structure of the sludge and to decrease the mass of sludge to promote fermentation of gases and soluble salts and finally to utilise the organic matter as the fertiliser. Methane fermentation is one of the processes wherein sludge is separated from sewage and treated. The mineralisation of salts of acetic and butyric acid liberates methane as follows:

$$2CH_3COOK + H_2O \rightarrow K_2CO_3 + CO_2 + 2CH_4 \uparrow$$

$$(CH_3CH_2CH_2COO)_2\ Ca + 3H_2O \rightarrow CaCO_3 + 2CO_2 + CH_4 \uparrow$$

Such conversion is rapid in the absence of oxygen in anaerobic condition. Other methane producing reactions are described in the following reactions:

$$\underset{\text{Ethanol}}{2C_2H_5OH} + CO_2 \rightarrow CH_4 \uparrow + 2CH_3COOH$$

$$\underset{\text{Butanol}}{2CH_3CH_2CH_2CH_2\text{–}OH} + CO_2 \rightarrow CH_4 \uparrow + 2CH_3CH_2CH_2COOH$$

$$\underset{\text{Isopropylalcohol}}{4CH_3CHOHCH_3} + CO_2 \rightarrow CH_4 \uparrow + 4CH_3CH_2CH_2COOH + 3H_2O$$

$$\underset{\text{Acetic acid}}{CH_3COOH} + CO_2 \rightarrow CH_4 \uparrow + 2CO_2$$

Thus in the above transformations, ethanol, butanol, isopropyl alcohol and acetic acid are converted to CH_4 and other useful products. The formic acid, formaldehyde and methanol also generates methane as

$$\underset{\text{Formic acid}}{4HCOOH} \rightarrow CH_4 \uparrow + 3CO_2 + 2H_2O$$

$$\underset{\text{Formic acid}}{HCOOH} + 3H_2 \rightarrow CH_4 \uparrow + 2H_2O$$

$$\underset{\text{Formaldehyde}}{HCHO} + 3H_2 \rightarrow CH_4 + H_2O$$

$$\underset{\text{Methanol}}{CH_3OH} \rightarrow + H_2 \rightarrow CH_4 + H_2O$$

Water acts as an oxidising agent in methane fermentation. Thus water acts upon enzymes then itself breaking up to form hydrogen and oxygen. The hydrogen is used by micro-organism to reduce CO_2 to CH_4, while oxygen oxidises the decarbonised radical to form acid. Thus the methane forming bacteria play an important role in circulation of substances and energy turnover in nature. They absorb CO, CO_2 and H_2 to give hydrocarbon, methane and synthesise its own cell substances. Table 19.4 gives composition of liquid sludge during fermentation. Table 19.5 gives composition of gas generated. It is worthwhile to compare them with study of effect of treatment and products generated, before and after treatment (Table 19.3).

The mechanism of conversion of organic matter to various gases and anion is shown below:

Table 19.4 Composition of liquid sludge

S. No.	*Parameter*	*Acid fermentation*	*Alkaline fermentation*
1.	pH	4.9	7.3
2.	Volatile fatty acid	64.2 mg/litre	16.0 mg/litre
3.	Nitrogen of ammonium salt	144 mg/litre	270 mg/litre
4.	Alkalinity	None	30 mg/eq/litre

Table 19.5 Composition of fermented gas

S. No.	*Gas*	*Acid fermentation*	*Basic fermentation*
1.	H_2S	46%	None
2.	CO_2	45%	26%
3.	CH_4	None	72%
4.	CH_3–CH_3	None	None
5.	H_2	7.5%	None
6.	N_2	None	None

$$\text{Organic–C} + O_2 \rightarrow CO_2 \quad ...(i)$$

$$\text{Organic–H} + O_2 \rightarrow H_2O \quad ...(ii)$$

$$\text{Organic–N} + O_2 \rightarrow NO_3^- \quad ...(iii)$$

$$\text{Organic–S} + O_2 \rightarrow SO_4^{2-} \quad ...(iv)$$

$$\text{Organic–P} + O_2 \rightarrow PO_4^{3-} \quad ...(v)$$

The bacteria are tiny automatic chemical reactors. They are self-adjusting and self-maintaining. This process is called mineralisation.

19.6 TERTIARY TREATMENT

This is used to improve the quality of an effluent after primary and secondary treatment. This is used to remove additional suspended solids including algae and further reduction in faecal coliform numbers. The suspended solid reduction consequently reduces BODs. Algae are 50% protein. The solids are removed by flocculation and sedimentation or floatation and microstaining with sand filtration. Alum is used to flocculate, while rapid sand filters are used for staining operation. The removal of faecal bacteria is achieved by disinfection. The faecal coliform can be removed by maturation of ponds. They are suitable for treatment of effluent from aerated lagoons. Chlorine is used for disinfection. The only difference in comparison with water is a large dose of chlorine, which is required due to presence of a large amount of organic matter (~10–25 mg/litre). The contact time of chlorine is also enhanced from half-an-hour to 2 hours. The excess chlorine can be removed by treatment with SO_2.

The practise of disposal of sewage into sea and septic tank are quite popular. Septic tank is used in rural sewage system. Other methods include sewage dehydration or freezing thermal treatment, mechanical dehydration by centrifugation and filtration. Sewage sludge is used for fertilisers, fortification of animal feed, as building materials and as biogas.

Table 19.6 gives information of effluent characteristics after tertiary treatment. It clearly shows reduction in pH, SS, BOD, COD and chloride contents. While Table 19.7 indicates efficiency of tertiary

Table 19.6 Nature of domestic effluent after tertiary treatment (mg/L)

S. No.	*Property*	*Solid mass*	*Liquid effluent*
1.	pH	6.9–7.2	7.2–7.9
2.	Total dissolved solids	400–600	400–600
3.	Suspended solids	70–100	–
4.	BOD	150–300	1–2
5.	COD	200–500	3–6
6.	Hardness ($CaCO_3$)	150–200	35–45
7.	Alkalinity	150–200	100–150
8.	Chlorides	60–80	60–80

Table 19.7 Efficiency of tertiary treatment (%)

S. No.	*Sub-treatment*	*Elimination of*	*Efficiency of process (%)*
1.	Anaerobic treatment	Nitrate–N	80–90
2.	Denitrification	Nitrate–N	80–95
3.	Algae treatment	Ammonia–N	50–90
4.	Ammonia stripping	Nitrogen–Phosphorus	80–95
5.	Ion exchange	Dissolved solids	80–92
6.	Electrodialysis	Organic matter	10–40
7.	Carbon adsorption	Dissolved solids	90–98
8.	Reverse osmosis	Dissolved solids	65–65
9.	Distillation	Surfactants	90–98

treatment in removal of pollutant by various sub-treatment procedures. This clearly shows that the highest efficiency of 80–95% is attainable with anaerobic treatment, carbon adsorption process and distillation, while electrodialysis, algae treatment removes a small percentage of the pollutant. Costwise, the reverse osmosis the ion-exchange and carbon adsorption are the most expensive processes, while the algae treatment and distillation are the cheapest processes, they have no practical applications.

19.7 ANALYSIS OF DOMESTIC EFFLUENT

In order to asses the efficiency of various treatment processes used for domestic effluent, it will be necessary to analyse samples of sewage from time to time. Depending upon results of analysis, one can step up the mode of treatment to mitigate the problems of hazards to health. Thus the tests are usually performed to evaluate the sewage treatment plant performance. Sewage as such is too complex, and its nature varies from place to place, hour to hour and day-to-day; so that determination of exact quantities of all the constituents is not possible. However, it is more important to adjudge the chemical characteristics of sewage. Although it is 90% waste water or what is termed as liquid effluent, it carries dissolved and solid matter which are most difficult to analyse.

It is not very difficult to analyse organic matter and mineral contents of sewage, but obnoxious carbonaceous and nitrogenous materials pose a serious problem during handling. Fresh sewage becomes stale in 120 minutes. The reduction of sludge volume during analysis is also of prime importance. The resulting humus has to be filtered and all efforts should be made to minimise sludge bulking. The efficiency is determined by the gas production and solid reduction process. The complex nature of sewage precludes its complete analysis. Since oxygen consumed can be easily measured, the concentration of organic matter is measured in terms of oxygen consumed for oxidation. This is expressed in terms of oxygen demand. Such oxygen demand is expressed in three ways.

(a) Oxygen Demand by Sewage

It is expressed in terms of ThOD; BOD and COD. Here ThOD represents theoretical oxygen demand; which is required to oxidise organic fraction of waste completely to CO_2 and H_2O. It is not possible to calculate theoretical oxygen demand. However, it is roughly equal to COD. The COD is obtained by oxidising waste with boiling dichromate solution. Such reaction proceeds to 95% completion. The chief merit of COD determination is rapid three-hour analysis, the only limitation being it does not say anything about the proportion of the waste that can be oxidised by bacteria. Also, no light is thrown on the rate of bio-oxidation. In BOD the amount of oxygen required for oxidation of waste by bacteria is known. It is a true measure of concentration of organic waste which is biodegradable, BOD is done for 5 days at 20°C. It is written as BOD_5, while the ultimate 20-day BOD is written as BOD_4. It is seen that ThOD > COD > BOD_4 > BOD_5. Further, as interesting relation exists between different stages of untreated domestic sewage. The approximate ratios are:

$$BOD_5 : COD = 0.5$$

$$BOD_4 : BOD_5 = 1.5$$

The presence of agricultural and industrial wastes may alter these ratios substantially. Table 19.8 shows average breakdown of daily per capita BOD_5 contributions.

BOD thus measures putrescibility of sewage. COD is used all over the world under different names, e.g. permanganate value (4 hour) oxygen consumed; oxygen absorbed; permanganate value (3 minutes); dichromate value or COD. The rate of COD/BOD is 2 : 1 for raw sewage which becomes 5 : 1 on secondary treatment during winter and is still higher in summer. The difference between two values

Table 19.8 Daily per capita BOD_5 contributions

S. No.	*Operations*	*Developed countries (gm per person)*	*Developing countries (gm per person)*
1.	Individual washings	9	5
2.	Utensil washing	6	8
3.	Solid waste disposal	31	0
4.	Cloth washing	9	5
5.	Toilet—Faeces	11	11
	Urine	10	10
	Others	2	1
6.	Total average per day per person	78	40

gives an idea about the type of waste. The DO analysis needs modification by Alsterberg in the presence of nitrite in sewage. If DO is high, putrefactive odour will be developed. Sodium azide modification to Winkler Alsterberg methods involves the following reactions.

$$NaN_3 + H_2SO_4 \rightarrow NaHSO_4 + HN_3$$

$$HNO_2 + HN_3 \rightarrow N_2 + N_2O + H_2O$$

(Thus HNO_2 will not react with KI in presence of N_3H)

The amount of DO is important in activated sludge process for aerobic oxidation. Sludge clarification is usually a serious problem.

(b) Temperature

Table 19.9 indicates level of dissolved oxygen in water. The measurement of temperature is of utmost importance. The temperature of sewage is 11–20°C. Initially the sewage temperature is higher than the atmospheric temperature. The high temperature causes problem of odour. This is measured by Beckmen thermometer.

Table 19.9 Solubility of oxygen in water at different temperature

Temperature, °C	*Solubility, (mg/litre)*
15	10.07
20	9.08
22	8.74
24	8.42
26	8.12
28	7.84
30	7.57
32	7.23

(c) pH of Sewage

This is determined by portable pH meter. The pH of fresh sewage is highly alkaline, while the pH of the stale or septic sewage is < 7.5. It is thus acidic. The pH goes on decreasing on standing. The degradation of the organic matter is pH dependent. pH meter is best for such analysis.

(d) Dissolved Chlorides

The sewage contains good amount of chlorides due to large consumption of salt. It will be fine if one has ion selective electrode for chloride to measure it by ion analyser. In the absence of such instrument, one has to take resort to Volhard's or Mohr's volumetric method involving precipitation titrations which is most dependable.

All operations in such a situation should be carried out by remote control. One must use rubber bulb for pipetting solutions in the pipette. The sample should be kept to 5°C and pH~2.0, before such analysis. The phosphate interferes in Mohr's method involving precipitation. Chloride level rises due to food and urine excretion. Chlorides are unaffected by biological action.

(e) Dissolved Chlorine Gas

For the purpose of disinfection of domestic effluent, chlorine gas is used. A large dose is needed (> 30 mg/litre) on account of presence of organic matter as well as ammonia in sewage. The excess of chlorine sometime is eliminated by treatment with sulphur dioxide. Chlorine in solution is present in form of HOCl, Cl^- and Cl_2 gas, it is also called free available chlorine. It is used to control odour, slime and insects in trickling filters in aerobic treatment and to prevent activated sludge bulking in anaerobic sludge treatment. If one has facility of amperometry, it gives best and quick results of analysis. The OTA method is an ideal one, which involves spectrophotometeric measurement.

(f) Suspended Solids

The tertiary treatment using rapid sand filters eliminates major portion of suspended solids. Table 19.10 gives the nature of refuse in principal cities in India, which contribute greatly to suspended solids. A perusal of Table 19.10 will show that refuse in Kolkata has a large percentage by weight of dust and ash; while sewage in Delhi has considerable amount of vegetable and fruits in the refuse; same is true of refuse in Chennai; refuse in Nagpur has a large volume of leaves and shell, while Pune follows pattern of Chennai and Delhi. The dissolved as well as suspended solids are likely to be large. They are weighed from known volume of sewage, and their percentage is calculated.

Table 19.10 Nature of refuse in major Indian cities (percentage by weight of wet refuse)

S. No.	*Nature*	*Kolkata*	*Delhi*	*Chennai*	*Nagpur*	*Pune*
1.	Fruit, vegetable	16.0	20.3	57	3.8	68
2.	Leaves, shell	18	–	–	12.9	4.6
3.	Paper, cardboard	3.2	5.9	4.8	2.0	8.7
4.	Glass, bones	7.4	0.6	0.5	0.8	0.6
5.	Metal, tins	0.7	0.6	0.6	0.3	0.7
6.	Dust, ash	41.6	6.0	–	–	–
7.	Wt. per person	0.51	0.31	0.56	0.35	0.3
8.	Calorific value	6300	6600	6700	5300	7100

(g) Nitrogen

In domestic effluent, it can be present in form of ammoniacal-N or nitrite-N or nitrate-N. The fresh sewage contains ammoniacal-N, which has a tendency to degrade to nitrite and finally to nitrate. Such valence change of nitrogen occurs as $NH_3 \rightarrow NO_2^- \rightarrow NO_3^-$. The valency change is brought about by living micro-organisms. Such change can be positive or negative, depending upon presence of oxygen. The treated sewage has high nitrate ion concentration and is said to be well nitrified. High nitrate and phosphate concentrations in sewage promotes eutrophication, thereby causing algae bloom. The ammonia is determined by Kjeldahl's method, while nitrite is estimated by Saltzman method and nitrate by Brucine method. Latter two are spectrophotometric methods. In view of easy availability of ion selective electrode for NH_3 and NO_3^-, such analysis has been simplified with use of ion analyser.

(h) Sludge Index

The bulking of sludge is a serious problem during treatment. Chlorine gas is injected in sludge to prevent bulking and promote contraction of humus. The settling characteristics of activated sludge have been expressed in different ways, viz. SVI and SDI. The sludge volume index (SVI) is defined as the volume occupied by one gram of activated sludge after settling the aerated liquid for at least 30 minutes. The sludge density index (SDI) is a factor obtained by dividing percentages of suspended solid (SS) by weight times 100 by sludge volume in the aeration tank settled in 30 minutes time. Both SVI and SDI are useful parameters for control of activated sludge aeration process. It can give warning in advance that the sludge is starting to bulk, which will lead to formation of poor effluent. The high percentage of water with no possibility of reduction to small volume is called sludge bulking.

The laboratory control of sludge digestion from analysis can be done with control of pH. At a pH 6.4, stoppage of liberation of methane may occur, while the excess accumulation of organic acid also drops pH value. The efficiency of digestion tank is measured by reduction in presence of volatile and solids.

It is not a general practice in domestic effluent analysis to investigate presence of micro-organism in terms of MPN count. Such information is made available from the value of BOD.

ADDITIONAL READINGS

19.1 L. Metclaf and H.P. Eddy, *Sewerage and Sewage Disposal*, Tata McGraw-Hill Book Co. (1972).

19.2 G.M. Fair, J.C. Gayer and D.A. Okun, *Water and Waste Water Engineering*, Vol. 3, John Wiley & Sons (1966).

19.3 N.L. Nemerow, *Liquid Waste of Industry; Theories, Practice and Treatment*, Addison-Wesley (1971).

19.4 E. Bosselover and M. Schwartz, *Treatment of Industrial Waste*, 2nd Ed., McGraw-Hill and Co. (1975).

19.5 P.W. Powers, *How to Dispose of Toxic Substances and Industrial Waste*, Noyes Data Co., New Jersey (1976).

19.6 R. Feacham, M. McGarry and D. Mara, *Sewage Treatment*, John Wiley & Sons, ELBS Series (1977).

19.7 D. Mara, *Sewage Treatment in Hot Climate*, John Wiley & Sons, ELBS Series (1977).

19.8 N.F. Vozaya, *Chemistry of Water and Microbiology*, Mir Publishers, Moscow (1979).

19.9 T.H.P. Veal, *The Disposal of Sewage*, Chapman and Hall Ltd., London (1956).

19.10 K. Imhoff, W.J. Muller and D.K.B. Thistlethwayte, *The Disposal of Sewage and other Waterborne Waste*, Butterworths, London (1971).

19.11 H.E. Babbit and E.R. Baumann, *Sewage and Waste Treatment*, Asia Publishing House, New Delhi (1958).

19.12 D. Barnes and F. Wilson, *Chemistry and Unit Operation in Sewage Treatment*, Applied Science Lab., London (1978).

CHAPTER 20

Land Pollution, Pesticides and Their Analysis

20.1 INTRODUCTION

India is agricultural country and more than 50% of its population resides in rural regions like villages and small towns. By and large developing countries have paid all attention on problem of urban pollution while rural or land pollution was totally neglected. No attempts were made to characterise agricultural land and to find out what kind of crops grow better in particular soil. As opposed to that several peasants from Europe and America have done land mapping by the remote control sensing in order to distinguish between barren and the fertile land. We have several agriculture universities, of course very few of them are carrying out excellent job of correlating crop growth with quality of soil. They have tailor made plans for getting maximum yields from specific soil in the country with production of particular crops or fruits.

Another alarming feature facing with our agricultural farming is the indiscriminate use of pesticides, insecticides and fungicides. In earlier days the concept of N-P-K balance was properly understood. The farmers maintained optimum ratio of nitrogen, phosphorus and potassium in soil to get maximum yield of crops. But with concept of fast growing food, with greed of taking two crops in year peasants, commensed using array of pesticides without much knowledge of ill effects of its excessive usage. Consequently in course of time it led to disaster of DDT, its use being banned by developed countries, although we still use it freely in fields.

Therefore in order to understand long term interrelationship between land and pesticide, a rigorous study on their applications as well as ill effects and knowledge of permissible dose for applications of manure is necessary. Agriculturist have to understand the intricate chemistry of the soil and define what constituents the agrictultural pollution and the role of micronutrients in soil? Pesticide is generic name is broad class of compounds comprising of insectides, herbicides and fungicides, but chemically we have a class of compounds as chlorinated hydrocarbons like DDT, organophosphorus compounds like malathion, carbamate pesticides like zectnon and new compounds like 2-4T, 2, 4, 5-T. Modern instrumental techniques like GLC or spectrophotometry and polarography provide a rapid tool for analysis. The knowledge from analysis should go longway to provide reliable formation on Lethal dose (LD_{50}) and tolerance limit by human beings as well as plants and animals.

The environmental scientists as told before have paid their attention towards pollution due to the industrial activity. As such this pollution was largely confined to the urban areas, in cities and towns.

Fortunately, it is possible to contain such pollution by effectively controlling the faulty industrial growth and related activities. Much worse than this industrial pollution is pollution of soil and pollution due to pesticide residues. As told earlier scant attention was paid towards this problem. As a matter of fact indiscriminate use of fertilisers and excessive use of pesticides has created a great menace. Both of these products find their way into food chain and they have direct implications on human health. The pollution of lithosphere due to fertilisers or pesticides is so widespread that in agricultural country like ours it is almost impossible to control it. It has much wider ramification. In future we will have to pay considerable attention to mitigate such soil pollution, in order to increase yield of crops, and food production and avoid ill effects on health.

There is clear evidence for crops being destroyed by industrial effluents, while very few grow well on them. Apart from consideration of role of nitrogen and phosphorus which promote eutrophication of effluent, the role of elements like potassium, copper, zinc, iron, manganese, molybdenum must be considered. Even boron is the micronutrient in soil. They promote growth of crops. As such the trace elements from soil, sediments are valuable asset. No doubt the pesticides are a gift to man, but also they are a curse if not properly used. The reason is, majority of these pesticides are not biodegradable even at elevated temperature. They leave behind a toxic residue in the environment. Such toxicity is determined by LD_{50} value for pesticides. In view of this it is absolutely essential to study so called agricultural pollution—a term which is attaining new dimension in developing countries. An endeavour is made in this chapter to explain soil pollution and analysis of pesticide residue. The deails of the chemical analysis are furnished in earlier chapters.

20.2 CHEMISTRY OF SOIL

The internal structure of the earth is shown in Fig. 20.1. The first 36 km is upper mantle, the next 40 km is transition zone, while the next 3000 km is lower mantle and then we have the core of lithosphere. Figure 20.2 represents lithosphere-biosphere interaction. The atmospheric carbon dioxide reacts with silica, carbon, calcium carbonate to return to lithosphere as the great atmospheric sink. This in turn returns these materials to hydrosphere through water vapour and micronutrients from the soil. Soil contains considerable amount of organic matter. Primary minerals such as silica, limestone and dolomites account for 70% of soil mass. Clay material is present in significant amounts. The humus is also present in soil in colloidal form, biological activity in soil depends upon form of water present in it. The micro-organisms present in soil are also of great relevance. The important microorganisms are algae, protozoa and actinomycetes. Such soils are productive. These micro-organisms play an important role in mineral and fertiliser breakdown in soil formation. A typical fungus produces readily citric acid in soil and then influences solubilisation of Si, Al, Fe and Mg in basalt. Some convert organic nitrogen to inorganic nitrogen. To understand the nature of relationship of the soil with knowledge of crop growth is important.

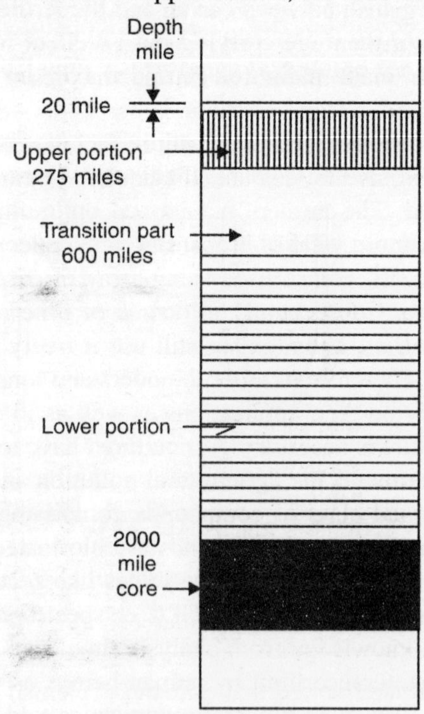

Fig. 20.1 Structure of the earth

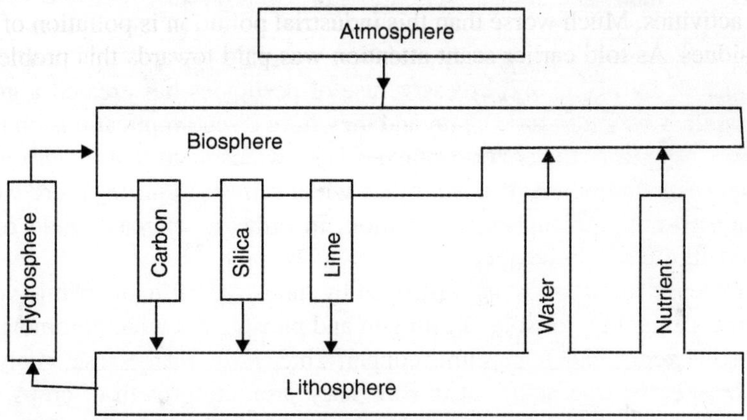

Fig. 20.2 Interaction of litho and biosphere

20.3 SOIL IRRIGATION BY EFFLUENTS

It is interesting to note the growth of crops on soil fed with industrial effluents. Table 20.1 gives typical interpretation of soil irrigated by effluents from different sources like rain water, well water, waste water from paper and pulp industries. In all such cases the soil is alkaline; with considerable amount of nitrate-nitrogen and phosphorus, which are of prime importance for the growth of plants. The contents of exchangeable calcium and magnesium indicates that water is generally hard in nature. Table 20.2 gives an idea about sustained efforts to grow crops on waste water for a period of six years. The crops grown were paddy, wheat and sugarcane in varying area of soil measuring 2–25 acre. There was rise in yield of paddy from 2500 to 3500 tons; while wheat growth did not rise and sugarcane showed marginal rise in growth in course of six years. This showed that all crops cannot grow well in industrial waste water. Table 20.3 lists certain varieties of trees like nilgiri, babul, rain tree, which had grown well on ash, as they contained useful micronutrients conducive for growth of plants.

Table 20.1 Analysis of quality of soil sample irrigated by different effluents (mg/L)

S. No.	*Irrigated by*	*pH*	*Conductivity*	*Organic matter*	*Nitrate-nitrogen*	*Phosphorus*	*Ca/Mg*
1.	Natural rain	7.5	0.3	0.3	6.3	9.6	24.0
2.	Well water	8.1	0.3	0.9	6.7	12.0	20.8
3.	Waste water from paper mills	8.2	0.5	0.7	7.8	12.0	30.0
4.	Effluent from pulp factory	8.1	0.6	1.2	7.3	20.0	28.8
5.	Irrigated both by paper and pulp effluent	8.5	0.5	1.2	8.1	9.6	22.4

It is observed that effluents replenish important nutrients of soil, and problem of exhaustion of the non-renewable resources of lithosphere is somewhat made up. The yield of crop is better in developed

Table 20.2 Sustained effort to grow crops on waste water

Year in succession	*Paddy yield (tons)*	*Wheat yield (tons)*	*Sugarcane yield (tons)*
1.	2300	–	70
2.	780	700	–
3.	2530	1200	65
4.	2600	1100	90
5.	3100	1400	80
6.	3500	1450	80

Table 20.3 Trees grown on ash

S. No.	*Botanical name*	*Popular name*
1.	Eucalyptus globulus	nilgiri
2.	Casuarina equisetifolia	saru tree
3.	Pithecellophis dulce	rain tree
4.	Leucaena leuciphalia	subabul tree
5.	Acacia auriculaformis	Australian acasia
6.	Iopmea, lantana, ziziphus	ipoma, tantani, beri

than in developing countries; e.g. the yield of wheat in the USA per acre is 1440 lbs and rice 2210 lbs, corresponding figures for India are wheat 760 lbs and rice 870 lbs. This is attributed to modern methods of farming in the USA. The per capita food growth has remained stationary in India. The remedy lies in making a success of green revolution. This is possible only by massive application of biofertilisers and judicious use of pesticides. Experiments to use waste water for better growth of plants is not that encouraging. Even if such efforts culminate in partial success, it is not the panacea for stagnating agricultural production.

20.4 AGRICULTURAL POLLUTION

Agricultural activity directly influences the chemistry of environment in several ways; like disposal of animal, plant or food processing waste; the contamination from applications of pesticides and fertilisers; depletion of natural soil nutrients and finally salt and mineral accumulation from irrigation. Thus agricultural waste is more serious than the solid or liquid waste generated due to industrial activity. The waste from food and milk industry poses a serious problem in India. The excessive growth of fruits like oranges in Vidharbha and apples in Kashmir may pose problem of disposal of solid waste. The food processing industry produces more than 60% of waste from fruit peels. No doubt it can be used for animal feed and as manure in soil. Cotton waste is useless as it contains a good amount of pesticide residue, which in the long run is harmful to animal or cattle. The mechanised meat processing industry will produce a liquid waste due to residual blood, flesh and bones which will be difficult to dispose. It will also add to the nuisance of flies and vultures.

There is no substantial evidence to prove that all fertilisers cause water pollution. However, it leads to excessive release of nitrogen and phosphorous in soil. Nitrate is the worst enemy of high-yield farming. Nitrate is reduced to nitrite and causes a disease called methaemoglobinemia (*i.e.* blue babies). Phosphate is also a wide spread pollutant from agriculture. It travels much slower in the soil as compared to nitrate, but it promotes eutrophication of effluents with algal bloom. Nitrogen and phosphorus pollution illustrate far reaching chemical consequences of man's agricultural activity. Agricultural activity removes these materials from lithosphere and drops them into the sea, which is a permanent sink for waste. This agricultural pollution represents a permanent and irreversible depletion of the resources of the lithosphere.

20.5 ROLE OF MICRONUTRIENTS IN SOIL

Mineral contents of the soil gets depleted due to rain, seepage, growth of plants and crops. The soil deficient in mineral contents is rendered unsuitable for further cultivation. As a matter of fact nitrogen, phosphorus and potassium are essential for growth of plants, but they are replenished through organo nitrogen and organo phosphorus compounds, which are dumped in soil as the sink by plant and vegetable kingdom. However, vitally more important than these are elements like iron, manganese, zinc, copper, molybdenum and boron, which are most essential for the healthy upkeep of soil. It is not true that each and every plant or crop needs all of these metals, but they thrive better on at least one of them. The plants have a tendency to extract one or more of these elements. Content of such an element is termed as 'available element' contents, while whatever available in total soil is termed as the 'total amount of a particular element' in the soil. Figure 20.3 histograms depicts the relationship of nutrient availability and pH of the soil. It will be noted from this figure that there is good availability in soil of molybdenum,

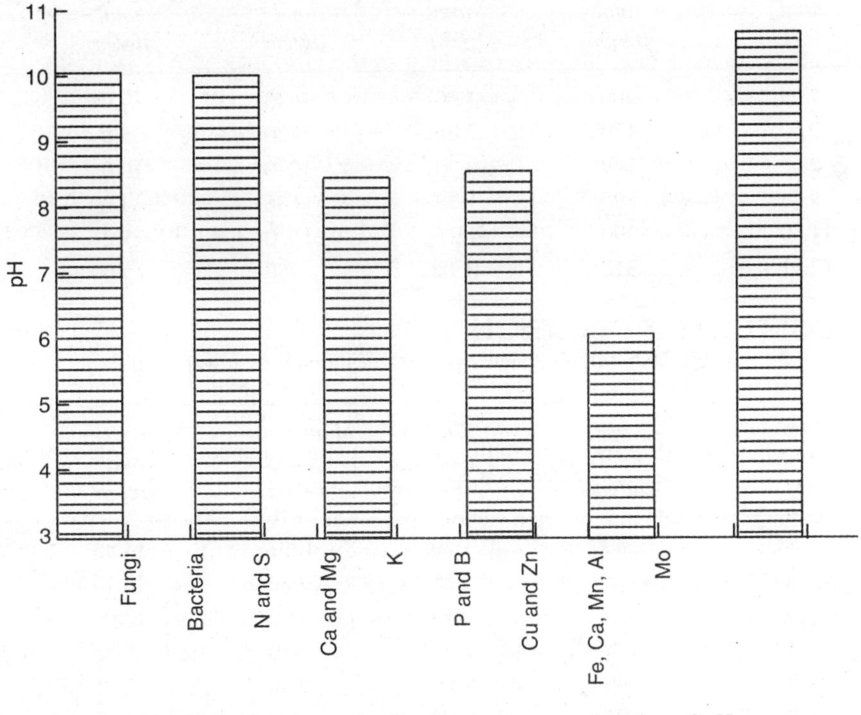

Fig. 20.3 Histogram of soil micronutrients as function of pH

copper, zinc and to a lesser exent iron, mangnese, cobalt and aluminium. No doubt calcium, potassium, magnesium, nitrogen, sulphur and phosphorus are otherwise present in substantial concentrations in the soil. The figure also shows beyond pH 10 is alkaline media, at which value the micronutrients are not available to plants. Tables 20.4 to 20.6 indicate the role of micronutrients in proper growth of soil. A systematic investigation on concentration of micronutrients in selected fields of West Maharashtra soil was undertaken for four years. As many as four stations in Mumbai, Pune and Rahuri universities were

Table 20.4 Micronutrients present in soil of sugarcane crop

Sample No.	*Iron total (%)*	*Copper total μg/gm*	*Zinc total (%)*	*Manganese total μg/gm*	*Molybdenum total μg/gm*	*Boron total (%)*
1.	7.95	380	0.90	1810	10.08	0.001
2.	6.02	203	1.10	1187	9.08	0.069
3.	3.51	161	0.51	653	15.10	0.032
4.	6.9	283	0.65	450	4.80	0.071
5.	8.10	150	0.70	750	12.10	0.011
6.	9.20	190	0.90	820	15.10	0.012

Table 20.5 Micronutrients present in the soil of paddy crop

Sample No.	*Iron total (%)*	*Copper total μg/gm*	*Zinc total (%)*	*Manganese total μg/gm*	*Molybdenum total μg/gm*	*Boron total (%)*
1.	7.99	200	1.00	645	7.01	0.084
2.	9.6	170	2.00	1120	5.0	0.062
3.	9.6	140	0.90	1230	10.0	0.010
4.	9.73	295	1.20	606	4.00	0.007
5.	11.50	290	1.21	1050	18.21	0.095
6.	12.01	310	0.980	950	19.00	0.09

Table 20.6 Micronutrients present in the soil of jowar crop

Sample No.	*Iron total (%)*	*Copper total μg/gm*	*Zinc total (%)*	*Manganese total μg/gm*	*Molybdenum total μg/gm*	*Boron total (%)*
1.	6.81	160	1.20	1291	15.85	0.090
2.	7.90	170	1.35	1083	16.10	0.091
3.	8.01	150	0.90	1045	12.00	0.061
4.	9.89	198	0.99	833	7.85	0.09
5.	11.01	140	0.69	940	12.01	0.095
6.	11.90	291	0.92	940	8.50	0.025

set up in Maharashtra and the analysis of available and total element content in fields of sugarcane, paddy and *jowar* crop was carried out using atomic absorption spectroscopy. The results in Table 20.4 indicate that high levels of manganese in soil, was conducive for growth of sugarcane (~ 1810 μg/gm of soil). Table 20.5 indicates high levels of manganese as well as copper showed that copper was the best micronutrient for growth of paddy. Table 20.6 represents micronutrient contents of *jowar* field. The study showed that amongst high level of manganese, copper, zinc and iron, was conducive for growth of *jowar.*

20.6 ANALYSIS OF MICRONUTRIENTS IN SOIL

Table 20.7 and 20.8 indicate available trace elements and total trace elements in soil. Before considering their analysis, it is worthwhile to consider the procedure for sampling of soil. The surface samples are collected from 10 cm horizon. The samples are ground in to fine powder (100 mesh). In order to determine total element content, they are digested in a mixture of nitric-perchloric acid. The resulting

Table 20.7 Available trace element contents of soil samples (μg/gm)

Soil sample No.	*Concentration of soil samples in water*				
	Mn	*Fe*	*Cu*	*Zn*	*Mo*
Reference	19.80	20.50	4.40	5.40	1.10
1.	32.70	44.60	2.00	6.70	0.08
2.	67.40	42.50	11.00	6.90	0.20
3.	72.00	58.30	10.80	6.30	0.10
4.	102.30	28.50	8.00	6.40	0.20
5.	115.30	19.20	6.00	6.20	0.20
6.	117.60	35.10	11.50	6.80	1.50
7.	120.00	28.20	7.00	6.30	0.10

Table 20.8 Total trace element contents of soil samples

Soil sample No.	*Concentration of soil samples in μg/gm*				
	Mn	*Fe*	*Cu*	*Zn*	*Mo*
Reference	1260.00	28,400.00	102.00	70.00	16.00
1.	1512.00	20,700.00	110.00	140.00	5.50
2.	1718.00	20,150.00	110.00	98.00	13.80
3.	2074.00	21,750.00	135.00	81.00	19.20
4.	2010.00	19,900.00	130.00	50.00	7.30
5.	2194.00	20,650.00	195.00	98.00	14.20
6.	2521.00	20,700.00	140.00	92.00	9.20
7.	2590.00	20,500.00	124.00	102.00	11.70

Above the **bar below** number indicates highest concentration.

residue is evaporated to dryness and then extracted with water and made to a known volume before analysis by AAS. For the analysis of available trace elements from soil, the samples must be extracted with suitable buffers for various elements as follows:

(*a*) Iron with ammonium acetate (pH 4.8) buffer
(*b*) Zinc with ammonium acetate buffer (pH 6.5)
(*c*) Manganese with ammonium acetate buffer (pH 7.0)
(*d*) Molybdenum with ammonium oxalate buffer (pH 3.3)
(*e*) Copper with ammonium citrate buffer (pH 8.5)
(*f*) Boron with sodium carbonate buffer (pH 6.0)

All these elements were extractable in appropriate buffers at specified pH. The organic contents from extract was destroyed by dry ashing with nitric perchloric acid treatment. The residue was laxivated with water and the available contents of element was analysed by AAS. The difference in the available and total contents gives an idea of extractability of micronutrients by plants.

20.7 TRACE ELEMENTS ANALYSIS IN SOIL

Sometimes it is necessary to determine trace elements from soil. A Van Vee Grab sampler is used to collect samples of sediments. They are dried at 100°C. The dried samples are dissolved in nitric perchloric acid mixture to destroy organic matter. The residue is dissolved in dilue hydrochloric acid and trace elements are analysed as usual. Thus Tables 20.7 and 20.8 give information on available trace elements in water and total trace elements in soil. Along with nine samples, a reference sample (from virgin land) was compared in both tables. The result showed high level of manganese in station No. 1. So also iron and zinc content was higher than reference source. While Table 20.8 shows high levels of manganese in station No. 7, copper was maximum at station No. 5.

An attempt was made to analyse trace elements from sediments in Thane Creek and Ulhas river estuary around Mumbai. Tables 20.9 and 20.10 present results of such an analysis. In these samples also, manganese and iron contents were high, while cadmium content was very low. The reason for such high concentration of iron, manganese is the location of small-scale industries around Thane belt. Water is alkaline in aquatic marine environment. Hence industrial soil sample suffered from iron and manganese toxicity. The high level of copper (Table 20.10) is due to effluents emerging from pesticide industry. Zinc was higher (Table 20.9) due to disposal of effluents from rayon industry. Lead was high in both samples, which was released by automobile emissions on Vashi bridge, which connects island

Table 20.9 Trace elements from sediments in Thane Creek

Station number	*Concentration in μg/gm*					
	Mn	*Fe*	*Cu*	*Zn*	*Cd*	*Pb*
Reference	658.50	61,170.00	70.50	102.00	1.20	30.50
1.	783.00	69,569.00	144.00	273.50	2.00	78.20
2.	722.50	68,369.00	104.40	209.50	1.70	56.80
3.	922.00	70,369.00	98.70	159.50	1.80	47.20
4.	1127.00	64,769.00	90.80	149.00	2.30	53.90
5.	1167.00	72,369.00	97.20	152.00	2.10	50.70
6.	1467.00	75,469.00	99.20	168.00	1.80	48.10

of Mumbai to main land. Traces of cadmium were detected due to disposal of effluent from nearby nuclear reactors, in which it is used as the cladding material. Not much work has been carried out on boron. The wavelength for analysis, detection limit and type of flame used in AAS for various micronurients is listed in Table 20.11.

Table 20.10 Trace elements in sediments in Ulhas river in Mumbai

Station number	*Concentration in µg/gm*							
	Mn	*Fe*	*Co*	*Ni*	*Cu*	*Zn*	*Cd*	*Pb*
1.	220.0	58,500.00	38.5	88.0	88.2	73.0	1.4	32.8
2.	198.0	61,200.00	41.0	87.5	89.5	125.5	1.3	36.4
3.	251.0	63,700.00	45.7	95.4	91.0	129.0	2.7	36.0
4.	252.0	64,800.00	43.5	91.7	82.9	130.0	2.7	35.6
5.	208.0	67,900.00	44.3	96.4	108.0	199.0	3.1	39.2
6.	212.0	68,200.00	43.8	94.5	97.5	180.0	2.3	33.4

Table 20.11 AAS determination of micronutrients

S. No.	*Element*	*AAS Wavelength λ max*	*Detection limit µg/ml*	*Types of flame*	*Spectrophotometry with chromogenic ligands*
1.	B	249.8	2.0	Nitrous-actylene	Caramine red
2.	Cd	228.8	0.002	Air acetylene	Dithizone
3.	Cu	324.8	0.004	Air acetylene	Cupferron
4.	Fe	248.3	0.03	Air acetylene	1-10 phenanthroline
5.	Mn	279.5	0.003	Air acetylene	Permanganate
6.	Mo	313.3	0.03	Nitrous-acetylene	Thiocyanate
7.	Zn	213.9	0.003	Air acetylene	Dithizone
8.	Pb	283.6	0.02	Air acetylene	Dithizone

As regards analysis of boron, it was carried out by spectrophotometric technique using caramine red as the chromogenic ligand. The only limitation in such analysis was the need to avoid glassware. It is essential to use only polythene, teflon or platinum ware. This is because, glass contains a substantial amount of boron and is bound to give erroneus results during determination of boron.

Boron is one of the micronutrients amongst non-metals. It is conducive for the growth of citrus fruits like oranges, lemon which grow in size, if the soil sample is fed with reasonable concentration of boron compounds like say boric acid. The amount of boron is 0.001–0.095% in western Maharashtra. On the contrary, soils in Vidharbha which is the courtyard of orchard of oranges, shows the high level of boron in the range 0.095–0.10%.

20.8 PESTICIDES AND POLLUTION

Pesticides are the blessing for peasants. It is worst enemy of man on this planet. It not only contaminates lithosphere, but also pollutes the atmosphere. Pesticides include herbicides, insecticides, fungicides or rodenticides. The details of methods for their analysis is presented in earlier chapters. They are persistent pollutants obtained during interaction of lithosphere and biosphere. Many are cumulative poisons in the body, which accumulate in food chain and are hazardous. Few of them undergo metabolic formation and biodegradation. The pesticides residue so formed by degradation is more dangerous than the parent compound. They break down in the environment, involving either of the following process, namely oxidation, dehydrogenation, reduction, conjugation like metal complex formation; hydrolytic reaction like cleavage, exchange reaction and isomerisation. The reaction in body involves interaction with iron porphyrins. Several residues are carcinogenic. The metabolic products are also dangerous. The arsenic pesticides render soil permanently infertile. They sometimes increase pest population, due to effect of immunisation. A pesticide like DDT changes fertility of mammals. Pesticides thus pose an alarming problem for mankind. Although DDT kills mosquitoes which spread malaria. Its use in developed countries has gone down due to its harmful effects. However, the production has remained same due to its consumption by developing countries. Its indiscriminate application poses a serious health problem.

Figure 20.4 presents data on production of pesticides in developing countries. Highest production was in 1960–1965, with somewhat less production in 1975. Amongst pesticides, highest production is of insecticides, followed by herbicides and finally fungicides. Figure 20.5 shows how pesticide residue enters the environment. A small fraction of pesticides reaches the target pest (3), some reaches the crop (1), water (5), and sediments (8 and 4), some part ends up in the atmosphere (6), some for instance like arsenical herbicide (35%) gets mixed up into air by volatisation. Thus lithosphere and atmosphere act as sinks for these pesticides, as chlorinated hydrocarbons are transported easily in air. DDT, the worst among pesticides, is lost to the soil by voltalisation, removal by harvesting of organic compounds, water run-off and chemical degradation. It is removed from atmosphere by rainfall, settling of aerosols, diffusion and chemical photodegradation.

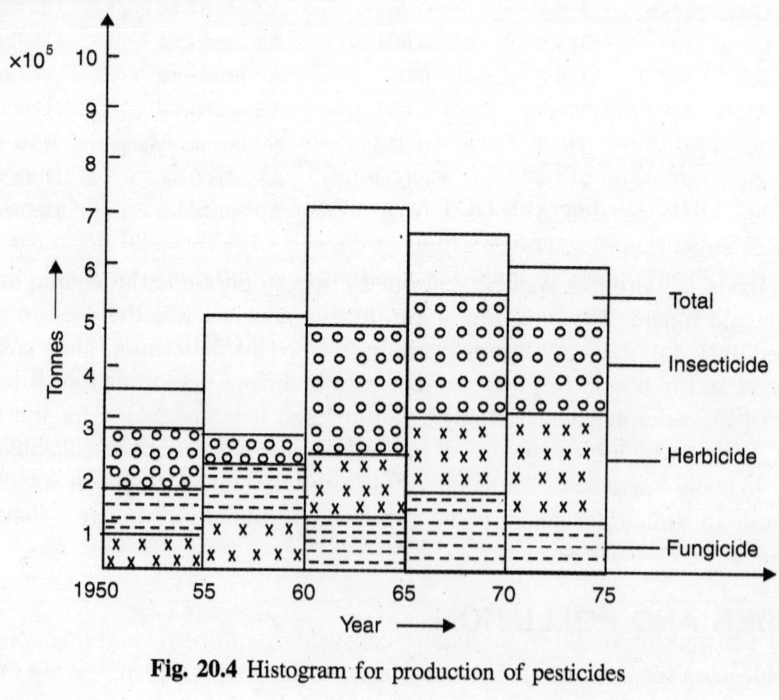

Fig. 20.4 Histogram for production of pesticides

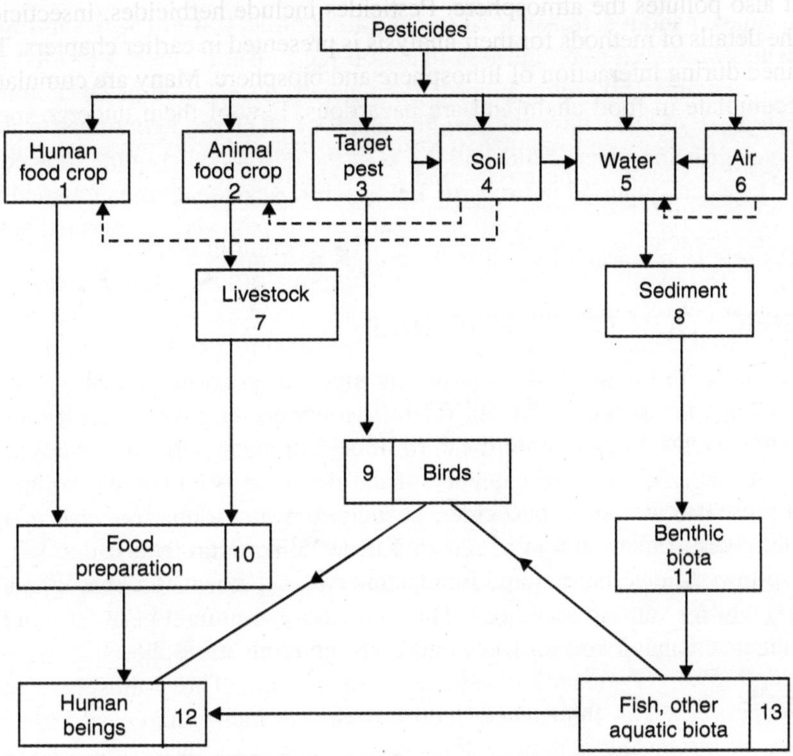

Fig. 20.5 Path of pesticide cycle in environment

20.9 DILEMA OF DDT

The chemical name of compound is 1,1,1 trichloro –2,2 bis (p-chlorophenyl) ethane. It converts to DDE, *i.e.* 1,1 dichloro 2,2 bis (p-chlorophenyl) ethylene. The average level of DDT in air is 0.015–0.149 μg/cm^2. According to survey conducted the per day intake of DDT is 0.178 mg in restaurant, 0.041 mg in college dinning hall, 0.184 mg in hospital, 0.026 mg/litre in basket of fruit, with corresponding figures, for DDE together with DDT in % are for restaurant 37%, college hall 39%, hospital 14% and market 40%. The body fat shows high levels of DDT. Table 20.12 shows levels of various chlorinated hydrocarbon pesticides in body fat in different countries in world.

Table 20.12 Pesticide in body fat (PPM)

S. No.	*Country*	*DDT*	*DDE*	*BHC*	*Dieldrin*	*Endrin*	*Heptachlor*
1.	USA	3.7	6.9	0.40	0.25	0.02	0.10
2.	France	3.1	6.5	1.19	–	–	–
3.	India	18	12	1.43	0.040	0	0
4.	UK	–	–	0.02	0.21	0.02	0.01

The mode of distribution of DDT in aquatic ecosystem is through bottom sediments, suspended solids and water. It goes from water to aquatic plants, to invertebrates, the carnivores and herbivores animals. Fish consumes it and it is transmitted to fish eating birds. If we make a survey of a typical river for its DDT contents, we would note that less than 500 mg/g of it is present in mud plankton or fish, while birds and fish contains 1500 ng/g. Figure 20.4 indicates production of DDT in developing countries. Even if we stop use of DDT now, the residual effect of DDT shall continue for the next 30 years. A survey of soil sites growing vines, fruits and crops in the developed country showed that the highest residue (11.4 g/M^2) was encountered in orchards fields, while vegetable crops retained the least amount (0.25 g/M^2). The conclusion is, DDT is very necessary to kill pest and save crops, but in the long run it would be worthwhile to replace it by a less hazardous pesticide.

20.10 CLASSIFICATION OF PESTICIDES

The detailed account on the quantitative analysis is given in previous chapters. They are classified largely into five groups as shown in Table 20.13. The important class is chlorinated hydrocarbons containing aldrin, BHC, DDT, lindane, chlordane, while organic phosphorus pesticides include malathion, parathion (ethyl and methyl); the third category of the pesticide is carbamate compounds containing zectrion and carbaryl. The inorganic pesticides include copper, nickel, lead and arsenic salts, while mercury containing pesticides are 2, 4 D or 2, 4, 5 T. Depending upon their use to kill insects, rodents, they are termed as insecticides, herbicides, fungicides, etc., as shown in Table 20.14. This table also lists LD_{50} dose (mg/kg) for various pesticides. This is the dose required to kill 50% of affected species. The chemical name and popular name of pesticides are given in Table 20.17.

Table 20.13 Classification of pesticides

S. No.	*Classification*	*Example*	*Used as*
1.	Chlorinated hydrocarbons	Aldrin, chlordane, BHC, DDT, dieldrin, lindane, heptachlor (PCB)	Insecticides
2.	Organophosphorus compounds	Malathion, parathion (methyl) parathion (ethyl)	Insecticides
3.	Carbamate compounds	Zectrion, carbaryl (sevin)	– – –
4.	Inorganic compounds	$CuSO_4$, $NiCl_2$, PbO_2, As_2O_3	Fungicides
5.	Miscellaneous compounds	2, 4 D, 2, 4, 5 T, carbaryl organic mercurials	Herbicides

Table 20.14 shows that Aldrin, PCB, carbaryl, DDT, endrin, heptachlor, lindane, malathion and parathion are insecticides, while 2, 4 D, 2, 4, 5 T containing organomercury compounds are herbicides, while inorganic compounds are fungicides. The most toxic amongst them is parathion with LD_{50} of 4 mg/kg, so also endrin is toxic with LD_{50} as 5 mg/kg; the least toxic among them is malathion and DDT with LD_{50} ~ 400 mg/kg, 2, 4, 5 T is dangerous with LD_{50} of 0.03 mg/kg, so also dioxane. Pharsenol is less injurious to health.

20.11 DEGRADATION OF PESTICIDES

The main problem associated with pesticide application is the presence of pesticide residue. These pesticide residues are formed from the main pesticide. Such residues are injurious to health in comparison with the parent compound, e.g. DDT degrades to DDE. Such degradation is possible by various transformation, such as oxidation (e.g. 2, 4 D undergo hydroxylation), side chain oxidation (DDT converts to OH-form), phorate undergoes to form sulphoxide, dehydrogenation and dehydrohalogenation

Table 20.14 LD_{50} dose and threshold limiting values of pesticides

Pesticides	*TLV* mg/m^3	LD_{50} *dose* *(mg/kg)*
Insecticides		
Aldrin	0.25	54–56
PCB	0.50	250
Carbaryl	–	340
DDT	1.0	420–800
Endrin	–	5–43
Heptachlor	0.5	90
Lindane	0.5	125–200
Malathion	10.0	480–1500
Parathion	0.1	4–30
Dieldrin	0.25	40
Herbicides		
2, 4 D	10.0	666
2, 4, 5 T	–	0.03
Pentachlorophenol	0.5	27–80
Dioxane	–	0.03
Fungicides		
$CuSO_4$	–	Very high
Pharsenol	–	825

(like DDT), conjugation or metal complex formation like amitrole; hydrolytic cleavage like malathion; exchange reactions like parathion and isomerisation of mostly organophosphorus compounds. Some compounds in this process become more toxic as indicated in Table 20.15. Such transformation involves no photosynthesis; no chlorophyll synthesis, absence of oxidephosphyration, etc. Table 20.16 shows percentage of degradation of pesticide at varying temperatures. It will be noted e.g. for DDT, if temperature is raised from 600 to 1000°C, degradation is complete, while pesticides like dieldrin hardly show any degradation; with rise of temperature by 400°C, it is only about 0.4%. These residue can remain in biosphere and lithosphere for more than 30 years, even if now we stop use of these pesticides. A large number of factors are responsible for degradation of pesticide residue in the environment. The rate of degradation depends upon the chemical structure and nature of the pesticide. Inspite of degradation they are not totally eliminated and as such degradation is incomplete or partial.

Table 20.15 Transformation of pesticides

S. No.	*Effect*	*Name of compounds*
1.	No photosynthesis	Trianzines
2.	No chlorophyl synthesis	Triazoles
3.	No oxidephosphorylane	Carbamats
4.	Harmonic analogue	2, 4 D, 2, 4 5 T
5.	No vit. B synthesis	Chlorihydrocarbon

Table 20.16 Extent of degradation at varying temperatures (%)

Pesticide	*600°C*	*800°C*	*1000°C*
Picloram	90.8	95.6	99.2
Atrazine	87.8	88.8	89.0
Malathion	95.3	96.3	96.7
Dieldrin	99.0	99.5	99.5
DDT	99.2	99.7	100.0
Parathion (mèthyl)	64.0	96.5	97.5
Sevin	88.7	88.8	90.0
Dapalon	64.3	67.8	97.0
DNBP	99.8	99.8	99.8
Nemagon	99.6	99.6	99.6

This is not end of manifestation of hazards. The agricultural pesticides do not hit target and rarely get degraded to harmless compounds. The pesticide residue in soil is taken up by plants and cause phytotoxicity. They are added to aquatic environment and enter in food web. The important pesticide in common usage with their popular name and chemical nomenclature are listed in Table 20.17.

Table 20.17 Important pesticides

S. No.	*Popular name*	*Chemical nomenclature of pesticide*
1.	Aldrin	1, 2, 3, 10, 10 hexachloro-1, 4, 4a, 5, 8, 8a hexahydro 1–4 endo-exo 5–8 dimethanonaphthalene
2.	BHC	1, 2, 3, 4, 5, 6 hexachlorocyclohexane (several isomers)
3.	Carbaryl (sevin)	1–naphthyl methyl carbamate
4.	2, 4, D	2–4 dichlorophenoxyacetic acid
5.	DDE	1, 1 dichloro 2, 2 bis (p-chlorophenyl) ethane
6.	DDT	1, 1, 1 trichloro 2, 2 bis (p-chlorophenyl) ethane
7.	Heptachlor (PCB)	1, 4, 5, 6, 7, 8, 8 heptachloro 3a, 4, 7, 7a tetrahydro 4, 7, methanoindene
8.	Lindane	1, 2, 3, 4, 5, 6 hexachlorocyclohexane
9.	Malathion	S[1, 2 bis (ethoxycarbonyl) ethyl] 0, 0 dimethyl phosphorodithioate
10.	Parathion	0, 0 diethyl 0–p nitrophenylphosphorothioate
11.	Phorate (Thimet)	0, 0 diethyl S-(ethylthio) methyl phosphorodiniate
12.	2, 4, 5 T	2, 4, 5 Trichlorophenoxyacetic acid
13.	Endrin	1, 2, 3, 4, 10, 10 hexachloro 6, 7 epoxy–1, 4, 4a, 5, 6, 7, 8, 8a octahydro 1–4 endoendo 5, 6 dimethano phthalene
14.	Dieldrin	1, 2, 3, 4, 10, 10 hexachloro 6, 7 epoxy–1, 4, 4a, 5, 6, 7, 8, 8a octahydro 1, 4 exoendo 5, 6 dimethanophthalene

Before considering analysis of pesticide, it is worthwhile to examine persistence of some common pesticides in soil. The unit for such measurement is time for 50% disappearance of the amount applied to start with; e.g. Aldrin—a chlorinated hydrocarbon needs two months for degradation, while carbaryl (sevin) and phorate need one month, while rest of pesticides like parathion, malathion mainly organophosphorus pesticide take 20 days for 50% degradation. All of them need as much as three months to reach residue level of 0.1 ppm, which corresponds to 0.3% of applied dose. Thus it will be noted that there is not a single pesticide which undergoes instantaneous degradation. That is one of the reasons why BOD of effluent does not indicate presence of pesticides residue. Only COD furnishes such information.

20.12 ANALYSIS OF PESTICIDES

For details of such analysis refer to the earlier chapter. Among agro-chemicals, pesticides comprise of one of the biggest and vitally important groups of compounds. Analytical methods are widely used for pesticide analysis involving study of pesticide residue, degradation products and new formulations. Largely chromatographic methods, polarographic technique and spectroscopic methods of analysis are used for their characterisation. The important chromatographic techniques include thin-layer chromatography (TLC), gas-liquid chromatography (GLC), and high performance liquid chromatography (HPLC). While the spectral methods of analysis cover UV–IR spectrophotometry, NMR spectrometry and GC–MS technique. Since persistent use of pesticide (e.g. chloroorgano compounds), the pesticide residue and degradation products accumulate in biota, thus needing analysis. GLC detects nanogram to picogram concentrations of pesticides. NMR helps to identify metabolites and degradation products of pesticides. They are extracted by extraction or adsorption on column. TLC is a cheap method and is useful for qualitative as well as quatitative analysis. The carbamates are analysed by TLC in preference to GLC. In GLC, organochlorine pesticides are detected with electron capture detector. While organophosphorus pesticides are best analysed with FID detector in GLC, pesticides containing oxidisable or reducible groups are determined by polarography. Pesticide residues are best analysed by IR spectroscopy. NMR technique has low sensitivity, but is useful for elucidation of the structure of a pesticide. Isomers can be best be analysed by NMR methods. Mass spectrometry is useful for identification of new pesticides. A brief review for analysis of typical pesticides by chromatography, spectroscopy and polarography is given in the earlier chapters.

20.13 CHROMATOGRAPHIC CHARACTERISATION OF PESTICIDES

The organochlorine pesticides are best detected by TLC, e.g. parathion or malathion with $AgNO_3$–bromophenol blue spraying reagent in TLC. Presence of 100–200 ηg of pesticide can be easily detected. The carbamate pesticides like zectran, carbaryl (sevin) and organophosphorus pesticides like phorate, malathion, are best characterised by TLC technique. Gas chromatography is useful for the analysis of volatile pesticides. The residue analysis of chlorinated pesticide is carried out by GLC. Acetonitrile is used as the solvent. Hexane is also used as the solvent for organophosphorus pesticides. From drinking water, chlorinated pesticide can be absorbed on a column packed with resin Amberlite XAD-2 and eluted. It is also analysed by GLC, e.g. lindane, DDT, DDE, aldrin. Polychlorinated biphenyl (PCB) is characterised by GC–MS with glass capillary column. The HPLC is a new invention for the analysis, it is mainly useful for analysis of new formulations which are not volatile. The photometric, fluorometric detectors are also used in HPLC technique.

20.14 POLAROGRAPHIC ANALYSIS OF PESTICIDES

Chlorine containing pesticides are analysed, e.g. hexachloro cyclohexane, endrin, aldrin and dieldrin are analysed on dropping mercury electrode in LiCl in isopropyl alcohol. The half-wave potential ($E_{1/2}$) of nitrogroup containing pesticides is available in literature. The supporting electrolytes used are nitric acid (0.1 M), acetate buffer, soldium hydroxide (0.1 N) and ammonia-ammonium chloride buffer. Half wave potential varies from –0.32 V to –0.72 V for parathion, e.g. –0.26 V in HNO_3, –0.62 V in acetate buffer and +0.74 V in ammonia buffer. Voltammetric methods can also be used. The sensitivity of the method is however not satisfactory. The chief advantage of polarography in the pesticide analysis is the advantage of identifying a particular pesticide from the knowledge of half-wave potential in a sample of industrial effluent. This is followed by a quantitative analysis.

20.15 SPECTROSCOPIC ANALYSIS OF PESTICIDES

The UV-spectrophotometry is used for the analysis of formulations. The standardised adsorbents are required for clean up before analysis. The direct reflectance spectometry is very useful, e.g. analysis to triazine herbicides. IR methods are used for analysis of several isomers of heptachlor or BHC. IR is less sensitive than UV and not suitable for residue analysis; the only advantage is that IR does not need rigorous clean up and offers a choice of identifying peaks. NMR methods are useful for identification and characterisation of pesticide compounds. Proton NMR is best for such purposes; DDT and DDE are best analysed by NMR. Mass spectrometry, which is the most reliable and definite tool for the analysis of pesticides belonging to the class of carbamates, chlorinated hydrocarbons and metabolites. Chlorophenols can be studied by mass spectrometry. GC–MS hyphenated method provides reliable information. For Lindane 2, 4, 7, 8 T and PCB were therefore best analysed by GC–MS. The development of several ionisation spectral techniques will lead to further enhancement of sensitiviy and reliability in elucidation of the structures of pesticides. The hyphenated GC–MS or GC–HPLC is likely to indicate spectacular results in future years.

ADDITIONAL READINGS

20.1 P.R. Hesse, *A Textbook of Soil Chemical Analysis*, John Murray, London (1971).

20.2 O.A. Black, *Methods of Soil Analysis,* American Society of Agronomy, USA (1965).

20.3 M.S. Cresser, *Solvent Extraction in Flame Spectroscopic Analysis,* Butterworth (1978).

20.4 I.L. Marr and M.S. Cresser, *Environmental Chemical Analysis,* International Textbook Co., New York (1983).

20.5 R.A. Horn, *The Chemistry of Our Environment,* John Wiley & Sons, New York (1978).

20.6 A. Volborth, *Elemental Analysis in Geochemistry*, Elsevier, New York (1969).

20.7 M.A.Q. Khan, *Pesticides in Aquatic Environment,* Plenum Press, New York (1977).

20.8 P.L. McEwen and G.R. Stepenson, *Use and Significance of Pesticides in the Environment,* Wiley, New York (1979).

20.9 C.A. Edwards, *Environmental Pollution by Pesticides,* Plenum Press, London (1973).

20.10 K.G. Das, *Pesticide Analysis,* Marcel Dekker, New York (1981).

Chapter 21

Noise Pollution and Its Measurement

21.1 INTRODUCTION

A melodious music makes the man happy. It may be either a vocal recital or instrumental music emanating from sitar, sarod or violin or a singer. However, there is another aspect of noise or sound which is not pleasant which causes headache and pain. This latter category of sound covers several sources of noise such as traffic, industries and construction activity. One cannot tolerate impulsive noise generated by pounding a pillar at construction or hammering site. Such noise is more harmful specially to children, senior citizens and the sick patients in home or in hospitals having no silence zones for the locality.

This problem led to need of understanding phenomena of noise. Such study consisted of identifying the source of noise, its adverse effects on human health, what is called as impulsive noise, the origin of the sonic boom? The most important aspect is the measurement of the noise. Such measurement (on particular scale) provided a person to know its intensity and to compare one kind of noise from that of other. It has provided a means to know the quality and intensity of sound. The unit of measurement that is accepted by the environmentalist is decibel. Later on there was discovery of pascal scale for such measurement, which is related to decibel by relation 35 dB = 960 μPa, *i.e.*, 1dB = 27 μPa (approx.)

With the availability of unit of measurement of noise it greatly facilitated person to devise certain standards of acceptance or rejection of the level of noise. It also facilitated the mapping of noise in industrial establishment. Number of sound level meters (SLM) were designed and standardised. There was development of anechoic chamber or sound absorbing chambers in hospitals, auditorium and in public places like temples.

Therefore an endeavour is made in this chapter to provide significance of noise, origin of noise, its effects on human health, kind of noise, meaning of sonic boom and impulsive noise and finally the various methods utilised to measure exactly the intensity of noise. This in turn leads to noise mapping or measurement and it facilitates to set uniform standards in fixing levels of noise or sound in home, garden, streets, offices and the industrial houses.

In our country with so many states only Maharashtra was the first state to pass water pollution act in 1968. An air pollution Bill was enacted by central government in 1984, while a Comprehensive Environmental Protection Bill was approved by the parliament in 1989. Until then there was no act which could prevent an individual from generating noise. The control and abatement of noise is still under jurisdiction of police in most of the states. With the growth of industries and specially with noise associated with traffic, there is a rising awareness of noise pollution in public.

21.2 WHAT IS NOISE?

Sound is a form of energy which requires a material medium for propagation. Sound can propagate through solids and liquids but not through vacuum. Any vibrations are capable of producing sound. In loudspeaker, the diaphragm moves forward and backward to generate sound. On account of compression, gas molecules move back and forth progressively. These oscillating molecules with definite frequency are responsible for creation of sound pressure. If the pressure varies regularly and repeatedly, then time between successive high and low pressure is called the period. The frequency is the reciprocal of period. Sound is defined as the energy in the frequency range 0.02–20 kHz or 20–20,000 Hz (hertz is the unit of frequency representing cycles per second) which is the typical range of hearing of human ears. The lowest frequency which is audible is 20 Hz, while the highest audible frequency is 12,000 Hz. This region is called frequency band, which defines a range of frequencies between the specified or implied limits. It is used specially in photoacoustic spectrometer for quantitative analysis.

The intensity of sound can be measured with precision and is related to the amount of sound energy received per second from the source of sound. The intensity level is ten times the logarithm of intensity ratio. For instance if I_L represents intensity level, then $I_L = 10 \log (I/I_0)$, if I = measured intensity, I_0 = reference intensity. Since $I_L = (P)^2 = 2 \times 10 \log (P/P_0) = 20 \log (P/P_0)$ = SPL. Thus there are practical difficulties in measuring the intensity, however, the variation in pressure can be easily measured. As intensity is directly proportional to the square of the pressure, the sound pressure level (SPL) in dB is defined by SPL = $20 \log (P/P_0)$ dB, while P_0 (reference pressure) = 2×10^{-5} Nm^{-2} where P is the measured pressure in Nm^{-2}. This value is used since it is the nearest whole number corresponding to the reference intensity.

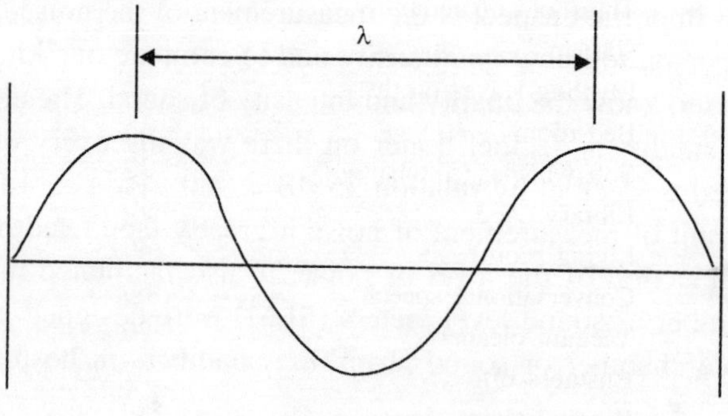

(a) Low pitch of sound

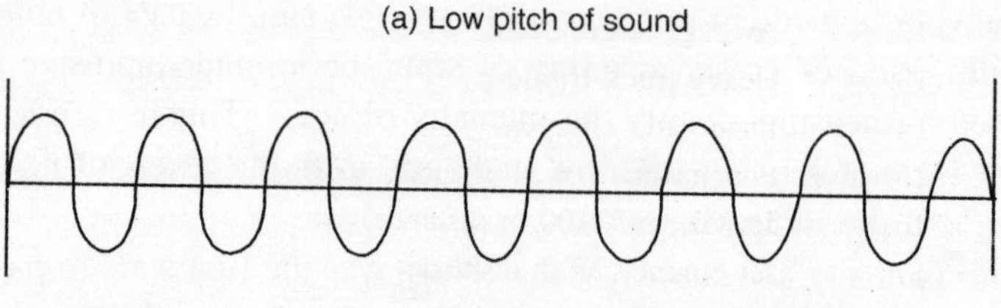

(b) High pitch of sound

Fig. 21.1 Pitch of sound

The sound vary in pitch, which is defined as the relative high or low pitch of a sound. Pitch is measured by the frequency of sound wave—the rapidity with which it vibrates. A high pitched sound with its fast vibration is more penetrating than one with low pitch Fig. 21.1. The intensity of sound is measured in decibels (dB). This scale was devised to measure the smallest difference in sound which is detectable by the human ears. Its gradations move up not in simple arithmetic progression, but in multiple progression based on logarithmic scale, *i.e.* each increase of one dB represents a large change of intensity of sound.

21.3 SOURCE OF NOISE

The sound which is pleasing to the ears, which is melodious to hear and which gives comfort like musical notes is defined as sound. While one which is penetrating and uncomfortable, which creates psychological stress is called as noise. The source of sound or noise in domestic as well as industrial environment is shown in Table 21.1 which lists domestic sound in decibels. It is interesting to note that threshold of audibility is zero dB. If a pin falls on a hard floor, it gives 10 dB, sound up to 40 dB is encountered in a garden, a bedroom, or libraries. Whisper (30 dB) is considered to be pleasant. An average conversation produces 60 dB. Anything between 50 and 90 dB is considered as uncomfortable noise, e.g. vacuum cleaner, office, alarm clock, street traffic and pneumatic chipper, while one beyond 90–120 dB is considered to be a health hazard. Such noise is produced by aircraft, pop-music, heavy truck traffic, etc. level of noise at road junctions, in Mumbai are presented in Table 21.2, and that at suburban railway stations in Table 21.3.

Apart from domestic source of noise pollution, industrial activity creates enormous noise. Table 21.4 lists source of industrial noise. Alarming noise commences from 70 dB and can extend up to

Table 21.1 Source for domestic noise

S. No.	*Source of noise*	*SPL in dB*
1.	Threshold of audibility	0
2.	Garden	20
3.	Rustling leaves on three	20
4.	Bed room	25
5.	Soft whisper (5′ away)	30
6.	Library	30
7.	Living room	40
8.	Conversational speech	60
9.	Vacuum cleaner	70
10.	Business office	67–70
11.	Ringing alarm clock	80
12.	Average street traffic	80–85
13.	Heavy truck traffic	90
14.	Pneumatic chipper	100
15.	Pop music	110
16.	Jet take-off (100 m distance)	130
17.	Jet engine (25 m distance)	140
18.	Threshold pain	140

Table 21.2 Noise level at road junction in Mumbai in the evening

S. No.	*Road junction (New name)*	*Noise level (dBA)*
1.	Flora Fountain (Hutama Chowk)	78
2.	Bombay VT (Bori Bunder)	81
3.	Byculla Bridge	74
4.	Lalbaugh Junction	76
5.	Parel (Dr. Ambedkar Marg)	79
6.	Dadar (Khodad Circle)	80
7.	King Circle (Maheshwari Udyan)	75
8.	Sion Circle (Laxmibai Chowk)	78

Table 21.3 Noise at railway suburban stations (Central Rly.) in Mumbai in the evening peak hours

S. No.	*Suburban station*	*Noise levels (dBA)*
1.	Bombay VT	84
2.	Masjid	75
3.	Sandhurst Road	76
4.	Byculla	81
5.	Chinchpokli	73
6.	Curry Road	82
7.	Parel	74
8.	Dadar	82
9.	Matunga	76
10.	Sion	78
11.	Kurla	76

140 dB. It includes noise generated by air compressors, concrete mixers, scrapper, dozer, payer, powersaw, generators, rockdrill and pile foundation operations, riveting process. Such noise is bound to cause ill effect on health. Another source of noise is traffic, which is discussed separately in a subsequent section.

21.4 NOISE AND HEALTH

Noise causes several kinds of health hazards. Such effects may be physical or mental, which would have long term consequences on communications, working efficiency, personal comfort and can lead to industrial accidents. The health problems may be acute and chronic. The acute effects depends upon frequency. 120–150 dB noise can cause permanent deafness. If a person is exposed to 90 dB noise for 40 hours week for say 30 years, he is bound to be deaf at the end of his career. A low level noise cause nausea, loss of physical control and physiological changes caused by stress. The important chronic effect is loss of hearing. The loss starts at a frequency of 4000 Hz, with more damage done by sudden impulsive noises. The noise causes prolonged loss of sleep. The mental effect causes lack of concentration and mental disorientation. Regarding communications, noise interferes communicability and causes proneness to commit error in judgement. The personal comfort no doubt gets affected, destroying peaceful life and enjoyment of leisure. It contracts blood vessel, thereby supplying less blood to heart and brain. It causes digestive disorder.

Table 21.4 Source of industrial noise

S. No.	*Source of noise*	*SPL in dB*
1.	Vacuum cleaner	75
2.	Electric generator	76
3.	Garbage collector	85
4.	Air compressor	85
5.	Concrete mixer	85
6.	Dozer	87
7.	Jack hammer	88
8.	Scrapper	88
9.	Dumb truck	88
10.	Paver	89
11.	Powersaw	90
12.	Rockdrill	98
13.	Noisy factory	100
14.	Pile driver	101
15.	Riveter	115

21.5 IMPULSIVE NOISE

An impulse is much more dangerous than a continuous noise. A sudden noise generated with high pitch or intensity but with a life-time of less than one second is called as impulse. Table 21.5 lists impulses caused by an exploding bomb (190 dB), naval gunshooting, firing crackers, pile driving and metal beating, which are capable of producing noise to the extent of 140 dB. An unexpected thud of sound with a short life has high impulse and is dangerous. Ultrasound or ultrasonics is a science which deals with sound vibrating at frequencies too high to be heard by human ear (> 20,000 Hz), but ultrasound can be heard by dogs, horses or dolphins in the sea. The infrasound is sound vibrating at too low frequencies (< 18 Hz), which cannot be heard by human ear. Such sound can be detected by electronic

Table 21.5 Sources of impulses

S. No.	*Nature*	*SPL, dB*
1.	Exploding bomb	190
2.	Naval gun shooting	160–180
3.	Firing hand gun	140–170
4.	Diwali crackers	125–160
5.	Pile driving	110–130
6.	Metal heating, drop forging	120–140

gadgets. The vibrations which occur because of natural resonance are known as sympathetic vibrations, they cause accoustic fatigue. The construction activities produces impulses exceeding 85–100 dB.

21.6 TYPES OF NOISE

We had considered the sources of noise. There are a variety of noises such as domestic noise, traffic noise, aircraft noise, construction noise and industrial noise, all leading to discomfort to human beings. In Tables 21.1 and 21.4 we have considered noise from domestic and industrial sources houses, in cities. Traffic noise makes our lives most uncomfortable. To a greater extent domestic noise can be easily controlled, by rational use of radio, T.V., through use of earphones; noise in household can also be minimised by using carpets, drapery and upholstered furniture. The noise of sirens can be also modulated. An attempt was made by traffic police in Mumbai to create silent zones around educational institutes, hospitals, but unfortunately it did not work although it was most successful in Britain. Traffic noise is due to cars, motorcycles, trucks and buses. For instance, a car gives 60 dB, a bus 80 dB, while a truck creates 85 dB, while motorcycles, scooters, mopeds, TVs, produce noise of the level of 85–90 dB. These noises are bounced on hard surfaces of buildings and are reflected back to the ears. The noise is produced by air turbulence around the bodywork of a moving car; tyres are an usual source of noise specially on cement roads, super express highways if a car is travelling at a high speed. In cold countries snow tires make more noise. The motorcycles are the worst culprits, wherein the noise originates from exhaust and engine. It can give 110–120 dB noise. Mufflers are designed in motorbikes to reduce noise. Trees and bushes are planted on highways to reduce noise by dispersing it on leaves and branches by process of absorption. In good town planning, the bedrooms in building are located far away from roads to produce peace and tranquility.

21.7 SONIC BOOM

Aircraft noise is the worst gift to metropolitan cities like Mumbai and Kolkata. Fortunately, in India concords are not allowed, which would have made life more miserable. The Jumbo Jets like Al-777 are the main culprits, which causes sonic boom—a loud noise produced when a moving object crosses the speed of sound. An aircraft flying at supersonic speed builds up a shock wave of compressed air, which surrounds the plane and trails behind it (Fig. 21.2). When the trailing edge of the shock wave reaches

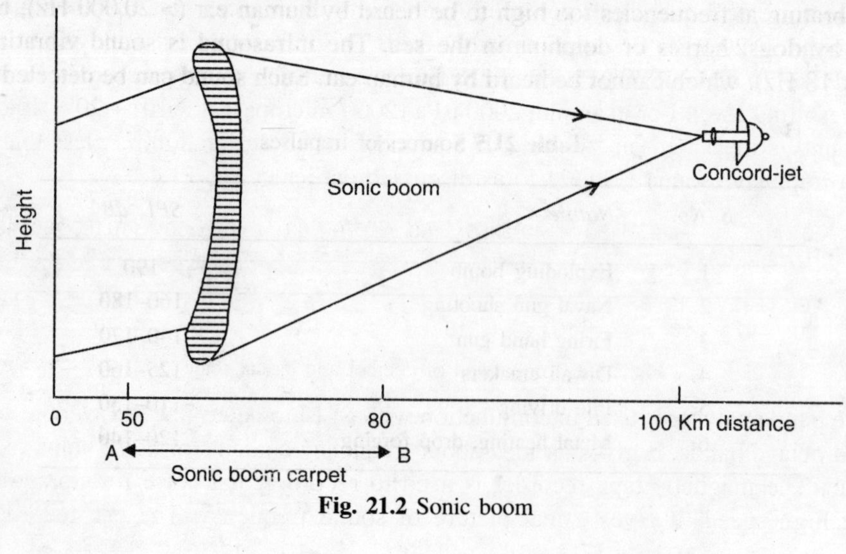

Fig. 21.2 Sonic boom

the ground, it produces sonic boom. As a matter of fact, aircraft noise is measured in terms of NNI (*i.e.* noise number index). It specifies a certain number. An aircraft noise in terms of NNI in busy airport like Chicago in USA is 50, while in our airports like Mumbai with comparatively low aircraft density, it is 35–40 NNI. If an aircraft lands at an angle of 60°, it would generate minimum noise. Noise abatement in airports in Narita, Japan is most rigorous. The residents around the Haneda old airport in Japan had to be paid compensation for noise nuisance. The Japanese had therefore shifted their International airport to Narita from Haneda, which is about 80 km outside the city limits of Tokyo, with very little activity of domestic traffic at Haneda airport, which is located in the heart of the city. Night flights are virtually banned in western countries. The stacking of aircraft can be done to minimise noise at airports, this is done in Germany at Frankfurt or Los Angeles in USA where large number of planes land. In United States the most busy airport is O'Hare at Chicago and Los Angeles in Western America. In spite of enacting stricter legislation, the life surrounding these airports is noisy. In England at Heathrow airport in London, the frequency of landing is high. This makes the airport noisy. In India, the problem is growing with concern. The Santacruz and Sahar airports in Mumbai and Palam in Delhi airport are somewhat busy, while Dumdum at Kolkata and Meenambhakom at Chennai are relatively less noisy and less populated. The new airports at Hyderabad and Bangalore are relatively noise free. Similarly, proposed new airport for Mumbai in outskirts of Mumbai at Panvel is expected to be noise free.

21.8 NOISE MEASUREMENT

There are several devices to measure sound or noise. There are sound level meters (SLM), octave band analyser, magnetic tape recorder, pen recorder, statistical analyser, noise average meter and noise analysers. The sound level meter (SLM) is used for evaluation of sound pressure on linear or weighted scale. The dynamic characteristics of meters is possibly of slow, fast and impulsive sound measurements. There are two grades of sound level meter, one with limited accuracy (IEC–123), while the other with (IEC–179) high degree of accuracy. Former are cheaper and robust while the latter are expensive but accurate.

The chief objective of sound measurement is to reduce its level, so as to enjoy it as music; carry out scientific analysis to reduce its annoying effect, to ascertain its capability to cause hearing damage and to serve as powerful diagnostic tool to reduce it. Sound travels at the rate of 750 miles per hour 3–4 seconds per kilometre or 2–5 seconds per mile. Apart from dB, one can measure sound in pascal units, e.g. weakest sound is 20 μPa, *i.e.* approximately 0 dB (Fig. 21.3). Pascal scale is not practical, e.g. 35 dB = 960 μPa or 70 dB = 1920 μPa, etc. Each time we multiply sound pressure in Pa by 10, we add 20 dB to the resulting level. For example, 2000 μPa (2000 micropascal × 10) + 20 = 40 dB. As shown in Fig. 21.4 sound is measured on 'A' scale. If a sound is of short duration, *i.e.* less than one second, it is termed an impulsive sound (Fig. 21.5) as discussed earlier in Sec 21.4.

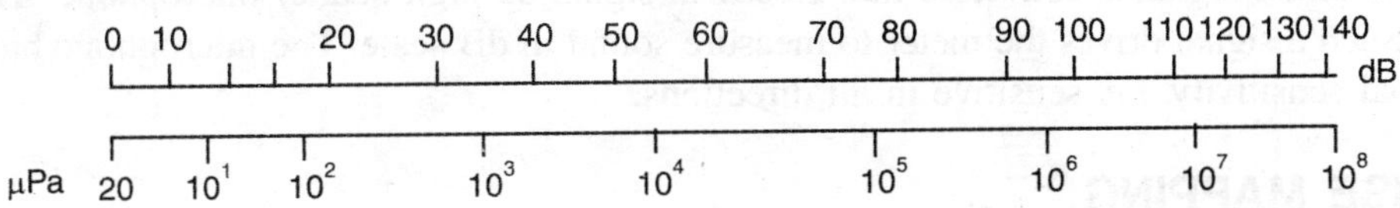

Fig. 21.3 Comparison of decibel and pascal scale

The octave band analyser is used in conjunction with SLM to determine the average sound pressure of the standard octave bands. It gives an idea about frequency contents of noise helps to decide abatement technique. The magnetic tape recorder is used to record actual noise for storage and latter for evaluation. At high speed, it gives a real picture of sound frequency. The pen recorder can be fed

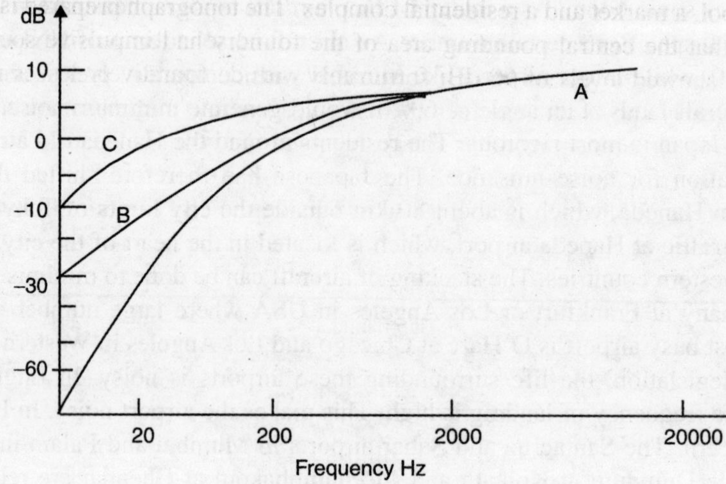

Fig. 21.4 Measurement of sound pressure

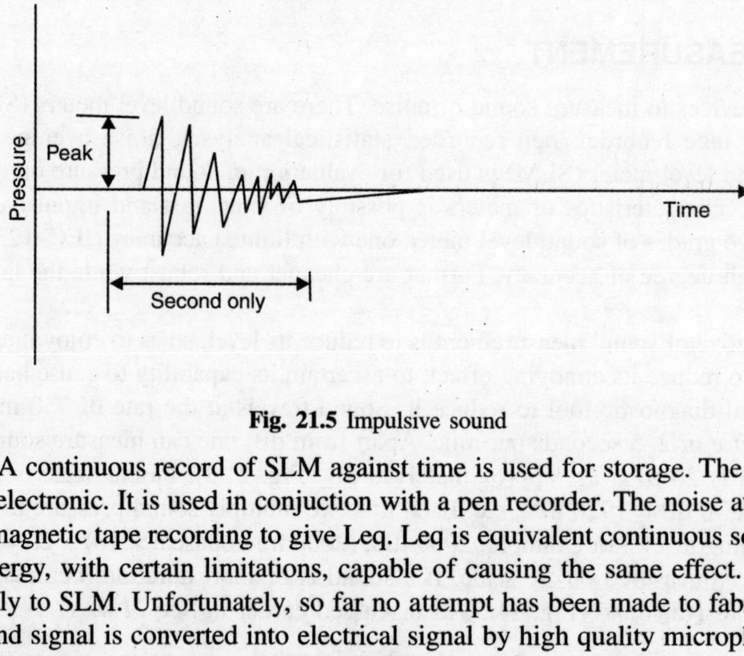

Fig. 21.5 Impulsive sound

directly to SLM. A continuous record of SLM against time is used for storage. The statistical analyser is mechanical or electronic. It is used in conjuction with a pen recorder. The noise average meter is fed directly or form magnetic tape recording to give Leq. Leq is equivalent continuous sound pressure level or amounts of energy, with certain limitations, capable of causing the same effect. Finally noise analyser is fed directly to SLM. Unfortunately, so far no attempt has been made to fabricate noise meters in India. The sound signal is converted into electrical signal by high quality microphone. The signal is amplified. Such a signal drives the meter to measure sound in dB scale. The microphone has to be Om in directional sensitivity, *i.e.* sensitive in all directions.

21.9 NOISE MAPPING

In order to assess the impact of noise on human health, it is necessary to undertake a survey of noise levels in different parts of a locality, hence sound maps are prepared. These are also called as topographs. The areas which are sound prone, like dwellings around industrial establishments, airports, railway stations, cinema houses or commercial plaza are selected for sound mapping. A typical survey was undertaken for an industrial establishment in Greater Mumbai. The said industry was surrounded by drama theatre, school, a market and a residential complex. The topograph prepared is shown in Fig. 21.6. The figure shows that the central pounding area of the foundry had impulsive sound of 105–110 dB, while factory shed showed levels of 90 dB; fortunately outside foundry premises noise level was less than 70 dB. This clearly shows that the factory is not very serious health hazard to the locality. Noise topograph is obtained by connecting lines drawn between the points of equal sound levels to exhibit sound distribution patterns. A topograph indicates zones of noise and is a first step to mitigate or abate sound nuisance. A topograph with thick lines would indicate need of putting waxed ear plugs or ear muffets to prevent damage to hearing.

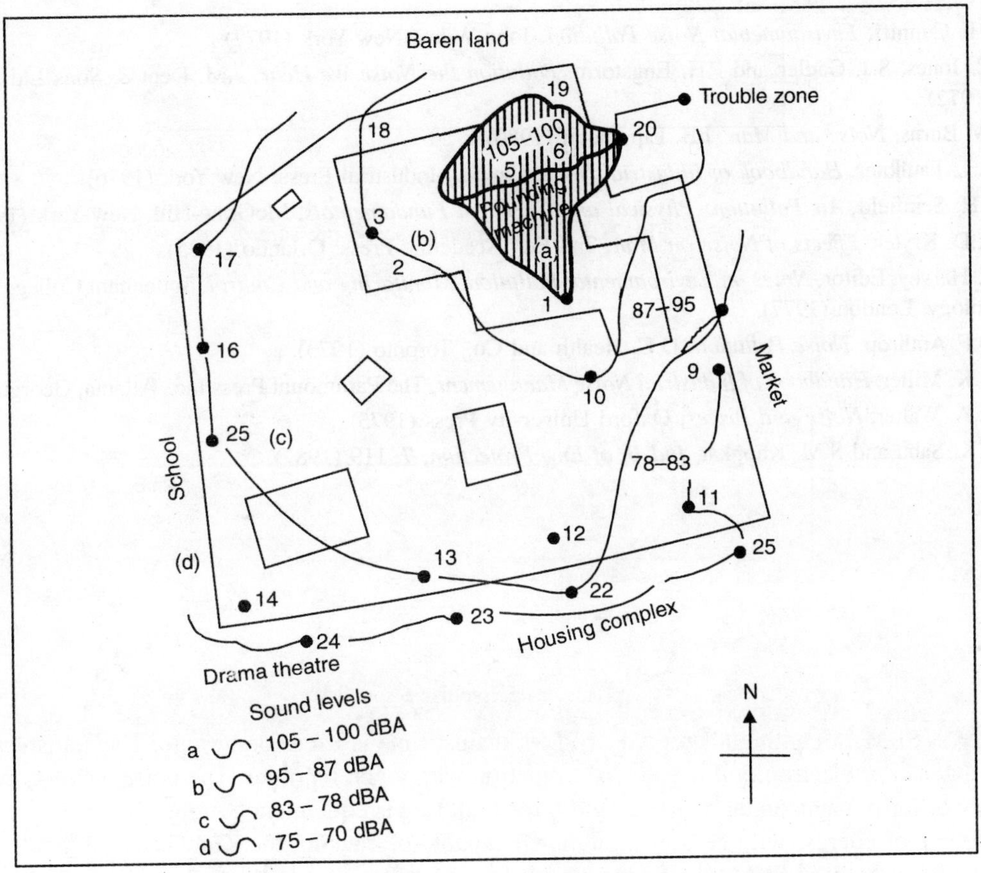

a - pounding zone; b - external zone; c - periphery and d - outside factory

Fig. 21.6 Topograph

21.10 ANECHOIC CHAMBERS

Further it is possible to control such noise danger zone by constructing Anechoic chambers or sound absorbing enclosures. After measuring sound in field free of reflecting objects, an anechoic chamber is constructed. The roof, sidewalls, ceiling, floor are covered by highly absorbing materials to eliminate reflection. The sound reflecting chambers are called reverberating chambers. These are opposite of anechoic chambers. For example, in Dome of Gol Ghumat in Bijapur, Karnataka, one notices whispering galleries. Such reverberating chambers are useful for noise investigations of machines.

21.11 ENVIRONMENT AND NOISE MEASUREMENTS

The wind blowing around noise meter would give wrong results, hence a ball of porous polyurethane sponge should be used over a microphone. The humidity up to 90% does not affect sound level meter (SLM), but rain and snow does. SLM operates in –10° to 50°C range. A variation in pressure of about ± 10% will not affect measurement. Vibration should be avoided during sound measurements. Foam rubber pads dampens vibration effectively. The magnetic field influence sound level measurements.

ADDITIONAL READINGS

21.1 P.F. Cunniff, *Environmental Noise Pollution*, John Wiley, New York (1977).

21.2 C. Jones, S.J. Gadler and P.H. Engstorm, *Pollution the Noise We Hear*, J.M. Dent & Sons Ltd., London (1972).

21.3 W. Burns, *Noise and Man*, J.B. Lippincott (1980).

21.4 L.L. Faulkner, *Handbook of Industrial Noise Control*, Industrial Press, New York (1976).

21.5 J.H. Seinfield, *Air Pollution: Physical and Chemical Fundamentals*, McGraw-Hill, New York (1975).

21.6 K.D. Kryter, *Effects of Noise on Man*, 2nd Ed., Academic Press, Orlando (1985).

21.7 T. Harvey, Editor, *Notes on Environmental Pollution, Monitoring and Control*, Tottenham College of Technology, London (1977).

21.8 D.F. Anthrop, *Noise Pollution*, D.C. Health and Co., Toronto (1973).

21.9 R.K. Miller, *Handbook of Industrial Noise Management*, The Paramount Press Inc., Atlanta, Georgia (1976).

21.10 A.A. Walter, *Noise and Prices*, Oxford University Press (1975).

21.11 S.K. Sahu and S.M. Khopkar, *Ind. J. of Env. Protection*, **7**, 119 (1987).

CHAPTER 22

Odour and Its Measurement

22.1 INTRODUCTION

Although there has been a spectacular development in the field of chemical instrumentation, no attempt has so far been made successful for the fabrication of design and development of odour meter or odour measuring instrument. Therefore we have to depend heavily on our nose as human being who can easily distinguish between the various tonnes of odour. So the practise of sensing and differentiation of odour continues to depend upon the reliability of ones nose. The odour originates not only from air, water but also from industrial effluents and largely sewage and sludge. If one looks into the chemistry of odour and specially its origin, one would normally confronts with the presence of organic compounds. No doubt several inorganic compounds also produce odour. Amongst organic compounds, musk like compounds as mercaptans, esters, phenols and halogenated hydrocarbons like chloroform produce odour. Within the domain of the inorganic compounds we have largely sulphides of metals like lead, copper, arsenic which produce bad smell. The sampling of odour is as important as the sampling of air. Many times we get faint smell and it is necessary to concentrate it in ratio to get perceptible odour. Only in few cases like pungent smell, we have to resort to dilution. The unit of odour is Threshold Odour Number (TON). This broadly indicates the limit beyond which if dilution is continued with diluent odour free, the odour would disappear. As regards scale of measurement of diluent we have what is called as the hedonic scale which rates odour on basis of likes and dislikes of person sensing odour.

There is good relationship between chemical constitution of a compound and odour. As a rule the compounds of nitrogen or sulphur no matter from where they are inorganic or organic have greater tendency to emanate odour. So also the micro-organism on decaying also produce offensive odour. The anerobic and aerobic bacteria play significant part in generating odour. Therefore this chapter dwells on source, sampling, dilution or concentration, measurement, ascertaining quality and intensity and quantitative analysis of odour. This originates from air, water, waste water or sewage.

Odour is defined as perception of smell referring to the experience, or one that is smelled referring to stimulus. Such perception is possible through the receptors of olfactory epithelium, in an area of an average one inch within the nasal of olfactory epithelium, in an area of an average one inch within the nasal cavity. A psychologically pure odorant is sensed by olfactory receptors. The odours like, chemical smell is termed as impure odour. Such chemical odours have a sense like pain, heat or cold.

The airborne gases, vapours or aerosols have a definite odour. They originate from solid, liquid or concentrated gas. Such origin is called as the odour source. The dispersion of odour in atmosphere is accompanied by dilution. The dispersion is also possible by a process of evaporation, desorption or condensation. Most gases and vapours like those of P_2, N_2, H_2O, CO_2 and noble gases are odorous in nature, while some gases like CO, H_2 have no odour.

22.2 SOURCE OF ODOUR

If a storage tank is filled with a liquid component, it discharges odour to the atmosphere. A volume of air is saturated with liquid component at prevailing temperature of the liquid. Such odour originates from a confined source. The odour is confined when its rate of discharge to the atmosphere can be measured. The characterisation of odour should include knowledge of rate of discharge, temperature, moisture content of air, location and elevation of the source and finally the description of the state of aggregation of the discharge. A direct human sensory characterisation of a confined odour source is not always possible, due to intolerable human exposure. An example of unconfined source of odour is drainage, sewage line, ditch discharge, etc. The rate of such discharge is affected by wind direction and wind velocity. Other examples of unconfined source of odour are garbage dump, settling lagoons, chemical storage areas and oxidation pond.

22.3 SAMPLING OF ODOUR

One can directly have sense of odour on sight, but many a times it is necessary to dilute or concentrate odour. It is necessary to dilute odorant or concentrate it, *i.e.* before human exposure for its measurement. There are four methods of sampling of odour, like grab sampling, sampling by dilution or concentration, and sampling by adsorption and absorption.

(a) Grab Sampling

It involves sampling without change in concentration of odorant. The containers for grab sampling should be inert and should have a large volume-to-surface ratio. The odour should not condense on cooling. Grab sampling is carried out by allowing the air to fill in an evacuated container such as stainless steel, glass container, plastic, mylar, teflon or polymeric containers with a known volume.

(b) Sampling with Dilution

At a time it is necessary to dilute concentrated odour in order to render it suitable for human exposure during measurements. We define

$$\text{Dilution ratio} = \frac{\text{Vol. of sample after dilution}}{\text{Vol. of sample before dilution}} \text{ (at constant pressure and temperature)}$$

It is also expressed in terms of pressure rather than volume

$$\text{Dilution ratio} = \frac{\text{Total pressure of sample after dilution}}{\text{Partial pressure of odorant in container}}$$

Container for dilution should be a stainless steel vessel and it should itself be odourless.

(c) Sampling with Concentration

Sometimes a faint odorant will have to be concentrated before characterisation by chemical techniques or their analysis. In such a case, ratio of concentrations between the odour source and the odorant outdoors should be maintained in the range $10^3 - 10^5$. Then infrared or gas chromatography can be used for characterisation of odour. Most commonly employed methods for concentration are:

(*a*) The odorant air stream is passed through a cold trap, which freezes and holds the odorous substance, but does not retain the major non-odorous components. The liquid air with b.p. –196°C is used for the purpose of cooling. Oxygen should not be trapped and cooled, as it creates a problem during analysis. Such a sample is good, because interactions among the concentrated components are minimised at low temperature.

(*b*) The odorant air steam is passed through a bed of activated carbon, silica gel, or some suitable adsorbent to concentrate odorous material. The best method for such purpose is to pass on activated carbon, as it absorbs organic matter. However, recovery from charcoal is somewhat difficult, but desorption in vacuum is comparatively easy. The solvent extraction process is a less drastic method but it yields poor results.

(*c*) It is preferable to concentrate and hold an odour sample easily with minimum use of equipment. Thus mineral odour-free oil or cotton can be used for adsorption of odour.

22.4 MEASUREMENT OF ODOUR

An odour can be characterised by its intensity, quality, threshold, effective tone or pleasant or unpleasant dimensions. Odour reduction is done by ventilation or absorption. The measurement is based upon consideration of the parameters like intensity, quality, dilution ratio and acceptability by human being.

(a) Intensity of Odour

Is the magnitude of the stimulus produced is measured when a person is exposed to an odorant, on the intensity scale, where reference points consists of standard odorants for the purpose of comparison. The odour standards need not always match the quality of the odours being measured. It is also possible to determine the intensity ratios that are based on certain reference point. This is called as 'ratio scales'. Some intensities are expressed in absolute terms as 'slight', moderate, strong and extreme odours. Intensity is similar to sound pitch.

(b) Quality of Odour

The quality of odour is its comparison in terms of resemblance to some other odour producing material. Thus fragrant, putrid, musky, phenolic are common words. Unfortunately no conceptual system has yet been devised which accurately predicts odour quality from the chemical composition of the odorant or explain mechanism of odour perception.

(c) Threshold Measurement of Odour

The odour threshold is the bare minimum concentration of odour, at which an odorous substance can be distinguished from odour-free air. It is also called as the detection threshold. This is the limit at which its quality can be recognised and is termed as recognisation threshold. The odour threshold level depends upon the nature of the substance as well as on the sensitivity of the judges. The threshold is the concentration at which the odour can be detected or recognised by definite population of judges. Threshold data can be used to predict conditions under which a given substance will be odorous or odourless. It can also decide the degree of dilution or ventilation to deodorise or dilute a given odour source. It represents the proportion of odorant that must be eliminated from a space, or it represents the quantity of a substance that must be injected into space to odorise. In fact this is the volume of air to be odorised by a given quantity of substance. Threshold concentrations are not reliable prediction of odour intensity.

The threshold level (C_t) is measured by quantitative dilution of a given volume (V) of odorant of known concentration (C). Thus $C = W/V$ and $C_t = W/V_t$, if W = quantity of odorant and V_t is the volume of sample after it is diluted to threshold level, V is volume of odorant and C is threshold concentration. Thus $C_t = CV/V_t$. The threshold dilution ratio is $V_t/V = C/C_t$. This is similar to reduced Beer-Lambert's law stated earlier. This is called threshold odour number (TON) or the odour pervasiveness.

(d) The Acceptibility of Odour Level

It depends largely upon aesthetic sense of a person. It can be rated either on a hedonic (like, dislike) scale or on action basis. A typical scale is numbered on like-dislike scale as follows:

1. Like extremely
2. Like very much
3. Like moderately
4. Like slightly
5. Neither like nor dislike
6. Dislike slightly
7. Dislike moderately
8. Dislike very much
9. Dislike extremely

This is called hedonic scale. This scale is not at all full proof and it certainly involves personal error during measurements.

22.5 JUDGEMENT FOR ODOUR MEASUREMENT BY PERSON

The following are the criteria for selection of judges during measurement of odour:

(a) The Criteria of Selection

The judges should be able to discriminate among different odours at low or moderate levels of intensity. He should be able to focus this attention on more than one odour quality or characterise a given odour sensation. His sensory ability should be good. The specific items to which he should be attendtive is speed, interest level, domination, independence and honesty. It is presumed that judge does not suffer from cold, cough or any pulmonary problems. He should not be allergic to smell. The person should become familiar with the conduct and execution of sensory testing. If not he should be given proper training to distinguish between various odours.

(b) Conditions for Measurement of Odour

The space where odour is measured should be itself free of competing distractions. The area should be provided with some means of space odour control, air cleaning can be done by local exhaust. The hard surfaced wall panels and mosaic or asphaltic tile flooring are low in background odour. No smoking, drinking, eating should be done in the area of measurement. Person should be discouraged from using perfumes or perfumed cosmetics before such measurement. There should be a gap of at least 30 minutes after drinking of coffee or hard drinks.

22.6 CHEMICAL CONSTITUTION AND ODOUR

The correlations between odorous substance and sensory measurement was not always possible, though odorants like other substances are amenable to chemical analysis. Thus a chemical analysis is not a final measurement of odour, though it may offer some clue on structural aspects. There is no specific recommended procedure for correlating chemical analysis with sensory measurements or the properties of odorants. The chemical analysis and sensory evaluation should be carried out side-by-side. GC–MS can be used with advantage for analysis and one can smell the fraction as soon as it comes out of chromatography chamber. The establishment of correlations between chemical and sensory measurements of complex mixtures in still demanding time consuming mixture of science and art.

22.7 ODOUR—NITROGEN AND SULPHUR CYCLE

An attempt was made to trace the origin of odour. Those of the compounds containing either nitrogen or sulphur are likely to produce odour. There is no hard and fast rule, but generally organic compounds containing sulphur or nitrogen emit odours. It is worthwhile therefore to consider the nitrogen cycle (Fig. 22.1) and sulphur cycle (Fig. 22.2).

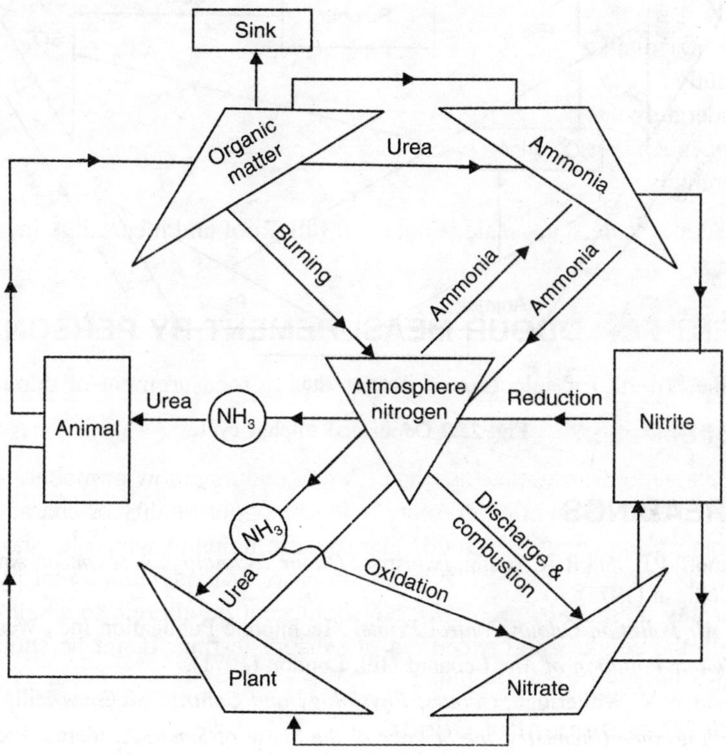

Fig. 22.1 Odour and nitrogen cycle

The nitrogen cycle indicates that odour is generated due to oxidation of nitrosomonas or nitrobacteria. The dead organic matter generated from animal or plant protein also gives rise to obnoxious odour. Amongst inorganic compounds containing nitrogen, ammonia, is well known example of odorous material.

As regards sulphur cycle (Fig. 22.2) the principal source of odour generation are sulphites, sulphide and animal and plant proteins. Such odour originates from animal faeces and urine. Unfortunately, there is no evidence to prove or disprove this argument.

Finally, it is interesting to note in the environmental pollution analysis that in spite of greater advancement in computers, microprocessor and IC chips, we have failed to produce a machine to replace olfactory sensory measuring sense of person. The stimulus is recognised by brain and so far science and technology has failed to devise a brain, though we have mechanical robots.

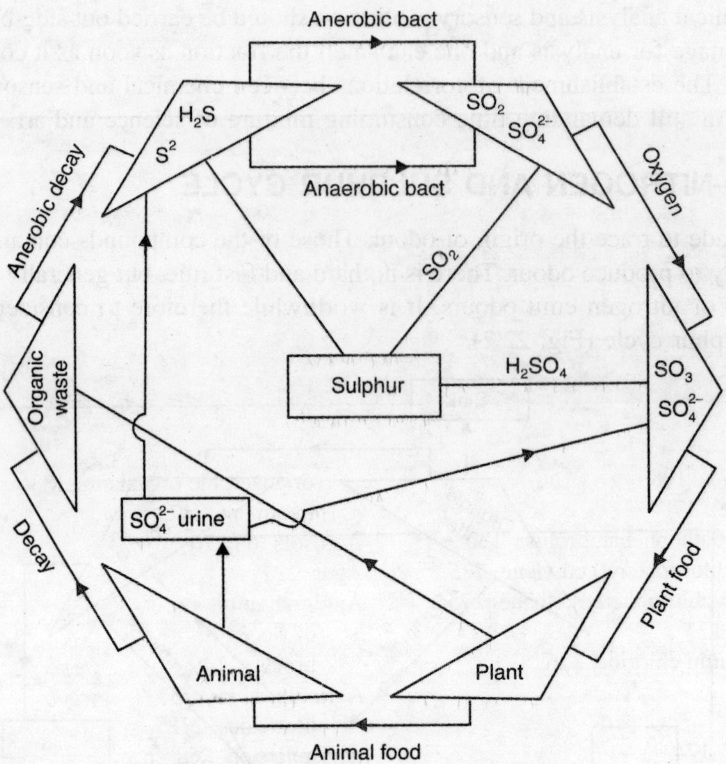

Fig. 22.2 Odour and sulphur cycle

ADDITIONAL READINGS

22.1 P.N. Cheremisinoff, P.E. and R.A. Young, *Industrial Odour Technology Assessment*, Ann Arbor Science, Inc., Ann Arbor, Michigan (1975).

22.2 F.L. Cross Jr., *Air Pollution Odour Control Primer*, Technomic Publication Inc., Westport, Conn. (U.S.A.)

22.3 W. Summer, *Odour Pollution of Air*, Leonard Hill, London (1971).

22.4 C.P. McCord and W.N. Whiteridge, *Odours, Physiology and Control*, McGraw-Hill, New York (1949).

22.5 E.T. Theimer, *Fragrance Chemistry, the Science of the Sense of Smell*, Academic Press, New York (1982).

22.6 S.M. Khopkar, *Environmental Pollution, Monitoring and Control*, New Age International Publishers (P) Ltd., New Delhi (2007).

Index

B

C

N

Q

R

S

T

U

V